VOLUME FIFTY TWO

ADVANCES IN INSECT PHYSIOLOGY

VOLUME FIFTY TWO

ADVANCES IN INSECT PHYSIOLOGY

Insect Immunity

Edited by

PETROS LIGOXYGAKIS

Laboratory of Cell Biology, Development and Genetics, University of Oxford, Oxford, United Kingdom

ELSEVIER

ACADEMIC PRESS

An imprint of Elsevier

Academic Press is an imprint of Elsevier
125 London Wall, London, EC2Y 5AS, United Kingdom
The Boulevard, Langford Lane, Kidlington, Oxford OX5 1GB, United Kingdom
525 B Street, Suite 1800, San Diego, CA 92101-4495, United States
50 Hampshire Street, 5th Floor, Cambridge, MA 02139, United States

First edition 2017

ISBN: 978-0-12-811775-0
ISSN: 0065-2806

For information on all Academic Press publications
visit our website at https://www.elsevier.com/books-and-journals

Publisher: Zoe Kruze
Acquisition Editor: Kirsten Shankland
Editorial Project Manager: Alina Cleju
Production Project Manager: Vignesh Tamil
Cover Designer: Victoria Pearson

Typeset by SPi Global, India

CONTENTS

Contributors	*ix*
Preface	*xi*
1. Insect Antimicrobial Defences: A Brief History, Recent Findings, Biases, and a Way Forward in Evolutionary Studies	**1**
Naomi L.P. Keehnen, Jens Rolff, Ulrich Theopold, and Christopher W. Wheat	
1. Introduction	2
2. The Canonical Immune Genes: A Brief Review of the Components	6
3. From Flies to General Insights Across Insecta	9
4. Conclusion	22
Acknowledgements	22
References	22
2. Phagocytosis in Insect Immunity	**35**
Ashley E. Nazario-Toole and Louisa P. Wu	
1. Introduction	35
2. Insect Blood Cells	36
3. Phagocytic Receptors in Insects	37
4. Regulation of Signalling During Phagocytosis	65
5. Phagosome Maturation	68
6. Conclusion	73
References	73
3. The Melanization Response in Insect Immunity	**83**
Johnny Nakhleh, Layla El Moussawi, and Mike A. Osta	
1. Introduction	84
2. Melanin Biosynthesis Pathways in Insects	86
3. PPO Activation Pathways in Model Insects	87
4. The Contribution of the Melanization Response to Insect Immune Defence	93
5. Crosstalk Between Melanization and Other Immune Pathways	96
6. Concluding Remarks	101
References	102

4. Microbiota, Gut Physiology, and Insect Immunity **111**

Ji-Hoon Lee, Kyung-Ah Lee, and Won-Jae Lee

1. Introduction 112
2. Gut Structure and Gut Microbes 113
3. Gut Immunity 119
4. Gut Renewal 129
5. Concluding Remarks 132

Acknowledgements 133
References 133

5. Intestinal Stem Cells: A Decade of Intensive Research in *Drosophila* and the Road Ahead **139**

Yiorgos Apidianakis, Vasilia Tamamouna, Savvas Teloni, and Chrysoula Pitsouli

1. Introduction 140
2. The Evolutionary Origin of ISCs 141
3. Primary Reports on *Drosophila* ISCs 143
4. The Molecular Characterization of *Drosophila* ISCs 144
5. Other SC Lineages in the *Drosophila* Gastrointestinal Tract: SCs of the Malpighian Tubules, the Hindgut, the Gastric Region and Stomach, and the Midgut Copper Cell Region 147
6. The Developmental Origin of *Drosophila* ISCs 149
7. The ISC Niche (Autocrine, Paracrine) and Hormonal (Endocrine) Signals Regulate ISC Activity 152
8. Ageing and Oxidative Stress Modulate ISC Activity and Vice Versa 153
9. Epithelial Regeneration (Regenerative Inflammation) as a Mechanism of Host Defence 155
10. Conserved Signalling Pathways Controlling Midgut ISC Proliferation (Notch,Wg/Wnt, PVR/PDGFR-VEGFR, JAK/STAT, JNK, p38, Hpo, EGFR, TOR, BMP, Slit-Robo, Bursicon-dLGR2, Hh, Ca^{2+}) 157
11. ISCs and Intestinal Tumours 160
12. Dietary Control of ISCs 162
13. Molecularly Defined Midgut Compartmentalization and ISCs 164
14. Drug Testing and *Drosophila* ISCs 165
15. How to Kill ISCs 167
16. Sexual Identity of ISCs 168
17. Conclusions 169

References 169

6. Insect Symbiosis and Immunity: The Bean Bug–*Burkholderia* Interaction as a Case Study **179**

Jiyeun K. Kim and Bok L. Lee

1. Introduction 180
2. The Bean Bug–*Burkholderia* Symbiosis 183
3. The Effect of the *Burkholderia* Symbiont on the Immunity of Bean Bug 185
4. The Effect of the Immunity of Bean Bug on *Burkholderia* Symbionts 187
5. Conclusions 193

Acknowledgements 193
References 194

7. Exploiting Innate Immunity for Biological Pest Control **199**

Fei Liu, Wuren Huang, Kai Wu, Zhongying Qiu, Yuan Huang, and Erjun Ling

1. Biological Pest Control 200
2. Insect Innate Immunity 201
3. Agents for Biological Pest Control and Their Relationship With Host Immune Responses 205
4. The DUOX Pathway in the Gut Is Crucial in Protecting Host Development 212
5. Additional Factors Involved in Gut Protection 213
6. Immunodeficiency in Insects 214
7. Prospects of Increasing Host Susceptibility to Pathogens Through Immune Suppression 215
8. Outlook 218

Acknowledgements 219
References 219

8. Immunology of Insect Vectors: Midgut Interactions of Sandflies and Tsetse with Kinetoplastid Parasites as a Paradigm for Establishing Infection **231**

Megan A. Sloan and Petros Ligoxygakis

1. Introduction 232
2. Midgut Establishment 233
3. The Peritrophic Matrix: A Structural Barrier 234
4. Digestive Enzymes 235
5. Tsetse EP Proteins 236
6. Pathogen Recognition 237
7. Antimicrobial Peptides 238
8. Reactive Oxygen Species 239

9. Gut Microbiota and Symbionts 240
10. Using the Microbiota as a Transmission Blocking Strategy 243
Acknowledgements 244
References 244

CONTRIBUTORS

Yiorgos Apidianakis
University of Cyprus, Nicosia, Cyprus

Layla El Moussawi
American University of Beirut, Beirut, Lebanon

Wuren Huang
Key Laboratory of Insect Developmental and Evolutionary Biology, Institute of Plant Physiology and Ecology, Shanghai Institutes for Biological Sciences, Chinese Academy of Sciences, Shanghai, China

Yuan Huang
College of Life Sciences, Shaanxi Normal University, Xi'an, Shaanxi, China

Naomi L.P. Keehnen
Stockholm University, Stockholm, Sweden

Jiyeun K. Kim
Kosin University College of Medicine, Busan, South Korea

Bok L. Lee
Global Research Laboratory of Insect Symbiosis, College of Pharmacy, Pusan National University, Busan, South Korea

Ji-Hoon Lee
School of Biological Science, Seoul National University and National Creative Research Initiative Center for hologenomics, Seoul, South Korea

Kyung-Ah Lee
School of Biological Science, Seoul National University and National Creative Research Initiative Center for hologenomics, Seoul, South Korea

Won-Jae Lee
School of Biological Science, Seoul National University and National Creative Research Initiative Center for hologenomics, Seoul, South Korea

Petros Ligoxygakis
Laboratory of Cell Biology, Development and Genetics, University of Oxford, Oxford, United Kingdom

Erjun Ling
Key Laboratory of Insect Developmental and Evolutionary Biology, Institute of Plant Physiology and Ecology, Shanghai Institutes for Biological Sciences, Chinese Academy of Sciences, Shanghai, China

Fei Liu
College of Life Sciences, Shaanxi Normal University; College of Life Sciences and Food Engineering, Shaanxi Xueqian Normal University, Xi'an, Shaanxi, China

Johnny Nakhleh
American University of Beirut, Beirut, Lebanon

Ashley E. Nazario-Toole
Institute for Bioscience and Biotechnology Research, University of Maryland, College Park, MD, United States

Mike A. Osta
American University of Beirut, Beirut, Lebanon

Chrysoula Pitsouli
University of Cyprus, Nicosia, Cyprus

Zhongying Qiu
College of Life Sciences, Shaanxi Normal University, Xi'an, Shaanxi, China

Jens Rolff
Freie Universität Berlin; Berlin-Brandenburg Institute of Advanced Biodiversity Research (BBIB), Berlin, Germany

Megan A. Sloan
Laboratory of Cell Biology, Development and Genetics, University of Oxford, Oxford, United Kingdom

Vasilia Tamamouna
University of Cyprus, Nicosia, Cyprus

Savvas Teloni
University of Cyprus, Nicosia, Cyprus

Ulrich Theopold
Wenner-Gren Institute, Stockholm University, Stockholm, Sweden

Christopher W. Wheat
Stockholm University, Stockholm, Sweden

Kai Wu
School of Life Science, Shangrao Normal University, Shangrao, Jiangxi, China

Louisa P. Wu
Institute for Bioscience and Biotechnology Research, University of Maryland, College Park, MD, United States

PREFACE

As a field of research, insect immunity had its beginnings over 130 years ago when Kowalevsky in St. Petersburg (Kowalevsky, 1887) and Cuénot in Nancy (Cuénot, 1896) studied the role of phagocytic blood cells in clearing invading bacteria from the haemolymph. The field was further advanced with the studies of a research team headed by Sergei Ivanovich Metalnikov at the Pasteur Institute in the 1920s. The group included émigrés Chorine, Toumanoff and Zernoff, who had followed Metalnikov to Paris (see Metalnikov and Chorine, 1930; Metalnikow, 1920; Toumanoff, 1927; Zernoff, 1928). In parallel, Paillot in Lyon initiated his studies on the interaction of *Drosophila* with grapevine cultures (Paillot, 1920), which culminated in his classic treatise on insect immunity (Paillot, 1933).

As the centrepiece of their findings, both the Institute Pasteur team and Paillot conclusively established that the injection of attenuated cultures of bacteria into lepidopteran larvae conferred protection against subsequent injection of normally lethal doses. These results were in keeping with a report by Glaser where blood from grasshoppers was found to have antibacterial activity following injection of bacterial cultures (Glaser, 1918). Studies by Metalnikov and Paillot continued for a decade ending in two partially conflicting monographs (Metalnikov, 1927; Paillot, 1933). The former insisted that the antibacterial defence of lepidopteran larvae was mainly sustained by cellular reactions such as phagocytosis and encapsulation (see also Metalnikov 1933), while the latter supported the view that humoral soluble factors produced by cells were the main tenants of the immune reaction.

As is frequently the case, reality was a combination of the two. However, it would take another 30 years for the field to characterise in more detail the cellular and humoral components of insect immunity. A summary of research on cellular responses can be found in the classic monograph of Salt (1970), which formed the inspiration for the next 20 years of research. The breakthrough for the humoral side of the response was made with the studies of Rasmuson and Boman in Umeå, when they used the more established model of *Drosophila* to identify the humoral nature of insect immunity. Widely accepted as a seminal piece of work in the field, Boman's paper (Boman et al., 1972) settled that the humoral side of the response was inducible and lacking in specificity. Biochemical characterisation of the anti-microbial components was mainly done in the diapausing pupae of the moth

Hyalophora cecropia spanning the next 15 years (reviewed in Boman and Hultmark, 1987). This work opened the molecular era for the field and enabled the return to *Drosophila* for the genetic analysis of signalling pathways triggering immune responses.

The baton for this passed on to the Hoffmann lab in Strasbourg where over the next 15 years the genetic potential of the *Drosophila* model was harnessed to outline the main recognition and signalling pathways that governed insect immunity in a series of landmark publications (for examples of primary papers, see Lemaitre et al., 1995, 1996; for a review of that period, see Hoffmann, 2003). This culminated in an extremely important moment for the field when the Nobel Prize for Physiology or Medicine was awarded to Jules Hoffmann (as well as Bruce Beutler and Ralph Steinman) for work, showing that this first line host defence studied in insects had a very important evolutionary conserved component in mammals and probably most metazoans.

So, where are we now? To answer this question, I am happy to present the current volume of *Advances in Insect Physiology* that reviews and integrates some of the best research on insect immunity in the last decade. The volume is testament to the point where the field has arrived. As an editor of this volume, I have been struck with how the historical work described above has been integrated in terms of evolution, ecology, symbiosis, tissue physiology and gut microbiota in studies of my contemporaries. We start with an article by Keehnen and coauthors, which places the work on humoral antimicrobial defences in insects under a comparative genomic light. With this, the authors extend the traditional signalling patterns and pathways to include an evolutionary perspective. In addition, these authors review the clotting response as an integral and important part of insect immunity. In Chapter 2, Nazario-Toole and Wu present the current view of the cellular arm of the immune response with an impressively detailed analysis of recognition mechanisms and intracellular signalling, showing the advance of the field when compared, for example, with the Salt monograph.

Coupled to the two articles above is the review paper (Chapter 3) by Osta and coworkers. It sums up the state of play in melanisation, a combination of humoral components that require, however, crystal blood cells. This uniquely insect reaction is nevertheless very important if we consider that a strain of the malaria vector *Anopheles gambiae* mosquito has been shown as refractory to the parasite due to its heighted encapsulation and melanisation levels (Collins et al., 1986). Osta and coauthors describe the current state of knowledge connecting pathogen recognition to the melanisation cascade and its permutations in model insects.

One relatively novel aspect in the field is the view of immune responses from the epithelium and especially the gut epithelium. There, a tripartite relationship between signalling, cell biology and gut microbiota has created opportunities for exciting discovery. In Chapter 4, Lee and colleagues review the current state of play and offer a holistic view on how gut microbiota interact with epithelial responses and how the host is able to distinguish between commensal bacteria and pathogenic ones.

Further integration of insect immunity with gut physiology is described in Chapter 5, where Pitsouli and coworkers put intestinal stem cells (ISCs) centre stage. These authors take the concept of ISCs from their discovery to the present day, analysing the different signalling pathways that activate ISC proliferation following various environmental or endogenous cues. The work is centred on *Drosophila*, but it places a framework for similar discoveries in other, difficult to "unlock" but medically or agriculturally important insects.

Integration of insect immunity with ecology and physiology is exemplified in Chapter 6 where Lee and coworkers demonstrate the importance of the interaction between endosymbionts and their insect hosts. They choose to focus on the bean bug–*Burkholderia* relationship and outline the changes in both the host and the symbiont due to this relationship.

Can we harness insect immunity for agricultural benefit? Let's remind ourselves that Paillot started working on *Drosophila* as a vineyard pest. Chapter 7, an article by Ling and coworkers, is testament to the enormous progress of the field. These authors outline the exciting possibilities and important opportunities opening up for using our knowledge on insect immunity to produce biological pest control.

Finally, in Chapter 8, Sloan and Ligoxygakis discuss establishment of infection in tsetse flies (transmitting parasites responsible for sleeping sickness in humans and nagana in cattle) and sand flies (vector of *Leishmania* parasites). This is an example of two important (but difficult to study) insect vectors that can be "opened up" through our current knowledge of insect immunity.

I would like to thank all the colleagues who contributed to this volume and hope that their excellent reviews will provide a road map for the future of insect immunity.

Petros Ligoxygakis
Laboratory of Cell Biology, Development and Genetics,
University of Oxford, Oxford, United Kingdom

REFERENCES

Boman, H.G., Hultmark, D., 1987. Cell-free immunity in insects. Annu. Rev. Microbiol. 41, 103–126.

Boman, H.G., Nilsson, I., Rasmuson, B., 1972. Inducible antibacterial defence in *Drosophila*. Nature 237, 232–235.

Collins, F.H., Sakai, R.K., Vernick, K.D., Paskewitz, S., Seeley, D.C., et al., 1986. Genetic selection of a Plasmodium-refractory strain of the malaria vector Anopheles gambiae. Science 234, 607–610.

Cuénot, L., 1896. Etudes physiologiques sur les orthopteres. Arch. Biol. 14, 293–341.

Glaser, R.W., 1918. On the existence of immunity principles in insects. Psyche 25, 39–46.

Hoffmann, J.A., 2003. The immune response of *Drosophila*. Nature 426, 33–38.

Kowalevsky, A., 1887. Ein beiträge zur kenntniss des exkretions organen. Biol. Zentralbl. 6, 125–144.

Lemaitre, B., Kromer-Metzger, E., Michaut, L., Nicolas, E., Meister, M., et al., 1995. A recessive mutation, immune deficiency (imd), defines two distinct control pathways in the *Drosophila* host defence. Proc. Natl. Acad. Sci. U.S.A. 92, 9465–9469.

Lemaitre, B., Nicolas, E., Michaut, L., Reichhart, J.M., Hoffmann, J.A., 1996. The dorsoventral regulatory gene cassette spaetzle/Toll/cactus controls the potent antifungal response in *Drosophila* adults. Cell 20, 973–983.

Metalnikov, S., 1927. L'Infection Microbienne et I'Immunité chez la Mite des Abeilles *Galleria melonella*. Monogr. Institut Pasteur, Masson & cie, Paris.

Metalnikov, S., 1933. Immunité chez les insects. In: Proc. Intern. Congr. Entomol. 5th Congr. Paris 1932, pp. 209–220.

Metalnikov, S., Chorine, V., 1930. Étude sur l'immunité naturelle et acquise des *Pyrausta nubilalis*. Ann. Inst. Pasteur 44, 273–278.

Metalnikow, S., 1920. Immunité naturelle ou acquise des chenilles de *Galleria mellonella*. C. R. Acad. Sci. Paris 83, 817–820.

Paillot, A., 1920. L'immunité acquise chez les insects. C. R. Acad. Sci. Paris 83, 278–280.

Paillot, A., 1933. L'Infection Chez Les Insects. Impremerie de Trevoux, Paris.

Salt, G., 1970. The cellular defence reactions of insects: Cambridge Monographs in Experimental Biology No. 16. Cambridge University Press.

Toumanoff, K., 1927. Essais sur l'immunisation des abeilles. C. R. Acad. Sci. 185, 1078–1080.

Zernoff, V., 1928. Sur la specificité de l' immunité passive chez les chenilles de *Galleria mellonela*. C. R. Soc. Biol. 98, 1500–1502.

CHAPTER ONE

Insect Antimicrobial Defences: A Brief History, Recent Findings, Biases, and a Way Forward in Evolutionary Studies

Naomi L.P. Keehnen*, Jens Rolff†,‡, Ulrich Theopold§, Christopher W. Wheat*

*Stockholm University, Stockholm, Sweden
†Freie Universität Berlin, Berlin, Germany
‡Berlin-Brandenburg Institute of Advanced Biodiversity Research (BBIB), Berlin, Germany
§Wenner-Gren Institute, Stockholm University, Stockholm, Sweden

Contents

1. Introduction 2
 1.1 Goals of This Review 2
 1.2 A Brief History of Insect Antimicrobial Defences 3
2. The Canonical Immune Genes: A Brief Review of the Components 6
 2.1 Recognition 7
 2.2 Signalling 7
 2.3 Antimicrobial Peptides 8
3. From Flies to General Insights Across Insecta 9
 3.1 Species Diversity 9
 3.2 Timescale 10
 3.3 Bottom-up Studies on Candidate Genes 10
 3.4 Candidate Gene Biases 12
 3.5 Example for Possible New Insights 13
 3.6 Epistasis: On the Importance of Genetic Backgrounds 13
 3.7 Genome-Wide Top-Down Studies Reveal Novel Insights 15
 3.8 Antibacterial Function of the Cuticle and Peritrophic Matrix 17
 3.9 Haemocyte Insights: The Early Start of Insect Immunology 18
 3.10 The Immune Function of Haemolymph Clots 20
4. Conclusion 22
Acknowledgements 22
References 22

Advances in Insect Physiology, Volume 52
ISSN 0065-2806
http://dx.doi.org/10.1016/bs.aiip.2017.02.003

Abstract

We propose that an evolutionary and phenotype-driven approach, harnessing current technological developments, has much to offer for our understanding of insect immunity. After briefly reviewing the history of the discovery of canonical immune system, the current understanding of its components is reviewed and then we argue that the current paradigm of research may be biassed due to (a) its limited taxonomic perspective, (b) the evolutionary time scale being studied, and (c) a focus primarily if not exclusively, upon the canonical, humoural gene set. For the rest of the review, we then discuss the importance of a phenotype down approach as an understudied perspective, exemplified by the need for understanding the basis of cellular responses and wounding as a source of selection on immunity in the wild. We propose that research on those topics almost certainly will provide new insights into the evolution of the insect immune system.

1. INTRODUCTION

1.1 Goals of This Review

Discovery of the insect immune system began with the study of humoural responses and cell morphology, followed by identifying proteins induced by infections. Together with the advent of the genomics era, these findings led to the discovery of immune-specific genes and their regulatory pathways. Today, evolutionary studies of insect immunology are dominated by a gene-specific focus, at both the level of individual gene products and their interaction, or studies of the entire recognized set of immune genes (i.e. the recognition, signalling, and effector molecules), hereafter referred to as the 'canonical set'. While the study of the canonical set, via gene expression analysis following induction or sequence level studies focused upon the signatures of evolutionary dynamics, has taught us much about insect antimicrobial defences, we propose that such approaches are also inherently biassed in several ways.

After briefly reviewing the history of the canonical set's discovery, we review the current understanding of its components. We then argue that the current paradigm of research into the evolutionary dynamics acting on the immune system may be biassed due to its limited taxonomic perspective, the evolutionary time scale being studied, and a focus primarily if not exclusively, upon the canonical, humoural gene set. For the rest of the review, we then discuss the importance of a phenotype down approach as an understudied perspective, exemplified by the need for understanding the basis of cellular responses and wounding as a source of selection on immunity in the wild. We propose that research on those topics almost certainly will provide new insights into the evolution of the insect immune system.

1.2 A Brief History of Insect Antimicrobial Defences

Insect humoural immunity had been known to exist since the 19th century but the first antimicrobial peptides (AMPs) were not isolated until the 1970s by Hans Boman and his coworkers, who immunized larvae with bacteria and identified several immune-induced proteins, most of them AMPs (Boman et al., 1972, 1974; Faye et al., 1975, summarized in Faye and Lindberg, 2016). One of the first AMPs characterized molecularly was named cecropin, after its source, the moth *Hyalophora cecropia* (Hultmark et al., 1980; Steiner et al., 1981). A cecropin homologue was discovered afterwards in *Drosophila melanogaster* (Samakovlis et al., 1990), and AMPs have since been found across all multicellular life and are suggested to have ancient origins (Zasloff, 2002). The Toll pathway, known for its function in establishing dorso-ventral polarity in fly embryos (Anderson and Nusslein-Volhard, 1984; Nusslein-Volhard et al., 1987; Steward, 1987), was shown to activate AMPs, such as the antifungal peptide Drosomycin (Faye and Lindberg, 2016; Lemaitre et al., 1996). Toll activation leads to the nuclear translocation of the NF-κB-like transcription factors Dorsal or DIF (dorsal-related immunity factor, Ip et al., 1993; Lemaitre et al., 1995a), which in turn activates transcription of AMPs and additional immune-related genes. Both fungal (Gottar et al., 2006) and bacterial (Bischoff et al., 2004; Buchon et al., 2009; Gobert et al., 2003; Michel et al., 2001) activators converge on the proteolytic activation of the procytokine Spätzle (Schneider et al., 1994), which enables it to bind to Toll. Activation of Spätzle is regulated positively and negatively (El Chamy et al., 2008; Jang et al., 2006; Kambris et al., 2006; Ligoxygakis et al., 2002; Robertson et al., 2003) and can be activated by damage signals created by fungi that invade the host (Gottar et al., 2006; Ligoxygakis et al., 2002; Ming et al., 2014, summarized in Lindsay and Wasserman, 2014). The discovery of *Drosophila* Toll was soon followed by the discovery of Toll-like receptors in vertebrates (Medzhitov and Janeway, 1998; Poltorak et al., 1998), suggesting similar ancient origin. In insects, a second pathway (immunodeficiency (IMD) pathway) mediates the induction of an even larger set of AMPs. The IMD pathway was originally discovered due to its defects in AMP induction (Lemaitre et al., 1995b), followed by the characterization of its intra- and extracellular components (Choe et al., 2002; Gottar et al., 2002; Lu et al., 2001; Rämet et al., 2002; Rutschmann et al., 2000; Silverman et al., 2000; reviewed in Kleino and Silverman, 2014), which include a third NF-κB-like transcription factor (Relish: Dushay et al., 1996; Hedengren et al., 1999). Activation of both pathways involves recognition of microbial elicitors via pattern-recognition receptors (PRRs) such as peptidoglycan-recognition proteins (PGRPs) or

Gram-negative-binding proteins (GNBPs, see also later). Canonically, Toll signalling is activated by fungal β-1,3-glucans or Lys(lysine)-type peptidoglycan, which is present on most Gram+ bacteria and IMD signalling by DAP(meso-diaminopimelic acid)-type peptidoglycan, which is typical for Gram− and some Gram+ bacteria.

Transcriptome studies of insects immunized with different elicitors showed that in addition to AMPs, dozens of other genes were induced upon immunization (De Gregorio et al., 2001; Irving et al., 2001). Microarrays and RNA sequencing on RNA obtained from *Drosophila* adults at various times postinfection revealed induction of members from several gene families. Importantly, a large fraction of genes not previously implicated in immunity were found. The Toll and IMD pathways showed distinct but partially overlapping patterns of induction (De Gregorio et al., 2002). Similarly, comparison between responses to different microbes showed overlapping gene sets, along with sets of genes that were induced due to specific infections. Further refinements of transcriptome studies included the use of bacteria that are natural pathogens or more closely related to them such as *Erwinia carotovora carotovora*, *Enterococcus faecalis*, *Serratia marcescens*, and *Providencia rettgeri* (Lazzaro et al., 2004; Muniz et al., 2007; Sackton et al., 2010; Vodovar et al., 2005). These studies led to increased resolution of pathogen-specific induction patterns and thereby gene sets and pathways. In some cases, variation in responses could even be related to polymorphisms in recognition molecules such as PGRP-SD (Sackton et al., 2010). Detailed understanding of these correlations identified in transcriptome studies relied mostly on the use of loss-of-function or gain-of-function mutants or knockdown experiments, thus establishing the canonical pathways (Lemaitre and Hoffmann, 2007).

Genome studies have greatly facilitated our understanding of the evolutionary dynamics acting upon genes related to the insect immune system. Microevolutionary insights came from detailed association studies between natural variation in *D. melanogaster* and immune performance (Lazzaro et al. 2004), with macroevolutionary insights gained via the completion of the genomes of several insect species, most notably the 12 *Drosophila* genomes publication (Clark et al., 2007). To date, the immune genes of diverse insects have been extensively characterized: 32 Diptera, 4 Hymenoptera, 3 Hemiptera, 2 Lepidoptera, 1 Coleoptera (Fig. 1 and Table 1). The increasing diversity of insect species with omic (i.e. transcriptome or genome) datasets now facilitates the study of immune function at both micro- and

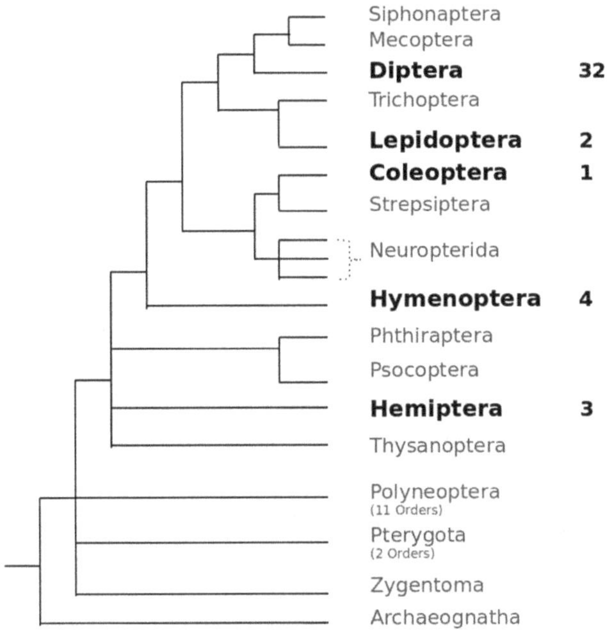

Fig. 1 Phylogenetic tree depicting the amount of species of which the canonical immune gene set has been identified. While many more transcriptome data are available, these have not been used for comparative studies of the canonical genes (perhaps due to their inherent biases). Tree based on Misof et al. (2014).

macroevolutionary scales. Importantly, these advances allow investigations to be driven by biological questions rather than feasibility alone, for applied (e.g. pests and vectors), model (*D. melanogaster*), as well as basic research questions in nonmodel species (e.g. ecological model species).

One of the most important contributions of the omics era to the study of the canonical gene set has been genome-wide analyses of gene family dynamics. Along with the influential *Drosophila* 12 genomes study (Clark et al., 2007), several insect species have been subject to studies on specific gene families, for example, the parasitic wasp *Nasonia vitripennis* and the lepidopteran *Manduca sexta* have their AMP genes reported (He et al., 2015; Tian et al., 2010). Together, these analyses show that there is considerable variation in the size and diversity of immune-related gene families across insect lineages. Immune genes are diverging as a result of expansions and contractions of gene families, and the host-pathogen arms race driven by positive selection, including both directional and balancing selection

Table 1 Insect Species on Which Genome-Wide Analysis on Immune Genes Have Been Performed

Insect Order	Species	References
Diptera	*Drosophila melanogaster*	Irving et al. (2001)
Diptera	12 *Drosophila* spp.	Sackton et al. (2007)
Diptera	*Musca domestica*	Scott et al. (2014)
Diptera	*Anopheles gambiae*	Christophides et al. (2002)
Diptera	16 *Anopheles* spp.	Waterhouse et al. (2007)
Diptera	*Aedes aegypti*	Neafsey et al. (2015)
Diptera	*Culex quinquefasciatus*	Bartholomay et al. (2010)
Lepidoptera	*Bombyx mori*	Tanaka et al. (2008)
Lepidoptera	*Plutella xylostella*	Xia et al. (2015)
Coleoptera	*Tribolium castaneum*	Zou et al. (2007)
Hymenoptera	*Apis mellifera*	Evans et al. (2006)
Hymenoptera	*Atta cephalotes*	Suen et al. (2011)
Hymenoptera	*Bombus impatiens*	Barribeau et al. (2015)
Hymenoptera	*Bombus terrestris*	Barribeau et al. (2015)
Hemiptera	*Acyrthosiphon pisum*	Gerardo et al. (2010)
Hemiptera	*Nilaparvata lugens*	Bao et al. (2013)
Hemiptera	*Rhodnius prolixus*	Mesquita et al. (2015)

(Viljakainen, 2015). Furthermore, a comparison of the genomes of *Drosophila* species showed that many immune genes evolve at a faster rate than non-immune genes (Sackton et al., 2007). Lineage-specific genes encoding recognition and effector proteins have been identified across diverse species, suggesting the emergence of evolutionary novelties via duplication dynamics is a general feature of importance in this system (Sackton et al., 2007).

2. THE CANONICAL IMMUNE GENES: A BRIEF REVIEW OF THE COMPONENTS

While there have been recent reviews on the canonical immune system (e.g. Hillyer, 2016; Kounatidis and Ligoxygakis, 2012; Rolff and Schmid-Hempel, 2016), here we highlight additional issues.

2.1 Recognition

Activation of the insect immune system begins with the recognition through the activation of PRRs. A variety of both cell surface and secreted PRRs are involved in the humoural immunity, such as PGRPs, β-1,3-glucan recognition proteins (βGRPs), and GNBPs. The PRRs involved in the cellular immune response are Nimrod family members, scavenger receptor proteins, and thioester-containing proteins (TEPs, see the review by Nazario-Toole and Wu in this issue for details). Despite being conserved from insects to mammals and having diverse functions in antimicrobial defence, families of PRRs show dynamic variation, especially in copy number across insect taxa. For example, the Nimrod gene family is highly divergent in *Musco domestica* (17 proteins) and *D. melanogaster* (11 proteins), whereas *Acyrthosiphon pisum* (pea aphid) have no Nimrod proteins (Gerardo et al., 2010).

The evolutionary dynamics differ between PRRs gene families depending on whether they are involved with cellular or humoural immunity. PGRPs, commonly involved with humoural immunity, are under purifying selection and highly conserved in *D. melanogaster* (Jiggins and Hurst, 2003). A comparison of 12 *Drosophila* genomes confirmed that PGRPs and GNBPs show few signs of positive selection with the exception of PGRP-LC and PGRP-LB (Sackton et al., 2007). The PGRP-LC is also involved with the cellular response (Rämet et al., 2002). Conversely, PRRs involved with cellular immunity were significantly more likely to show positive selection than nonimmune-related genes, of the 10 recognition genes that were found to be under positive selection, 9 were shown to be involved with phagocytosis (Sackton et al., 2007). Additional PRRs involved with the cellular response, such as scavenger receptors and TEPs also show signatures of positive selection in *Drosophila* (Jiggins and Kim, 2006; Lazzaro, 2005), *Anopheles* (Little and Cobbe, 2005), and *Bombus* (Barribeau et al., 2015).

The general patterns reveal that PRRs show considerable genomic rearrangement and adaptive sequence divergence. While they are evolutionarily stable on a short time scale, with the exception of receptors required for the cellular response, over longer evolutionary timescales the PRR gene families undergo considerable duplication and deletion events. This results in copy number variation among insect species (Lazzaro, 2008).

2.2 Signalling

Signalling genes are mostly present as single copy orthologos and species or lineage-specific paralogs, most likely to maintain functionality of the pathways (Kafatos et al., 2009). Both Toll and IMD signalling pathways are conserved across a wide range of insects, with the exception of the pea aphid *A. pisum* which appears to be missing crucial components of the IMD signalling pathway (Gerardo et al., 2010). In contrast, the *Drosophila* serine protease Grass, which is involved with the Toll pathway, appear to be subjected to divergent evolution (El Chamy et al., 2008). The core genes are orthologous conserved across long evolutionary distances. However, genes in the signalling pathways appear to be capable of evolving rapidly and adaptively at the amino acid sequence level (Lazzaro, 2008). For example, in the *D. melanogaster* lineage, a number of genes in the IMD pathway show accelerated rates of evolution and have experienced positive selection (Sackton et al., 2007). These adaptive substitutions appear to be located in protein domains responsible for release of the Relish transcription factor. This type of adaptive evolution in the Relish complex also occurs in *Nasutitermes* termites (Bulmer and Crozier, 2006).

2.3 Antimicrobial Peptides

Among insects there are both widespread (e.g. Defensins) and taxon-specific (e.g. abaecin in *Apis mellifera*) AMP families (Mylonakis et al., 2016). Generally speaking, AMPs appear to evolve slower compared to other immune genes (Unckless and Lazzaro, 2016), have undergone gene and exon duplication events, show rapid gene birth–death dynamics, and can arise via horizontal gene transfer or by de novo creation from noncoding sequences (Tennessen, 2005; Tian et al., 2010). Furthermore, exon shuffling has been indicated to be the mechanism behind the variability of the AMP defensin family and could be a more common mechanism in the evolution of AMP families (Froy and Gurevitz, 2003).

Although insect AMPs have traditionally been regarded as being under relaxed evolutionary constraints with no typical signatures of positive selection, a naturally occurring polymorphism in the AMP Diptericin sheds new light on insect AMP evolution: it is highly predictive of resistance to bacterial infections and convergent substitutions appears to have occurred at least five times across the genus *Drosophila* (Unckless et al., 2016). Further investigation throughout the insect clade revealed that adaptive maintenance of polymorphisms and balancing selection occurs more often in insect AMP genes and is suggested to be widespread (Unckless and Lazzaro, 2016). This is in line with the view on vertebrate AMPs, which show sequence variation in particular ancient families such as defensins (Maxwell et al., 2003).

Balancing selection on AMPs may be common across insects, due to the diversity of pathogens as well as interactions among AMPs (Unckless & Lazzaro, 2016). While observations of positive selection acting on AMPs in *Bombus* possibly exemplify different selection pressures operating across insect taxa (Barribeau et al., 2015; Erler et al., 2014), such observations may also arise due to changing selection dynamics over time and the instability of long-term balancing selection (Charlesworth, 2006).

It has been postulated that social immunity in colony-forming nests may make some branches, like AMPs, of immunity obsolete. In line with this, a reduction in AMP repertoire has been observed in the bee genome (Evans et al., 2006). There are opposite examples though, which indicate that eusocial insects such as ants have diversified their AMP repertoire (Zhang and Zhu, 2012). One has to also keep in mind that AMPs may have functions other than acting against microbes, including immune regulatory and wound-healing activity as well as inhibition of proteases (Mangoni et al., 2016).

3. FROM FLIES TO GENERAL INSIGHTS ACROSS INSECTA

While insects are highly diverse, the majority of molecular insights is derived from detailed studies of few species and comparative work mostly within a single genus. The challenge is now to use this detailed knowledge from a few species to study the variation in immune function and the evolutionary scenarios in across Insecta.

3.1 Species Diversity

With over 1 million species spread over at least 30 orders, insects are the most diverse group of animals on Earth (Mora et al., 2011). Yet, we only have detailed insights of the immune system from a handful of species from Diptera and Lepidoptera, and lack such insights for the most diverse insect order, Coleoptera (Fig. 1; Hillyer, 2016). While this limitation has recently been reviewed, it is worth noting that although an increasing number of insect genomes will become available in the coming years, functional studies are likely to be very slow to follow. Additionally, it is reasonable to expect immune genes with high copy number variation to be in those regions of the genome most difficult to assemble accurately. Thus, the evolutionary importance of gene family dynamics to immune performance is likely to remain challenging for some time.

One way to overcome or interact positively with these biases is to use current insights for generating specific tests. If these are conducted across taxa in a way that samples a wide taxonomic breadth, the strength of general insights from our current understanding can be quantified. For example, recent comparative work has investigated specific components of the canonical set across the arthropods. Interestingly, several immune genes exhibited large copy number variation across species, suggesting a potential change in function (Palmer and Jiggins, 2015) and thereby generating a hypothesis that is now ripe for testing.

3.2 Timescale

Here, we draw attention to a macroevolutionary (among species) temporal bias in studies of evolutionary dynamics acting on immune genes. While most recent attention has fallen upon the AMP genes, several studies have focused across the entire canonical set, e.g. *Drosophila* (Sackton et al., 2007), *Anopheles* (Neafsey et al., 2015; Waterhouse et al., 2007), and *Apis* (Barribeau et al., 2015) on a macroevolutionary scale. These studies gain their power from patterns of variation and divergence in coding regions over longer evolutionary time. Unfortunately, in general, there have been few studies on short-term, microevolutionary dynamics, such as population differences and recent adaptation to local pathogens and environmental conditions. For example, a study in *Anopheles gambiae* revealed striking differences in immune system genes between different wild populations, hypothesized to be a result of recent ecological niche specialization (Crawford et al., 2012). While macroevolutionary studies detect differences in the immune system (Clark et al., 2007), these must manifest initially at the microevolutionary scale (e.g. Lazzaro et al., 2004). Thus, determining the spatial and temporal scale over which populations are sufficiently immunologically challenged for an evolutionary response to occur, resulting in local adaptation, is an open question.

3.3 Bottom-up Studies on Candidate Genes

One general way to characterize studies that investigate how organismal phenotypes evolve is to divide them in those that begin with candidate genes (bottom-up) vs those that start with the phenotypes (top-down) (Fig. 2; Rogers and Bernatchez, 2007). While the discovery of the insect immune

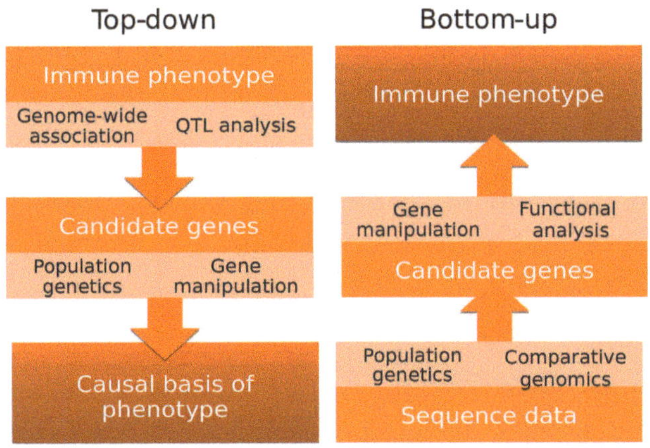

Fig. 2 Schematic of the top-down, phenotype first approach (*left*) and the bottom-up (*right*) candidate gene approach.

system began with *top-down* approaches (Boman et al., 1972), today the vast majority of studies employ a strong *bottom-up* approach given their primary focus upon variation in the canonical gene set and its potential evolutionary importance. With the obvious exception of *Drosophila* and *Anopheles*, these bottom-up studies often lack accompanying functional studies.

Bottom-up approaches, while potentially powerful, lend themselves to biassed insights for several reasons. First, sequence-based studies coupled with molecular tests of selection have a high false-positive rate coupled with inherently low power for realistic demographic scenarios in the wild (Pritchard and Di Rienzo, 2010). Entirely sequence-based approaches are therefore best considered hypothesis generators that warrant phenotypic and functional level studies. In the absence of performance differences in immunity between populations, sequence differences between populations in the canonical immune genes have questionable meaning at best. Conversely, difference in genes without immunity-related sequence annotations could nevertheless indicate a significant difference in immune performance between studied groups (Unckless et al., 2015), highlighting the inherent bias arising from annotation-based inferences. Second, there is a high likelihood that phenotypic variation in immune performance with evolutionary consequences may arise from variation outside of the canonical immune genes. To the degree that this is true, bottom-up studies on the canonical gene sets will fail to detect important components. In sum, while the bottom-up approach has been great for revealing and understanding the components of the canonical set and how they interact with one another, if our goal is to understand how immunity evolves on both a micro- and macroevolutionary time scale, more top-down studies are warranted. Ideally, a synergy will emerge between organismal performance and Darwinian fitness, biochemical performance and genome-wide inferences of selection dynamics of the corresponding genes (e.g. Ecological and Evolutionary Functional Genomics (Feder and Mitchell-Olds, 2003).

3.4 Candidate Gene Biases

At a fundamental level, the insect innate immune system consists of two mechanisms: the humoural and the cellular response. Humoural defences consist of the production of AMPs, which are produced through the signalling pathways mentioned earlier. Cellular defences of insects are mediated through their blood cells (haemocytes, Box 2) and consist of responses like the engulfment and digestion of the pathogen (phagocytosis), envelopment of parasites (encapsulation and nodulation), melanization (Box 1), and clotting (Lemaitre and Hoffmann, 2007). To date, the vast majority of studies investigating evolutionary dynamics of insect immunity have focused almost entirely on genes involved primarily in the humoural response (e.g. Sackton et al., 2007).

Several factors complicate studies of the evolutionary dynamics acting upon the cellular response. First, haemocytes are highly variable across taxa, as they play differing roles across species in the cellular response via recognition, modulation, and effectors. For example, lamellocytes are involved in melanization response for only specific lineages within *Drosophila* (Salazar-Jaramillo et al., 2014). Second, not only does this greatly complicate assignment of homology at the cellular level, but also for the genes uniquely functioning within these different haemocytes. For example, while a comparative study across 11 *Drosophila* species revealed that genes known to be associated with haematopoiesis are highly conserved, 11 novel genes were identified in the *Drosophila* lineages capable of lamellocyte-mediated encapsulation (Salazar-Jaramillo et al., 2014). This highlights that there are many unknown players in the cellular immune response that potentially experience significant selection pressures. Certainly such genes will be missed when working with only the canonical gene set, making it difficult to address if genes underlying the humoural and cellular response experience similar selection pressures.

BOX 1 Immune Response: Melanization

Besides the humoural and cellular immune response, insects also possess the capability to melanize their microbial invaders. Melanization is a major effector mechanism against pathogens, viruses, and parasites, either as part of humoural immunity or in haemolymph clots (Fig. 3), and is controlled by the enzyme phenoloxidase (PO) (Shelby and Popham, 2006; Valadez-Lira et al., 2011; Wyatt, 1961). Melanin works against pathogens in two ways: (1) it prevents growth and acquisition of nutrients by invaders and (2) during the production of melanin, highly reactive and cytotoxic intermediates are produced, such as phenols, quinones, and reactive oxygen species, which can directly kill pathogens (Cerenius and Söderhäll, 2004; Nappi and Christensen, 2005). Phenoloxidases are present as inactive enzyme precursors (zymogens) in haemocytes (proPO), and activation is achieved through PRRs, serine proteases, and serine protease inhibitors (Serpin) after pathogen detection (Cerenius et al., 2008). Quinones polymerize to form melanin, which is used in the immune response nodulation and encapsulation.

The prophenoloxidase cascade is activated by certain PGRPs. The bacterial PGN is recognized by PGRP-LE in *Drosophila*, but in both *Bombyx mori* and *Tenebrio molitor* the PGRP-S molecule activates this cascade. In *Holotrichia diomphalia*, the PGRP-1 also recognizes fungal β-glucan, activating the prophenoloxidase cascade (Lee, 2003).

All insect species have at least one proPO gene and can have as many as 10 (mosquitoes). Phylogenetic analyses of the known proPO genes revealed that there are three major insect proPO clades. Clade A contains conserved proPO genes found among a wide variety of insect orders, and proPO genes of clade B and C can only be found in *Lepidoptera* and *Diptera*, respectively (Lu et al., 2014).

3.5 Example for Possible New Insights

For the rest of the review, we present our perspective of a way to move forward in understanding the evolutionary dynamics acting upon the immune system of insects. We draw attention to emerging and less studied fields, as they highlight the importance and potential power of moving from the phenotype down to the genetic basis, which is likely to uncover important dynamics among canonical genes, as well as noncanonical genes that have impacts on immune performance.

3.6 Epistasis: On the Importance of Genetic Backgrounds

Broadly speaking, studies of the immune system have used a limited set of species (e.g. *D. melanogaster*), limited genetic backgrounds, and have focused primarily on functional analysis of knockdown lines and mutants in

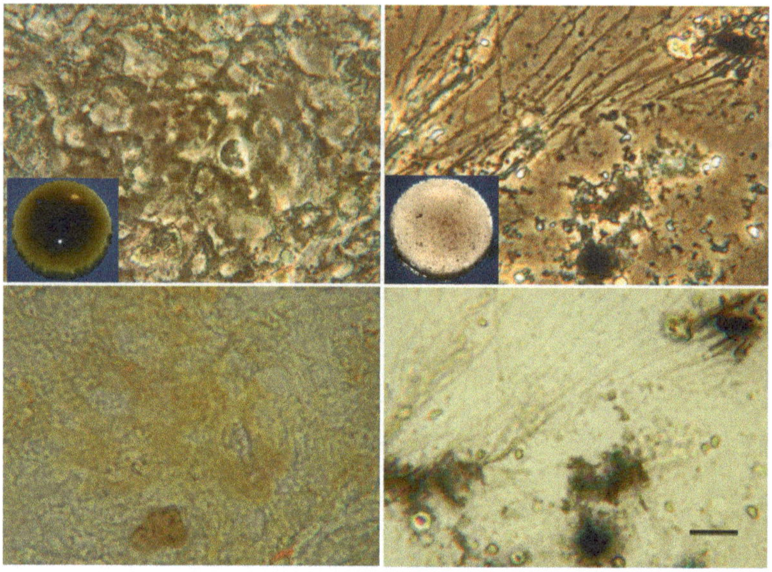

Fig. 3 Activation of PO in the Lepidoptera *Pararge aegeria* and Diptera *Anopheles* haemolymph clots. Haemolymph from *Pararge aegeria* (*left*) and *Anopheles gambiae* (*right*) was bled onto multiwell slides (a single well has a diameter of 4 mm), allowed to melanize for 30 min and either photographed directly (insets in *upper part*) or clots prepared using the hanging drop method and analysed microscopically either using phase contrast (*upper part*) or transmission light (*lower part*). Note the difference in clot structure as well as the strong but diffuse melanization in *Pararge aegeria* compared to the strongly localized melanin islands in *Anopheles gambiae*. Scale bar = 10 μm.

individual genes and pathways. This was facilitated by the fact that mutations in many immune genes do not show developmental phenotypes. While this has been a necessary and productive avenue of research for uncovering the primary components of the innate immune system, such experiments face challenges when trying to understand the role and impact of genetic variation in the evolutionary dynamics generally acting upon the immune system. This is partly due to the pleiotropic function of many 'immune' genes and to an increasingly recognized interaction of immunity with metabolic and physiological pathways (Owusu-Ansah and Perrimon, 2014).

Gene–gene interactions underlie complex phenotypes such as immune performance, and are highly influenced by epistatic interactions (Mackay, 2014). Epistasis between two genes occurs when variation at one gene affects the function of a second gene. Therefore, phenotypic effects of variation at one locus depend upon the variation at other loci. Since this variation can differ among individuals and populations, understanding and predicting genotype–phenotypic connections becomes very complex. Evidence for a significant role of epistasis in the function of the immune system is observed in the nonrepeatability of results across different lab strains, over time, or among populations.

The relevance of physiological epistasis for studies of natural populations is proportional to the allele frequencies of the interacting loci. If all the variants are rare, the effects are likely to be minimal since they will rarely occur within the same individual. However, if interacting alleles are at higher frequencies, their effects could substantially affect the evolution of allelic variation, and our ability to accurately quantify such effects (Sackton and Hartl, 2016).

While the majority of functional studies have focused upon specific genetic lines in a laboratory setting, evolutionary studies necessarily utilize the patterns of genetic variation in wild populations. Only recently have genome-wide studies of immune system performance been conducted across diverse genetic backgrounds (Sleiman et al., 2015; Unckless et al., 2015), thereby providing an unbiassed view of genes that could affect immunity.

3.7 Genome-Wide Top-Down Studies Reveal Novel Insights

Recently, two genome-wide scans of immune performance using the Drosophila Genome Reference Panel (DGRP) have found noncanonical genes with large effects, as well as evidence of epistasis in immune performance (Sleiman et al., 2015; Unckless et al., 2015). Genome-wide association analysis across the 140 lines assessed survival after postenteric infection by the entomopathogenic bacterium *Pseudomonas entomophila* (Sleiman et al., 2015). The most significant QTLs affecting immune performance were not part of the canonical immune response pathways. A nonsynonymous SNP in Neurospecific receptor kinase (*Nrk*) explained nearly 14% the phenotypic variance in survival and appears to affect the IMD pathway. The frequency of this minor allele is 15% within the panel, suggesting it is not a rare allele in the population from which the lines were derived. Another locus, Gyc76C, explained nearly 15% of the phenotypic variance and may induce the IMD pathway via its capability to activate and translocate Relish, which is an IMD transcription factor (Overend et al., 2012). Further indications of noncanonical gene effects were found in their RNA-Seq analysis of four susceptible and four resistant lines. PCA analysis on the 2000 genes with the highest expression changes between these groups and their controls identified 24 expression modules, only 1 of which had Gene Ontology (GO) annotations for the canonical immune pathways (Sleiman et al., 2015).

Beyond finding strong evidence for noncanonical genes of large effect, Sleiman et al. (2015) found that genetic background determines the response as well as outcome of infection. While lethality of *P. entomophila* infection was previously found to arise by translation inhibition in the gut reducing epithelial renewals, this was not observed here (Sleiman et al., 2015). Using these findings, the authors highlight the potential limitations of lab-based studies of specific strains to make general predictions in wild strains, indicating that epistasis degrades the predictive ability of immune performance from such genotypic information. In addition, the authors conclude that looking at higher levels of biological organization might be more informative, as the observed transcriptome phenotype was already different preinfection, for individuals that would go on to have resistant or susceptible phenotypes, again pointing to a very different set of genes involved in immune performance (ROS metabolism) than the canonical gene set.

Another DGRP study conducting an unbiassed genome-wide association investigation found that there were between 6 and 9 genes with SNPs showing diet-specific responses to infection (Unckless et al., 2015). Overall, bacterial infection by a natural pathogen (*P. rettgeri*) was more severe when reared on a high-glucose diet. While there were strong genome-wide associations between pathogen loads and variation in canonical genes (Diptericin, AMPs, thioester-containing protein 2), there were also substantial noncanonical findings. Importantly, several of these noncanonical observations were validated by RNAi knockdown that reduced pathogen loads (e.g. *multiplexin*, loss of function result in smaller fat bodies than wild-type flies, *defective proboscis extension response 6*, sensory perception of chemical stimulus, including gustatory perception of food), suggesting they impact immune responses via pleiotropic interactions. Top-down approaches also uncovered novel dynamics not previously seen, resulting in a more comprehensive story for some canonical genes. For example, Diptericin had a specific response to the pathogen used *P. rettgeri*, while previous studies did not find any AMP with specific responses to a range of previously tested bacterial strains (Unckless et al., 2015). Finally, the observed diet-dependent effects highlight the importance of studying organisms under a range of ecologically relevant conditions in order to understand the potential roles that existing genetic variation might have in relation to immune performance (Unckless et al., 2015).

Thus, working from phenotypes down to causal genotypes can provide more clarity than the reverse approach when working with natural variation. We highlight phenotypes that warrant more attention, moving away from the common humoural focus to that of the cellular immune response. We draw attention to the cuticle and wounding, which are at the interface between hosts and potential parasites.

3.8 Antibacterial Function of the Cuticle and Peritrophic Matrix

We argue that variation in the cuticle and the peritrophic membrane is likely to have consequences upon the most common routes of infection occurring in the wild. Studies focusing solely upon the canonical immune system may therefore miss both phenotypic and genetic variation in these components, which serve as a higher level of organization within which the canonical immune system functions. This issue is exacerbated by the common technique of injecting pathogens, which is practical and relates to some natural situations (wounding, see later), but circumvents the surface of the insect.

The insect cuticle, as well as the peritrophic matrix that covers the gut epithelium, provide protection against injuries and a large range of infections (Brey et al., 1993). To overcome these barriers, pathogens and parasites employ both mechanical tools, such as buccal teeth of entomopathogenic nematodes (Eleftherianos et al., 2010) and the ovipositor of parasitoid wasps (Schmidt et al., 2001), as well as enzymatic activities such as proteases and chitinases (Vodovar et al., 2005). The study by Vodovar and coworkers also shows that in addition to the peritrophic matrix, the underlying mucus layer may have a protective function and is targeted by the insect pathogenic *P. entomophila*. Several mucins are induced upon *P. entomophila* infection but their individual contribution to protection against gut pathogens remains to be determined (Buchon et al., 2013).

Upon infection, many hosts induce genes that code for components of both barriers including peritrophins and cuticular proteins. A recent example is the set of genes that are induced by the trypanosomal gut parasite *Crithidia bombi* in its host, the bumblebee *Bombus terrestris* (Riddell et al., 2014) or after infection with insect-pathogenic nematodes (Arefin et al., 2014; Kučerová et al., 2015). In addition to providing mechanical protection, the cuticle may also be covered with antibacterial secretions, for example, in social hymenopteran insects, either to protect themselves or as part of social immunity (Baracchi et al., 2012). In the latter case, the antibacterial secretions are also used to impregnate nests contributing to social immunity (Otti et al., 2014).

These observations of infection inducement of expression in components of both barriers suggest the host's attempt to buffer expected damage, which if successful could result in a successful immune response without potential involvement of the canonical gene set. Often the breaching of the cuticle results in a localized immune response (Haine et al., 2007; Scherfer et al., 2004). It has been speculated that this is related to the immune system reading the combination of MAMPs and danger signals (Lazzaro and Rolff, 2011). Once the cuticle has been breached, haemocytes form the first line of defence against infection.

3.9 Haemocyte Insights: The Early Start of Insect Immunology

With modern techniques, the versatile deployment of immune effector mechanisms reveals itself at the molecular level, but already decades ago the distinct morphologies of insect immune cells collectively called haemocytes indicated functional specialization (Gupta, 1984). Some of this variability is due to the multifunctional nature of haemocytes but even the same immune function appeared to be accomplished in different ways and by morphologically different cells (Gupta, 1984). Due to the fact that haemocytes were initially characterized based on morphological criteria, the nomenclature is not always consistent. One example for this is presented in Box 2 where we focus on lepidopteran and *Drosophila* haemocytes: the enzyme phenoloxidase, which is the key component of melanization during immune reactions (see Box 1 and Fig. 3) is harboured in crystalline form in *Drosophila* crystal cells (hence their name, Bidla et al., 2007; Lemaitre and Hoffmann, 2007) and without formation of visible crystals in lepidopteran haemocytes called oenocytoids (Gupta, 1984). Another example is the clotting of haemolymph after wounding, which has been shown to involve different cell types in different species and as a consequence the haemolymph clots look very distinct leading to their being grouped into different phenotypic classes (Gregoire, 1974). In part due to these difficulties, *Drosophila*, classification based on haemocyte morphology has been largely replaced by the use of molecular markers, in particular a set of monoclonal antibodies developed in the Ando group (Kurucz et al., 2007).

BOX 2 Insect Haemocyte Classes

The major classes of haemocytes are shown for *D. melanogaster* and for the order Lepidoptera. Immune functions for haemocytes comprise (1) uptake of objects by a single cell (phagocytosis, note that the frequency of phagocytic cells differs between species) (Lavine and Strand, 2002), (2) binding of several haemocytes to larger objects such as parasitoid eggs and nematodes (encapsulation), and (3) binding of several haemocytes to bacterial aggregates (nodulation). Coagulation is likely performed by the cells indicated although morphologically different coagulocytes have been described in some species (Gupta, 1984; Lavine and Strand, 2002). Melanization (see Box 1) occurs to varying degrees in nodules, capsules, and clots and involves release and activation of proPO (Bidla et al., 2009). Note the different nomenclature for lepidopterans and *Drosophila* (Ribeiro and Brehélin, 2006), for example, oenocytoids and crystal cells perform similar functions and *Drosophila* plasmatocytes are functionally related to both lepidopteran plasmatocytes, but also granulocytes. Encapsulation competence has been shown to correlate with lamellocyte counts both at the species (Fors et al., 2014; Salazar-Jaramillo et al., 2014) and population level (Fors et al., 2016). Phagocytosis has been shown to be regulated by the circadian clock with peak levels during night-time (Stone et al., 2012). Of note, components of the circadian clock are strongly affected during diapause both at the gene and transcriptional level (Kucerova et al., 2016).

	Haemocyte Class	Function
Lepidoptera	Prohaemocyte	Haematopoietic stem cells
	Spherule cell	Transport (cuticular components)
	Oenocytoid	Contain proPO
	Granular cell	Adhesive, phagocytosis, nodulation, encapsulation, coagulation
	Plasmatocyte	Adhesive, phagocytosis, nodulation, encapsulation, coagulation
Drosophila melanogaster	Prohaemocyte	Haematopoietic stem cells
	Plasmatocytes	Phagocytosis
	Lamellocytes	Encapsulation
	Crystal cells	Contain proPO

3.10 The Immune Function of Haemolymph Clots

To the best of our knowledge, the frequency of wounding has not been examined rigorously in the wild. Anecdotal evidence suggests that wounding is rather frequent. Males of many species, for example, inflict wounds on females during copulation. This includes obvious examples such as traumatic insemination (reviewed in Reinhardt et al., 2015), but also more cryptic internal wounding in the *D. melanogaster* group (Kamimura, 2007). While not directly reporting wounding, extensive studies on host resistance against parasitoids and melanized parasitoid eggs inside adult flies suggest that parasitoid attack is a frequent source of wounding (Kraaijeveld et al., 1998). Attack by predators either directly or indirectly additionally contributes to wounding, for example, spider webs or bird attacks may well be a source of frequent wounds for insects in the wild. A study on adults of the damselfly *Platycnemis pennipes* (Martens, 1992) revealed that adults have on average five legs in the wild. Taken together, these examples indicate that successfully responding to wounding events contributes to selection on immune function in insects.

In the event that cuticular barriers are breached a series of reactions at the biochemical, cellular, and transcriptome level are set in motion with the ultimate goal to restore epithelial integrity. Initially to prevent bleeding, a haemolymph clot forms at the wound site. This is followed by cellular rearrangements leading to the reestablishment of a functional epithelium and ultimately wound healing. When insect clots started to be studied at the molecular level, the observed phenotypic variability among species (Fig. 3; Gregoire, 1955) appeared to be corroborated: some clot components were found specific for the *Drosophila* lineage, others such as hemolectin were shared among insects or arthropods (phenoloxidase, Theopold et al., 2004). Only the cross-linking enzyme transglutaminase has been found to contribute to clot formation across all animal phyla, including humans, where clotting factor XIIIa helps in hardening of the blood clot (Wang et al., 2010).

In both insects and humans, transglutaminase serves an immune function by blocking entry of microbes or keeping them at the wound site thus preventing septicemia (Loof et al., 2011a,b; Theopold et al., 2014). Bacteria in *Drosophila* clots die at around 1 h after clot formation but the underlying mechanism is unknown (Bidla et al., 2009). One possibility is that, similar to blood clotting, proteolytic activation of clot formation in insects leads to the release of peptide fragments with bactericidal activity (Kasetty et al., 2011).

Interestingly, a protease (CG11313), which is activated during clotting (Karlsson et al., 2004) has also been found induced upon infestation by a parasitoid wasp (Wertheim et al., 2005) and lacks a Clip domain in a *Drosophila* species that is incapable of encapsulation, which is the standard immune reaction against wasp eggs (Salazar-Jaramillo et al., 2014). This lends molecular support to the old observation that clotting, encapsulation of larger foreign objects, and nodule formation against larger number of bacteria bear many similarities at the histological level (Ratcliffe and Gagen, 1977).

Recent work on chrysomelid beetles, collected from a variety of field sites in Sweden, identified a haemocyte type with similar fragility as the previously described coagulocytes (Fors et al., 2014). With the knowledge obtained from vertebrate models one may speculate that this 'coagulocyte' uses similar mechanisms as neutrophils upon the release of neutrophil extracellular traps, as has been suggested (Altincicek et al., 2008). To restore epithelial integrity, haemolymph clots are replaced with a new epithelial layer within one or a few days. Due to their genetic tractability and transparent nature, studies of wound healing and subsequent regenerative processes in insect have provided insights into the mechanisms that restore epithelia (Razzell et al., 2011; Stramer and Dionne, 2014).

Regeneration of both the cuticular epidermis and the gut were studied in detail in *Drosophila*. The cuticular epidermis reacts initially by cellular rearrangements, which include the formation of an actin-based purse string-like structure which helps to contract the wound and/or the formation of filopodia and lamellipodia, which extend into the wound and aid in wound closure (Razzell et al., 2011). This is followed by differential transcriptional induction in the cells at wound edges and neighbouring cells (Stramer and Dionne, 2014). In contrast, if the gut epithelium is damaged chemically, mechanically, or due to infection, the loss of enterocytes leads to proliferation of stem cells to replenish the epithelial cellular pool (Buchon et al., 2013; Erkosar and Leulier, 2014; Guo et al., 2016; You et al., 2014). Both regeneration of the cuticular and the gut epithelia restore their protective function. In addition, the gut epithelium serves to preserve a healthy microbiota (Buchon et al., 2013; Erkosar and Leulier, 2014; Guo et al., 2016; You et al., 2014).

Some of the molecular mechanisms that underlie wound healing appear to be conserved, for example, both *Drosophila* and zebrafish rely on reactive oxygen species (hydrogen peroxide) as a signal that activates wound healing (Stramer and Dionne, 2014). There are differences even when one looks at different stages of the same animal, for example, *Drosophila* embryos heal wounds in the absence of a clot, while clots initially seal wounds in larvae (Razzell et al., 2011). This indicates that wound sealing and healing may be evolutionary malleable, although detailed comparative studies are still rare.

4. CONCLUSION

While we have a very good understanding of the molecular biology of immune defences in some model species such as *Drosophila* and *Anopheles*, we lack such insights for the vast majority of insect species. Moreover, knowledge about the nature of selection shaping the evolution of immune responses insects is very limited. Here, we argue that using a top-down, phenotype first approach has the potential to yield a much deeper understanding of the malleability of immune responses. Top-down approaches are feasible for many species, whether they are of interest from an applied or basic research perspective. One interesting, yet hardly studied phenotype is wound repair in the wild, where the existing thorough understanding of wound repair from *Drosophila* can serve as a basis for wider investigations. We would like to highlight that while we focused on areas here that we feel will result in new insights on the evolution of insect immune defences; there are many other very worthwhile approaches. These include the study of tolerance of infections (Louie et al., 2016), the interactions between reproduction and immune function (Schwenke et al., 2016), and the role of nutrition (Lee et al., 2006) to name but a few.

ACKNOWLEDGEMENTS

The authors' work is supported by the Swedish Research Council (VR-2010-5988 to U.T.), the Swedish Foundation for International Cooperation in Research and Higher Education (STINT, IG2011-2042 to U.T.), the Knut and Alice Wallenberg Foundation (KAW2012.0058) and the Swedish Cancer Foundation (CAN 2010/553 to U.T.), the European Research Council (EVORESIN to J.R.), and the Deutsche Forschungsgemeinschaft (SFB 973 to J.R.). We thank Thomas Hauling for preparing Fig. 3.

REFERENCES

Altincicek, B., Stotzel, S., Wygrecka, M., Preissner, K.T., Vilcinskas, A., 2008. Host-derived extracellular nucleic acids enhance innate immune responses, induce coagulation, and prolong survival upon infection in insects. J. Immunol. 181, 2705–2712.

Anderson, K.V., Nusslein-Volhard, C., 1984. Information for the dorsal-ventral pattern of the *Drosophila* embryo is stored as maternal mRNA. Nature 311, 223–227.

Arefin, B., Kucerova, L., Dobes, P., Markus, R., Strnad, H., Wang, Z., Hyrsl, P., Zurovec, M., Theopold, U., 2014. Genome-wide transcriptional analysis of Drosophila larvae infected by entomopathogenic nematodes shows involvement of complement, recognition and extracellular matrix proteins. J. Innate Immun. 6, 192–204.

Bao, Y.-Y., Qu, L.-Y., Zhao, D., Chen, L.-B., Jin, H.-Y., Xu, L.-M., Cheng, J.-A., Zhang, C.-X., 2013. The genome- and transcriptome-wide analysis of innate immunity in the brown planthopper, *Nilaparvata lugens*. BMC Genomics 14, 1–22.

Baracchi, D., Mazza, G., Turillazzi, S., 2012. From individual to collective immunity: the role of the venom as antimicrobial agent in the *Stenogastrinae* wasp societies. J. Insect Physiol. 58, 188–193.

Barribeau, S.M., Sadd, B.M., du Plessis, L., Brown, M.J.F., Buechel, S.D., Cappelle, K., Carolan, J.C., Christiaens, O., Colgan, T.J., Erler, S., Evans, J., Helbing, S., Karaus, E., Lattorff, H.M.G., Marxer, M., Meeus, I., Naepflin, K., Niu, J., Schmid-Hempel, R., Smagghe, G., Waterhouse, R.M., Yu, N., Zdobnov, E.M., Schmid-Hempel, P., 2015. A depauperate immune repertoire precedes evolution of sociality in bees. Genome Biol. 16 (1), 83.

Bartholomay, L.C., Waterhouse, R.M., Mayhew, G.F., Campbell, C.L., Michel, K., Zou, Z., Ramirez, J.L., Das, S., Alvarez, K., Arensburger, P., Bryant, B., Chapman, S.B., Dong, Y., Erickson, S.M., Karunaratne, S.H.P.P., Kokoza, V., Kodira, C.D., Pignatelli, P., Shin, S.W., Vanlandingham, D.L., Atkinson, P.W., Birren, B., Christophides, G.K., Clem, R.J., Hemingway, J., Higgs, S., Megy, K., Ranson, H., Zdobnov, E.M., Raikhel, A.S., Christensen, B.M., Dimopoulos, G., Muskavitch, M.A.T., 2010. Pathogenomics of *Culex quinquefasciatus* and meta-analysis of infection responses to diverse pathogens. Science 330, 88–90.

Bidla, G., Dushay, M.S., Theopold, U., 2007. Crystal cell rupture after injury in *Drosophila* requires the JNK pathway, small GTPases and the TNF homolog Eiger. J. Cell Sci. 120, 1209–1215.

Bidla, G., Hauling, T., Dushay, M.S., Theopold, U., 2009. Activation of insect phenoloxidase after injury: endogenous versus foreign elicitors. J. Innate Immun. 1, 301–308.

Bischoff, V., Vignal, C., Boneca, I.G., Michel, T., Hoffmann, J.A., Royet, J., 2004. Function of the *Drosophila* pattern-recognition receptor PGRP-SD in the detection of gram-positive bacteria. Nat. Immunol. 5, 1175–1180.

Boman, H.G., Nilsson, I., Rasmuson, B., 1972. Inducible antibacterial defence system in *Drosophila*. Nature 237, 232–235.

Boman, H.G., Nilsson-Faye, I., Paul, K., Rasmuson Jr., T., 1974. Insect immunity. I. Characteristics of an inducible cell-free antibacterial reaction in hemolymph of *Samia cynthia* pupae. Infect. Immun. 10, 136–145.

Brey, P.T., Lee, W.J., Yamakawa, M., Koizumi, Y., Perrot, S., Francois, M., Ashida, M., 1993. Role of the integument in insect immunity: epicuticular abrasion and induction of cecropin synthesis in cuticular epithelial cells. Proc. Natl. Acad. Sci. U. S. A. 90, 6275–6279.

Buchon, N., Poidevin, M., Kwon, H.M., Guillou, A., Sottas, V., Lee, B.L., Lemaitre, B., 2009. A single modular serine protease integrates signals from pattern-recognition receptors upstream of the *Drosophila* toll pathway. Proc. Natl. Acad. Sci. U. S. A. 106, 12442–12447.

Buchon, N., Broderick, N.A., Lemaitre, B., 2013. Gut homeostasis in a microbial world: insights from *Drosophila melanogaster*. Nat. Rev. Microbiol. 11, 615–626.

Bulmer, M.S., Crozier, R.H., 2006. Variation in positive selection in termite GNBPs and Relish. Mol. Biol. Evol. 23, 317–326.

Cerenius, L., Söderhäll, K., 2004. The prophenoloxidase-activating system in invertebrates. Immunol. Rev. 198, 116–126.

Cerenius, L., Lee, B.L., Söderhäll, K., 2008. The proPO-system: pros and cons for its role in invertebrate immunity. Trends Immunol. 29, 263–271.

Charlesworth, D., 2006. Balancing selection and its effects on sequences in nearby genome regions. PLoS Genet. 2, e64.

Choe, K.M., Werner, T., Stoven, S., Hultmark, D., Anderson, K.V., 2002. Requirement for a peptidoglycan recognition protein (PGRP) in relish activation and antibacterial immune responses in *Drosophila*. Science 296, 359–362.

Christophides, G.K., Zdobnov, E., Barillas-Mury, C., Birney, E., Blandin, S., Blass, C., Brey, P.T., Collins, F.H., Danielli, A., Dimopoulos, G., Hetru, C., Hoa, N.T., Hoffmann, J.A., Kanzok, S.M., Letunic, I., Levashina, E.A., Loukeris, T.G., Lycett, G., Meister, S., Michel, K., Moita, L.F., Müller, H.-M., Osta, M.A., Paskewitz, S.M., Reichhart, J.-M., Rzhetsky, A., Troxler, L., Vernick, K.D., Vlachou, D., Volz, J., von Mering, C., Xu, J., Zheng, L., Bork, P., Kafatos, F.C., 2002. Immunity-related genes and gene families in *Anopheles gambiae*. Science 298, 159–165.

Clark, A.G., Eisen, M.B., Smith, D.R., Bergman, C.M., Oliver, B., Markow, T.A., Kaufman, T.C., Kellis, M., Gelbart, W., Iyer, V.N., Pollard, D.A., Sackton, T.B., Larracuente, A.M., Singh, N.D., Abad, J.P., Abt, D.N., Adryan, B., Aguade, M., Akashi, H., Anderson, W.W., Aquadro, C.F., Ardell, D.H., Arguello, R., Artieri, C.G., Barbash, D.A., Barker, D., Barsanti, P., Batterham, P., Batzoglou, S., Begun, D., Bhutkar, A., Blanco, E., Bosak, S.A., Bradley, R.K., Brand, A.D., Brent, M.R., Brooks, A.N., Brown, R.H., Butlin, R.K., Caggese, C., Calvi, B.R., Bernardo de Carvalho, A., Caspi, A., Castrezana, S., Celniker, S.E., Chang, J.L., Chapple, C., Chatterji, S., Chinwalla, A., Civetta, A., Clifton, S.W., Comeron, J.M., Costello, J.C., Coyne, J.A., Daub, J., David, R.G., Delcher, A.L., Delehaunty, K., Do, C.B., Ebling, H., Edwards, K., Eickbush, T., Evans, J.D., Filipski, A., Findeiß, S., Freyhult, E., Fulton, L., Fulton, R., Garcia, A.C.L., Gardiner, A., Garfield, D.A., Garvin, B.E., Gibson, G., Gilbert, D., Gnerre, S., Godfrey, J., Good, R., Gotea, V., Gravely, B., Greenberg, A.J., Griffiths-Jones, S., Gross, S., Guigo, R., Gustafson, E.A., Haerty, W., Hahn, M.W., Halligan, D.L., Halpern, A.L., Halter, G.M., Han, M.V., Heger, A., Hillier, L., Hinrichs, A.S., Holmes, I., Hoskins, R.A., Hubisz, M.J., Hultmark, D., Huntley, M.A., Jaffe, D.B., Jagadeeshan, S., Jeck, W.R., Johnson, J., Jones, C.D., Jordan, W.C., Karpen, G.H., Kataoka, E., Keightley, P.D., Kheradpour, P., Kirkness, E.F., Koerich, L.B., Kristiansen, K., Kudrna, D., Kulathinal, R.J., Kumar, S., Kwok, R., Lander, E., Langley, C.H., Lapoint, R., Lazzaro, B.P., Lee, S.-J., Levesque, L., Li, R., Lin, C.-F., Lin, M.F., Lindblad-Toh, K., Llopart, A., Long, M., Low, L., Lozovsky, E., Lu, J., Luo, M., Machado, C.A., Makalowski, W., Marzo, M., Matsuda, M., Matzkin, L., McAllister, B., McBride, C.S., McKernan, B., McKernan, K., Mendez-Lago, M., Minx, P., Mollenhauer, M.U., Montooth, K., Mount, S.M., Mu, X., Myers, E., Negre, B., Newfeld, S., Nielsen, R., Noor, M.A.F., O'Grady, P., Pachter, L., Papaceit, M., Parisi, M.J., Parisi, M., Parts, L., Pedersen, J.S., Pesole, G., Phillippy, A.M., Ponting, C.P., Pop, M., Porcelli, D., Powell, J.R., Prohaska, S., Pruitt, K., Puig, M., Quesneville, H., Ravi Ram, K., Rand, D., Rasmussen, M.D., Reed, L.K., Reenan, R., Reily, A., Remington, K.A., Rieger, T.T., Ritchie, M.G., Robin, C., Rogers, Y.-H., Rohde, C., Rozas, J., Rubenfield, M.J., Ruiz, A., Russo, S., Salzberg, S.L., Sanchez-Gracia, A., Saranga, D.J., Sato, H., Schaeffer, S.W., Schatz, M.C., Schlenke, T., Schwartz, R., Segarra, C., Singh, R.S., Sirot, L., Sirota, M., Sisneros, N.B., Smith, C.D., Smith, T.F., Spieth, J., Stage, D.E., Stark, A., Stephan, W., Strausberg, R.L., Strempel, S., Sturgill, D., Sutton, G., Sutton, G.G., Tao, W., Teichmann, S., Tobari, Y.N., Tomimura, Y., Tsolas, J.M., Valente, V.L.S., Venter, E., Craig Venter, J., Vicario, S., Vieira, F.G., Vilella, A.J.,

Villasante, A., Walenz, B., Wang, J., Wasserman, M., Watts, T., Wilson, D., Wilson, R.K., Wing, R.A., Wolfner, M.F., Wong, A., Ka-Shu Wong, G., Wu, C.-I., Wu, G., Yamamoto, D., Yang, H.-P., Yang, S.-P., Yorke, J.A., Yoshida, K., Zdobnov, E., Zhang, P., Zhang, Y., Zimin, A.V., Baldwin, J., Abdouelleil, A., Abdulkadir, J., Abebe, A., Abera, B., Abreu, J., Christophe Acer, S., Aftuck, L., Alexander, A., An, P., Anderson, E., Anderson, S., Arachi, H., Azer, M., Bachantsang, P., Barry, A., Bayul, T., Berlin, A., Bessette, D., Bloom, T., Blye, J., Boguslavskiy, L., Bonnet, C., Boukhgalter, B., Bourzgui, I., Brown, A., Cahill, P., Channer, S., Cheshatsang, Y., Chuda, L., Citroen, M., Collymore, A., Cooke, P., Costello, M., D'Aco, K., Daza, R., De Haan, G., DeGray, S., DeMaso, C., Dhargay, N., Dooley, K., Dooley, E., Doricent, M., Dorje, P., Dorjee, K., Dupes, A., Elong, R., Falk, J., Farina, A., Faro, S., Ferguson, D., Fisher, S., Foley, C.D., Franke, A., Friedrich, D., Gadbois, L., Gearin, G., Gearin, C.R., Giannoukos, G., Goode, T., Graham, J., Grandbois, E., Grewal, S., Gyaltsen, K., Hafez, N., Hagos, B., Hall, J., Henson, C., Hollinger, A., Honan, T., Huard, M.D., Hughes, L., Hurhula, B., Erii Husby, M., Kamat, A., Kanga, B., Kashin, S., Khazanovich, D., Kisner, P., Lance, K., Lara, M., Lee, W., Lennon, N., Letendre, F., LeVine, R., Lipovsky, A., Liu, X., Liu, J., Liu, S., Lokyitsang, T., Lokyitsang, Y., Lubonja, R., Lui, A., MacDonald, P., Magnisalis, V., Maru, K., Matthews, C., McCusker, W., McDonough, S., Mehta, T., Meldrim, J., Meneus, L., Mihai, O., Mihalev, A., Mihova, T., Mittelman, R., Mlenga, V., Montmayeur, A., Mulrain, L., Navidi, A., Naylor, J., Negash, T., Nguyen, T., Nguyen, N., Nicol, R., Norbu, C., Norbu, N., Novod, N., O'Neill, B., Osman, S., Markiewicz, E., Oyono, O.L., Patti, C., Phunkhang, P., Pierre, F., Priest, M., Raghuraman, S., Rege, F., Reyes, R., Rise, C., Rogov, P., Ross, K., Ryan, E., Settipalli, S., Shea, T., Sherpa, N., Shi, L., Shih, D., Sparrow, T., Spaulding, J., Stalker, J., Stange-Thomann, N., Stavropoulos, S., Stone, C., Strader, C., Tesfaye, S., Thomson, T., Thoulutsang, Y., Thoulutsang, D., Topham, K., Topping, I., Tsamla, T., Vassiliev, H., Vo, A., Wangchuk, T., Wangdi, T., Weiand, M., Wilkinson, J., Wilson, A., Yadav, S., Young, G., Yu, Q., Zembek, L., Zhong, D., Zimmer, A., Zwirko, Z., Jaffe, D.B., Alvarez, P., Brockman, W., Butler, J., Chin, C., Gnerre, S., Grabherr, M., Kleber, M., Mauceli, E., MacCallum, I., 2007. Evolution of genes and genomes on the *Drosophila* phylogeny. Nature 450, 203–218.

Crawford, J.E., Bischoff, E., Garnier, T., Gneme, A., Eiglmeier, K., Holm, I., Riehle, M.M., Guelbeogo, W.M., Sagnon, N., Lazzaro, B.P., Vernick, K.D., 2012. Evidence for population-specific positive selection on immune genes of *Anopheles gambiae*. G3 (Bethesda) 2, 1505–1519.

De Gregorio, E., Spellman, P.T., Rubin, G.M., Lemaitre, B., 2001. Genome-wide analysis of the *Drosophila* immune response by using oligonucleotide microarrays. Proc. Natl. Acad. Sci. U. S. A. 98, 12590–12595.

De Gregorio, E., Spellman, P.T., Tzou, P., Rubin, G.M., Lemaitre, B., 2002. The Toll and IMD pathways are the major regulators of the immune response in *Drosophila*. EMBO J. 21, 2568–2579.

Dushay, M.S., Asling, B., Hultmark, D., 1996. Origins of immunity: relish, a compound Rel-like gene in the antibacterial defense of *Drosophila*. Proc. Natl. Acad. Sci. U. S. A. 93, 10343–10347.

El Chamy, L., Leclerc, V., Caldelari, I., Reichhart, J.-M., 2008. Sensing of 'danger signals' and pathogen-associated molecular patterns defines binary signaling pathways "upstream" of Toll. Nat. Immunol. 9, 1165–1170.

Eleftherianos, I., ffrench-Constant, R.H., Clarke, D.J., Dowling, A.J., Reynolds, S.E., 2010. Dissecting the immune response to the entomopathogen *Photorhabdus*. Trends Microbiol. 18, 552–560.

Erkosar, B., Leulier, F., 2014. Transient adult microbiota, gut homeostasis and longevity: novel insights from the *Drosophila* model. FEBS Lett. 588, 4250–4257.

Erler, S., Lhomme, P., Rasmont, P., Lattorff, H.M.G., 2014. Rapid evolution of antimicrobial peptide genes in an insect host–social parasite system. Infect. Genet. Evol. 23, 129–137.

Evans, J.D., Aronstein, K., Chen, Y.P., Hetru, C., Imler, J.L., Jiang, H., Kanost, M., Thompson, G.J., Zou, Z., Hultmark, D., 2006. Immune pathways and defence mechanisms in honey bees *Apis mellifera*. Insect Mol. Biol. 15, 645–656.

Faye, I., Lindberg, B.G., 2016. Towards a paradigm shift in innate immunity—seminal work by Hans G. Boman and co-workers. Philos. Trans. R. Soc. B 371. 20150303.

Faye, I., Pye, A., Rasmuson, T., Boman, H.G., Boman, I.A., 1975. Insect immunity. 11. Simultaneous induction of antibacterial activity and selection synthesis of some hemolymph proteins in diapausing pupae of *Hyalophora cecropia* and *Samia cynthia*. Infect. Immun. 12, 1426–1438.

Feder, M.E., Mitchell-Olds, T., 2003. Evolutionary and ecological functional genomics. Nat. Rev. Genet. 4, 651–657.

Fors, L., Markus, R., Theopold, U., Hambäck, P.A., 2014. Differences in cellular immune competence explain parasitoid resistance for two coleopteran species. PLoS One 9, e108795.

Fors, L., Markus, R., Theopold, U., Ericson, L., Hambäck, P.A., 2016. Geographic variation and trade-offs in parasitoid virulence. J. Anim. Ecol. 85, 1595–1604.

Froy, O., Gurevitz, M., 2003. Arthropod and mollusk defensins-evolution by exon-shuffling. Trends Genet. 19, 684–687.

Gerardo, N.M., Altincicek, B., Anselme, C., Atamian, H., Barribeau, S.M., de Vos, M., Duncan, E.J., Evans, J.D., Gabaldón, T., Ghanim, M., Heddi, A., Kaloshian, I., Latorre, A., Moya, A., Nakabachi, A., Parker, B.J., Pérez-Brocal, V., Pignatelli, M., Rahbé, Y., Ramsey, J.S., Spragg, C.J., Tamames, J., Tamarit, D., Tamborindeguy, C., Vincent-Monegat, C., Vilcinskas, A., 2010. Immunity and other defenses in pea aphids, *Acyrthosiphon pisum*. Genome Biol. 11, R21.

Gobert, V., Gottar, M., Matskevich, A.A., Rutschmann, S., Royet, J., Belvin, M., Hoffmann, J.A., Ferrandon, D., 2003. Dual activation of the *Drosophila* toll pathway by two pattern recognition receptors. Science 302, 2126–2130.

Gottar, M., Gobert, V., Michel, T., Belvin, M., Duyk, G., Hoffmann, J.A., Ferrandon, D., Royet, J., 2002. The *Drosophila* immune response against Gram-negative bacteria is mediated by a peptidoglycan recognition protein. Nature 416, 640–644.

Gottar, M., Gobert, V., Matskevich, A.A., Reichhart, J.M., Wang, C., Butt, T.M., Belvin, M., Hoffmann, J.A., Ferrandon, D., 2006. Dual detection of fungal infections in *Drosophila* via recognition of glucans and sensing of virulence factors. Cell 127, 1425–1437.

Gregoire, C., 1955. Blood coagulation in arthropods. V. Studies on haemolymph coagulation in 420 species of insects. Arch. Biol. 66, 103–148.

Gregoire, C., 1974. Hemolymph coagulation. In: Rockstein, M. (Ed.), ThePhysiology of Insecta, 5. Academic Press, New York, pp. 309–360.

Guo, Z., Lucchetta, E., Rafel, N., Ohlstein, B., 2016. Maintenance of the adult *Drosophila* intestine: all roads lead to homeostasis. Curr. Opin. Genet. Dev. 40, 81–86.

Gupta, A.P., 1984. Cellular elements in the hemolymph. In: Kerku, G.A., Gilbert, L.I. (Eds.), Comprehensive Insect Physiology, Biochemistry and Pharmacology. Pergamon Press, Oxford, pp. 401–451.

Haine, E.R., Rolff, J., Siva-Jothy, M.T., 2007. Functional consequences of blood clotting in insects. Dev. Comp. Immunol. 31, 456–464.

He, Y., Cao, X., Li, K., Hu, Y., Chen, Y.-R., Blissard, G., Kanost, M.R., Jiang, H., 2015. A genome-wide analysis of antimicrobial effector genes and their transcription patterns in *Manduca sexta*. Insect Biochem. Mol. Biol. 62, 23–37.

Hedengren, M., Asling, B., Dushay, M.S., Ando, I., Ekengren, S., Wihlborg, M., Hultmark, D., 1999. Relish, a central factor in the control of humoral but not cellular immunity in *Drosophila*. Mol. Cell 4, 827–837.
Hillyer, J.F., 2016. Insect immunology and hematopoiesis. Dev. Comp. Immunol. 58, 102–118.
Hultmark, D., Steiner, H., Rasmuson, T., Boman, H.G., 1980. Insect immunity. Purification and properties of three inducible bactericidal proteins from hemolymph of immunized pupae of *Hyalophora cecropia*. FEBS J. 106, 7–16.
Ip, T.Y., Reach, M., Engstroem, Y., Kadalayil, L., Cai, H., Gonzalez-Crespo, S., Tatei, K., Levine, M., 1993. DIF, a dorsal-related gene that mediates an immune response in *Drosophila*. Cell 75, 753–763.
Irving, P., Troxler, L., Heuer, T.S., Belvin, M., Kopczynski, C., Reichhart, J.M., Hoffmann, J.A., Hetru, C., 2001. A genome-wide analysis of immune responses in *Drosophila*. Proc. Natl. Acad. Sci. U. S. A. 98, 15119–15124.
Jang, I.H., Chosa, N., Kim, S.H., Nam, H.J., Lemaitre, B., Ochiai, M., Kambris, Z., Brun, S., Hashimoto, C., Ashida, M., Brey, P.T., Lee, W.J., 2006. A Spatzle-processing enzyme required for toll signaling activation in *Drosophila* innate immunity. Dev. Cell 10, 45–55.
Jiggins, F.M., Hurst, G., 2003. The evolution of parasite recognition genes in the innate immune system: purifying selection on *Drosophila melanogaster* peptidoglycan recognition proteins. J. Mol. Evol. 57, 598–605.
Jiggins, F.M., Kim, K.-W., 2006. Contrasting evolutionary patterns in *Drosophila* immune receptors. J. Mol. Evol. 63, 769–780.
Kafatos, F., Waterhouse, R., Zdobnov, E., Christophides, G., 2009. Comparative genomics of insect immunity. In: Rolff, J., Reynolds, S.E. (Eds.), Insect Infection and Immunity: Evolution, Ecology, and Mechanisms. Oxford University Press, Oxford, pp. 86–105.
Kambris, Z., Brun, S., Jang, I.H., Nam, H.J., Romeo, Y., Takahashi, K., Lee, W.J., Ueda, R., Lemaitre, B., 2006. *Drosophila* immunity: a large-scale in vivo RNAi screen identifies five serine proteases required for toll activation. Curr. Biol. 16, 808–813.
Kamimura, Y., 2007. Twin intromittent organs of *Drosophila* for traumatic insemination. Biol. Lett. 3, 401–404.
Karlsson, C., Korayem, A.M., Scherfer, C., Loseva, O., Dushay, M.S., Theopold, U., 2004. Proteomic analysis of the *Drosophila* larval hemolymph clot. J. Biol. Chem. 279, 52033–52041.
Kasetty, G., Papareddy, P., Kalle, M., Rydengård, V., Walse, B., Svensson, B., Mörgelin, M., Malmsten, M., Schmidtchen, A., 2011. The C-terminal sequence of several human serine proteases encodes host defense functions. J. Innate Immun. 3, 471–482.
Kleino, A., Silverman, N., 2014. The *Drosophila* IMD pathway in the activation of the humoral immune response. Dev. Comp. Immunol. 42, 25–35.
Kounatidis, I., Ligoxygakis, P., 2012. *Drosophila* as a model system to unravel the layers of innate immunity to infection. Open Biol. 2, 120075.
Kraaijeveld, A.R., Van Alphen, J.J., Godfray, H.C., 1998. The coevolution of host resistance and parasitoid virulence. Parasitology 116, S29–S45.
Kučerová, L., Broz, V., Arefin, B., Maaroufi, H.O., Hurychova, J., Strnad, H., Zurovec, M., Theopold, U., 2015. The *drosophila* Chitinase-like protein IDGF3 is involved in protection against nematodes and in wound healing. J. Innate Immun. 8, 199–210.
Kucerova, L., Kubrak, O.I., Bengtsson, J.M., Strnad, H., Nylin, S., Theopold, U., Nässel, D.R., 2016. Slowed aging during reproductive dormancy is reflected in genome-wide transcriptome changes in *Drosophila melanogaster*. BMC Genomics 17, 50.
Kurucz, É., Váczi, B., Markus, R., Laurinyecz, B., Vilmos, P., Zsámboki, J., Csorba, K., Gateff, E., Hultmark, D., Andó, I., 2007. Definition of *Drosophila* hemocyte subsets by cell-type specific antigens. Acta Biol. Hung. 58, 95–111.

Lavine, M.D., Strand, M.R., 2002. Insect hemocytes and their role in immunity. Insect Biochem. Mol. Biol. 32, 1295–1309.
Lazzaro, B.P., 2005. Elevated polymorphism and divergence in the class C scavenger receptors of *Drosophila melanogaster* and *D. Simulans*. Genetics 169, 2023–2034.
Lazzaro, B.P., 2008. Natural selection on the *Drosophila* antimicrobial immune system. Curr. Opin. Microbiol. 11, 284–289.
Lazzaro, B.P., Rolff, J., 2011. Danger, microbes, and homeostasis. Science 332, 43–44.
Lazzaro, B.P., Sceurman, B.K., Clark, A.G., 2004. Genetic basis of natural variation in *D. melanogaster* antibacterial immunity. Science 303, 1873–1876.
Lee, M.H., 2003. Peptidoglycan recognition proteins involved in 1,3-β-D-glucan-dependent prophenoloxidase activation system of insect. J. Biol. Chem. 279, 3218–3227.
Lee, K.P., Cory, J.S., Wilson, K., Raubenheimer, D., Simpson, S.J., 2006. Flexible diet choice offsets protein costs of pathogen resistance in a caterpillar. Proc. R. Soc. B 273, 823–829.
Lemaitre, B., Hoffmann, J., 2007. The host defense of *Drosophila melanogaster*. Annu. Rev. Immunol. 25, 697–743.
Lemaitre, B., Meister, M., Govind, S., Georgel, P., Steward, R., Reichhart, J.-M., Hoffmann, J.A., 1995a. Functional analysis and regulation of nuclear import of dorsal during the immune response in *Drosophila*. EMBO J. 14, 536–545.
Lemaitre, B., Kromer-Metzger, E., Michaut, L., Nicolas, E., Meister, M., Georgel, P., Reichhart, J.M., Hoffmann, J.A., 1995b. A recessive mutation, immune deficiency (IMD), defines two distinct control pathways in the *Drosophila* host defense. Proc. Natl. Acad. Sci. U. S. A. 92, 9465–9469.
Lemaitre, B., Nicolas, E., Michaut, L., Reichhart, J.-M., Hoffmann, J.A., 1996. The dorsoventral regulatory gene cassette spätzle/Toll/cactus controls the potent antifungal response in *Drosophila* adults. Cell 86, 973–983.
Ligoxygakis, P., Pelte, N., Hoffmann, J.A., Reichhart, J.M., 2002. Activation of *Drosophila* Toll during fungal infection by a blood serine protease. Science 297, 114–116.
Lindsay, S.A., Wasserman, S.A., 2014. Conventional and non-conventional *Drosophila* Toll signaling. Dev. Comp. Immunol. 42, 16–24.
Little, T.J., Cobbe, N., 2005. The evolution of immune-related genes from disease carrying mosquitoes: diversity in a peptidoglycan- and a thioester-recognizing protein. Insect Mol. Biol. 14, 599–605.
Loof, T.G., Mörgelin, M., Johansson, L., Oehmcke, S., Olin, A.I., Dickneite, G., Norrby-Teglund, A., Theopold, U., Herwald, H., 2011a. Coagulation, an ancestral serine protease cascade, exerts a novel function in early immune defense. Blood 118, 2589–2598.
Loof, T.G., Schmidt, O., Herwald, H., Theopold, U., 2011b. Coagulation systems of invertebrates and vertebrates and their roles in innate immunity: the same side of two coins? J. Innate Immun. 3, 34–40.
Louie, A., Song, K.H., Hotson, A., Thomas, T.A., Schneider, D.S., 2016. How many parameters does it take to describe disease tolerance? PLoS Biol. 14 (4), e1002435. http://dx.doi.org/10.1371/journal.pbio.1002435.
Lu, Y., Wu, L.P., Anderson, K.V., 2001. The antibacterial arm of the *Drosophila* innate immune response requires an IkappaB kinase. Genes Dev. 15, 104–110.
Lu, A., Zhang, Q., Zhang, J., Yang, B., Wu, K., Xie, W., Luan, Y.-X., Ling, E., 2014. Insect prophenoloxidase: the view beyond immunity. Front Physiol. 5, 252.1–252.15.
Mackay, T.F.C., 2014. Epistasis and quantitative traits: using model organisms to study gene-gene interactions. Nat. Rev. Genet. 15, 22–33.
Mangoni, M.L., McDermott, A.M., Zasloff, M., 2016. Antimicrobial peptides and wound healing: biological and therapeutic considerations. Exp. Dermatol. 25, 167–173.
Martens, A., 1992. Egg deposition rates and duration of oviposition in *Platycnemis pennipes* (Pallas) (Insects: Odonata). Hydrobiologia 230, 63–70.

Maxwell, A.I., Morrison, G.M., Dorin, J.R., 2003. Rapid sequence divergence in mammalian β-defensins by adaptive evolution. Mol. Immunol. 40, 413–421.

Medzhitov, R., Janeway, C.A., 1998. An ancient system of host defense. Curr. Opin. Immunol. 10, 12–15.

Mesquita, R.D., Vionette-Amaral, R.J., Lowenberger, C., Rivera-Pomar, R., Monteiro, F.A., Minx, P., Spieth, J., Bernardo Carvalho, A., Panzera, F., Lawson, D., Torres, A.Q., Ribeiro, J.M.C., Sorgine, M.H.F., Waterhouse, R.M., Montague, M.J., Abad-Franch, F., Alves-Bezerra, M., Amaral, L.R., Araujo, H.M., Araujo, R.N., Aravind, L., Atella, G.C., Azambuja, P., Berni, M., Bittencourt-Cunha, P.R., Braz, G.R.C., Calderon-Fernandez, G., Carareto, C.M.A., Christensen, M.B., Costa, I.R., Costa, S.G., Dansa, M., Daumas-Filho, C.R.O., De-Paula, I.F., Dias, F.A., Dimopoulos, G., Emrich, S.J., Esponda-Behrens, N., Fampa, P., Fernandez-Medina, R.D., da Fonseca, R.N., Fontenele, M., Fronick, C., Fulton, L.A., Gandara, A.C., Garcia, E.S., Genta, F.A., Giraldo-Calderon, G.I., Gomes, B., Gondim, K.C., Granzotto, A., Guarneri, A.A., Guigo, R., Harry, M., Hughes, D.S.T., Jablonka, W., Jacquin-Joly, E., Patricia Juarez, M., Koerich, L.B., Latorre-Estivalis, J.M., Lavore, A., Lawrence, G.G., Lazoski, C., Lazzari, C.R., Lopes, R.R., Lorenzo, M.G., Lugon, M.D., Majerowicz, D., Marcet, P.L., Mariotti, M., Masuda, H., Megy, K., Melo, A.C.A., Missirlis, F., Mota, T., Noriega, F.G., Nouzova, M., Nunes, R.D., Oliveira, R.L.L., Oliveira-Silveira, G., Ons, S., Pagola, L., Paiva-Silva, G.O., Pascual, A., Pavan, M.G., Pedrini, N., Peixoto, A.A., Pereira, M.H., Pike, A., Polycarpo, C., Prosdocimi, F., Ribeiro-Rodrigues, R., Robertson, H.M., Salerno, A.P., Salmon, D., Santesmasses, D., Schama, R., Seabra-Junior, E.S., Silva-Cardoso, L., Silva-Neto, M.A.C., Souza-Gomes, M., Sterkel, M., Taracena, M.L., Tojo, M., Tu, Z.J., Tubio, J.M.C., Ursic Bedoya, R., Venancio, T.M., Walter-Nuno, A.B., Wilson, D., Warren, W.C., Wilson, R.K., Huebner, E., Dotson, E.M., Oliveira, P.L., 2015. Genome of Rhodnius prolixus, an insect vector of Chagas disease, reveals unique adaptations to hematophagy and parasite infection. Proc. Natl. Acad. Sci. U. S. A. 112, 14936–14941. http://dx.doi.org/10.1073/pnas.1506226112.

Michel, T., Reichhart, J.M., Hoffmann, J.A., Royet, J., 2001. *Drosophila* Toll is activated by Gram-positive bacteria through a circulating peptidoglycan recognition protein. Nature 414, 756–759.

Ming, M., Obata, F., Kuranaga, E., Miura, M., 2014. Persephone/Spatzle pathogen sensors mediate the activation of Toll receptor signaling in response to endogenous danger signals in apoptosis-deficient *Drosophila*. J. Biol. Chem. 289, 7558–7568.

Misof, B., Liu, S., Meusemann, K., Peters, R.S., Donath, A., Mayer, C., Frandsen, P.B., Ware, J., Flouri, T., Beutel, R.G., Niehuis, O., Petersen, M., Izquierdo-Carrasco, F., Wappler, T., Rust, J., Aberer, A.J., Aspöck, U., Aspöck, H., Bartel, D., Blanke, A., Berger, S., Böhm, A., Buckley, T.R., Calcott, B., Chen, J., Friedrich, F., Fukui, M., Fujita, M., Greve, C., Grobe, P., Gu, S., Huang, Y., Jermiin, L.S., Kawahara, A.Y., Krogmann, L., Kubiak, M., Lanfear, R., Letsch, H., Li, Y., Li, Z., Li, J., Lu, H., Machida, R., Mashimo, Y., Kapli, P., McKenna, D.D., Meng, G., Nakagaki, Y., Navarrete-Heredia, J.L., Ott, M., Ou, Y., Pass, G., Podsiadlowski, L., Pohl, H., von Reumont, B.M., Schütte, K., Sekiya, K., Shimizu, S., Slipinski, A., Stamatakis, A., Song, W., Su, X., Szucsich, N.U., Tan, M., Tan, X., Tang, M., Tang, J., Timelthaler, G., Tomizuka, S., Trautwein, M., Tong, X., Uchifune, T., Walzl, M.G., Wiegmann, B.M., Wilbrandt, J., Wipfler, B., Wong, T.K.F., Wu, Q., Wu, G., Xie, Y., Yang, S., Yang, Q., Yeates, D.K., Yoshizawa, K., Zhang, Q., Zhang, R., Zhang, W., Zhang, Y., Zhao, J., Zhou, C., Zhou, L., Ziesmann, T., Zou, S., Li, Y., Xu, X., Zhang, Y., Yang, H., Wang, J., Wang, J., Kjer, K.M., Zhou, X., 2014. Phylogenomics resolves the timing and pattern of insect evolution. Science 346, 763–767.

Mora, C., Tittensor, D.P., Adl, S., Simpson, A.G.B., Worm, B., 2011. How many species are there on earth and in the ocean? PLoS Biol. 9, e1001127.

Muniz, C.A., Jaillard, D., Lemaitre, B., Boccard, F., 2007. *Erwinia carotovora* Evf antagonizes the elimination of bacteria in the gut of *Drosophila* larvae. Cell. Microbiol. 9, 106–119.

Mylonakis, E., Podsiadlowski, L., Muhammed, M., Vilcinskas, A., 2016. Diversity, evolution and medical applications of insect antimicrobial peptides. Philos. Trans. R. Soc. B 371, 20150290.

Nappi, A.J., Christensen, B.M., 2005. Melanogenesis and associated cytotoxic reactions: applications to insect innate immunity. Insect Biochem. Mol. Biol. 35, 443–459.

Neafsey, D.E., Waterhouse, R.M., Abai, M.R., Aganezov, S.S., Alekseyev, M.A., Allen, J.E., Amon, J., Arcà, B., Arensburger, P., Artemov, G., Assour, L.A., Basseri, H., Berlin, A., Birren, B.W., Blandin, S.A., Brockman, A.I., Burkot, T.R., Burt, A., Chan, C.S., Chauve, C., Chiu, J.C., Christensen, M., Costantini, C., Davidson, V.L.M., Deligianni, E., Dottorini, T., Dritsou, V., Gabriel, S.B., Guelbeogo, W.M., Hall, A.B., Han, M.V., Hlaing, T., Hughes, D.S.T., Jenkins, A.M., Jiang, X., Jungreis, I., Kakani, E.G., Kamali, M., Kemppainen, P., Kennedy, R.C., Kirmitzoglou, I.K., Koekemoer, L.L., Laban, N., Langridge, N., Lawniczak, M.K.N., Lirakis, M., Lobo, N.F., Lowy, E., MacCallum, R.M., Mao, C., Maslen, G., Mbogo, C., McCarthy, J., Michel, K., Mitchell, S.N., Moore, W., Murphy, K.A., Naumenko, A.N., Nolan, T., Novoa, E.M., O'Loughlin, S., Oringanje, C., Oshaghi, M.A., Pakpour, N., Papathanos, P.A., Peery, A.N., Povelones, M., Prakash, A., Price, D.P., Rajaraman, A., Reimer, L.J., Rinker, D.C., Rokas, A., Russell, T.L., Sagnon, N., Sharakhova, M.V., Shea, T., Simão, F.A., Simard, F., Slotman, M.A., Somboon, P., Stegniy, V., Struchiner, C.J., Thomas, G.W.C., Tojo, M., Topalis, P., Tubio, J.M.C., Unger, M.F., Vontas, J., Walton, C., Wilding, C.S., Willis, J.H., Wu, Y.-C., Yan, G., Zdobnov, E.M., Zhou, X., Catteruccia, F., Christophides, G.K., Collins, F.H., Cornman, R.S., Crisanti, A., Donnelly, M.J., Emrich, S.J., Fontaine, M.C., Gelbart, W., Hahn, M.W., Hansen, I.A., Howell, P.I., Kafatos, F.C., Kellis, M., Lawson, D., Louis, C., Luckhart, S., Muskavitch, M.A.T., Ribeiro, J.M., Riehle, M.A., Sharakhov, I.V., Tu, Z., Zwiebel, L.J., Besansky, N.J., 2015. Highly evolvable malaria vectors: the genomes of 16 *Anopheles* mosquitoes. Science 347, 1258522.1–1258522.8.

Nusslein-Volhard, C., Frohnhofer, H.G., Lehmann, R., 1987. Determination of anteroposterior polarity in *Drosophila*. Science 238, 1675–1681.

Otti, O., Tragust, S., Feldhaar, H., 2014. Unifying external and internal immune defenses. Trends Ecol. Evol. 29, 625–634.

Overend, G., Cabrero, P., Guo, A.X., Sebastian, S., Cundall, M., Armstrong, H., Mertens, I., Schoofs, L., Dow, J.A.T., Davies, S.-A., 2012. The receptor guanylate cyclase Gyc76C and a peptide ligand, NPLP1-VQQ, modulate the innate immune IMD pathway in response to salt stress. Peptides 34, 209–218.

Owusu-Ansah, E., Perrimon, N., 2014. Modeling metabolic homeostasis and nutrient sensing in Drosophila: implications for aging and metabolic diseases. Dis. Model. Mech. 7, 343–350.

Palmer, W.J., Jiggins, F.M., 2015. Comparative genomics reveals the origins and diversity of arthropod immune systems. Mol. Biol. Evol. 32, 2111–2129.

Poltorak, A., He, X., Smirnova, I., Liu, M.-Y., Van Huffel, C., Du, X., Birdwell, D., Alejos, E., Silva, M., Galanos, C., Freudenberg, M., Ricciardi-Castagnoli, P., Layton, B., Beutler, B., 1998. Defective LPS signaling in C3H/HeJ and C57BL/10ScCr mice: mutations in Tlr4 gene. Science 282, 2085–2088.

Pritchard, J.K., Di Rienzo, A., 2010. Adaptation—not by sweeps alone. Nat. Rev. Genet. 11, 665–667.

Rämet, M., Manfruelli, P., Pearson, A., Mathey-Prevot, B., Ezekowitz, R.A.B., 2002. Functional genomic analysis of phagocytosis and identification of a *Drosophila* receptor for *E. coli*. Nature 416, 644–648.

Ratcliffe, N.A., Gagen, S.J., 1977. Studies on the in vivo cellular reactions of insects: an ultrastructural analysis of nodule formation in *Galleria mellonella*. Tissue and Cell 9, 73–85.

Razzell, W., Wood, W., Martin, P., 2011. Swatting flies: modelling wound healing and inflammation in *Drosophila*. Dis. Model. Mech. 4, 569–574.

Reinhardt, K., Anthes, N., Lange, R., 2015. Copulatory wounding and traumatic insemination. Cold Spring Harb. Perspect. Biol. 7 (5), a017582.

Ribeiro, C., Brehélin, M., 2006. Insect haemocytes: what type of cell is that? J. Insect Physiol. 52, 417–429.

Riddell, C.E., Garces, J.D.L., Adams, S., Barribeau, S.M., Twell, D., Mallon, E.B., 2014. Differential gene expression and alternative splicing in insect immune specificity. BMC Genomics 15, 1031.

Robertson, A.S., Belorgey, D., Lilley, K.S., Lomas, D.A., Gubb, D., Dafforn, T.R., 2003. Characterization of the necrotic protein that regulates the Toll-mediated immune response in *Drosophila*. J. Biol. Chem. 278, 6175–6180.

Rogers, S.M., Bernatchez, L., 2007. The genetic architecture of ecological speciation and the association with signatures of selection in natural lake whitefish species pairs. Mol. Biol. Evol. 24, 1423–1438.

Rolff, J., Schmid-Hempel, P., 2016. Perspectives on the evolutionary ecology of arthropod antimicrobial peptides. Philos. Trans. R. Soc. B 371, 20150297.

Rutschmann, S., Jung, A.C., Zhou, R., Silverman, N., Hoffmann, J.A., Ferrandon, D., 2000. Role of *Drosophila* IKK gamma in a toll-independent antibacterial immune response. Nat. Immunol. 1, 342–347.

Sackton, T.B., Hartl, D.L., 2016. Genotypic context and epistasis in individuals and populations. Cell 166, 279–287.

Sackton, T.B., Lazzaro, B.P., Schlenke, T.A., Evans, J.D., Hultmark, D., Clark, A.G., 2007. Dynamic evolution of the innate immune system in *Drosophila*. Nat. Genet. 39, 1461–1468.

Sackton, T.B., Lazzaro, B.P., Clark, A.G., 2010. Genotype and gene expression associations with immune function in *Drosophila*. PLoS Genet. 6, e1000797.

Salazar-Jaramillo, L., Paspati, A., van de Zande, L., Vermeulen, C.J., Schwander, T., Wertheim, B., 2014. Evolution of a cellular immune response in *Drosophila*: a phenotypic and genomic comparative analysis. Genome Biol. Evol. 6, 273–289.

Samakovlis, C., Kimbrell, D.A., Kylsten, P., Engström, A., Hultmark, D., 1990. The immune response in *Drosophila*: pattern of cecropin expression and biological activity. EMBO J. 9, 2969–2976.

Scherfer, C., Karlsson, C., Loseva, O., Bidla, G., Goto, A., Havemann, J., Dushay, M.S., Theopold, U., 2004. Isolation and characterization of hemolymph clotting factors in *drosophila melanogaster* by a pullout method. Curr. Biol. 14, 625–629.

Schmidt, O., Theopold, U., Strand, M., 2001. Innate immunity and its evasion and suppression by hymenopteran endoparasitoids. Bioessays 23, 344–351.

Schneider, D.S., Jin, Y., Morisato, D., Anderson, K.V., 1994. A processed form of the Spatzle protein defines dorsal-ventral polarity in the Drosophila embryo. Development 120, 1243–1250.

Schwenke, R.A., Lazzaro, B.P., Wolfner, M.F., 2016. Reproduction–immunity trade-offs in insects. Annu. Rev. Entomol. 61, 239–256.

Scott, J.G., Warren, W.C., Beukeboom, L.W., Bopp, D., Clark, A.G., Giers, S.D., Hediger, M., Jones, A.K., Kasai, S., Leichter, C.A., Li, M., Meisel, R.P., Minx, P., Murphy, T.D., Nelson, D.R., Reid, W.R., Rinkevich, F.D., Robertson, H.M., Sackton, T.B., Sattelle, D.B., Thibaud-Nissen, F., Tomlinson, C., van de Zande, L., Walden, K.K.O., Wilson, R.K., Liu, N., 2014. Genome of the house fly, *Musca domestica*

L., a global vector of diseases with adaptations to a septic environment. Genome Biol. 15, 466.

Shelby, K.S., Popham, H.J.R., 2006. Plasma phenoloxidase of the larval tobacco budworm, *Heliothis virescens*, is virucidal. J. Insect Sci. 6, 1–12.

Silverman, N., Zhou, R., Stoven, S., Pandey, N., Hultmark, D., Maniatis, T., 2000. A *Drosophila* IkappaB kinase complex required for Relish cleavage and antibacterial immunity. Genes Dev. 14, 2461–2471.

Sleiman, M.S.B., Osman, D., Massouras, A., Hoffmann, A.A., Lemaitre, B., Deplancke, B., 2015. Genetic, molecular and physiological basis of variation in *Drosophila* gut immunocompetence. Nat. Commun. 6, 7829.

Steiner, H., Hultmark, D., Engström, A., Bennich, H., Boman, H.G., 1981. Sequence and specificity of two antibacterial proteins involved in insect immunity. Nature 292, 246–248.

Steward, R., 1987. Dorsal, an embryonic polarity gene in *Drosophila*, is homologous to the vertebrate proto-oncogene, c-rel. Science 238, 692–694.

Stone, E.F., Fulton, B.O., Ayres, J.S., Pham, L.N., Ziauddin, J., Shirasu-Hiza, M.M., 2012. The circadian clock protein timeless regulates phagocytosis of bacteria in *Drosophila*. PLoS Pathog. 8, e1002445–11.

Stramer, B.M., Dionne, M.S., 2014. Unraveling tissue repair immune responses in flies. Semin. Immunol. 26, 310–314.

Suen, G., Teiling, C., Li, L., Holt, C., Abouheif, E., Bornberg-Bauer, E., Bouffard, P., Caldera, E.J., Cash, E., Cavanaugh, A., Denas, O., Elhaik, E., Favé, M.-J., Gadau, J., Gibson, J.D., Graur, D., Grubbs, K.J., Hagen, D.E., Harkins, T.T., Helmkampf, M., Hu, H., Johnson, B.R., Kim, J., Marsh, S.E., Moeller, J.A., Muñoz-Torres, M.C., Murphy, M.C., Naughton, M.C., Nigam, S., Overson, R., Rajakumar, R., Reese, J.T., Scott, J.J., Smith, C.R., Tao, S., Tsutsui, N.D., Viljakainen, L., Wissler, L., Yandell, M.D., Zimmer, F., Taylor, J., Slater, S.C., Clifton, S.W., Warren, W.C., Elsik, C.G., Smith, C.D., Weinstock, G.M., Gerardo, N.M., Currie, C.R., 2011. The genome sequence of the leaf-cutter ant *Atta cephalotes* reveals insights into its obligate symbiotic lifestyle. PLoS Genet. 7, e1002007.

Tanaka, H., Ishibashi, J., Fujita, K., Nakajima, Y., Sagisaka, A., Tomimoto, K., Suzuki, N., Yoshiyama, M., Kaneko, Y., Iwasaki, T., 2008. A genome-wide analysis of genes and gene families involved in innate immunity of *Bombyx mori*. Insect Biochem. Mol. Biol. 38, 1087–1110.

Tennessen, J.A., 2005. Molecular evolution of animal antimicrobial peptides: widespread moderate positive selection. J. Evol. Biol. 18, 1387–1394.

Theopold, U., Schmidt, O., Söderhäll, K., Dushay, M.S., 2004. Coagulation in arthropods: defence, wound closure and healing. Trends Immunol. 25, 289–294.

Theopold, U., Krautz, R., Dushay, M.S., 2014. The *Drosophila* clotting system and its messages for mammals. Dev. Comp. Immunol. 42, 42–46.

Tian, C., Gao, B., Fang, Q., Ye, G., Zhu, S., 2010. Antimicrobial peptide-like genes in *Nasonia vitripennis*: a genomic perspective. BMC Genomics 11, 187.1–187.19.

Unckless, R.L., Lazzaro, B.P., 2016. The potential for adaptive maintenance of diversity in insect antimicrobial peptides. Philos. Trans. R. Soc. B 371, 20150291.

Unckless, R.L., Rottschaefer, S.M., Lazzaro, B.P., 2015. The complex contributions of genetics and nutrition to immunity in *Drosophila melanogaster*. PLoS Genet. 11, e1005030.

Unckless, R.L., Howick, V.M., Lazzaro, B.P., 2016. Convergent balancing selection on an antimicrobial peptide in *drosophila*. Curr. Biol. 26, 257–262.

Valadez-Lira, J.A., Alcocer-Gonzalez, J.M., Damas, G., Nuñez-Mejía, G., Oppert, B., Rodriguez-Padilla, C., Tamez-Guerra, P., 2011. Comparative evaluation of phenoloxidase activity in different larval stages of four lepidopteran pests after exposure to *Bacillus thuringiensis*. J. Insect Sci. 12, 80.
Viljakainen, L., 2015. Evolutionary genetics of insect innate immunity. Brief. Funct. Genomics 14, 407–412. https://doi.org/10.1093/bfgp/elv002.
Vodovar, N., Vinals, M., Liehl, P., Basset, A., Degrouard, J., Spellman, P., Boccard, F., Lemaitre, B., 2005. *Drosophila* host defense after oral infection by an entomopathogenic *Pseudomonas* species. Proc. Natl. Acad. Sci. U. S. A. 102, 11414–11419.
Wang, Z., Wilhelmsson, C., Hyrsl, P., Loof, T.G., Dobes, P., Klupp, M., Loseva, O., Mörgelin, M., Iklé, J., Cripps, R.M., Herwald, H., Theopold, U., 2010. Pathogen entrapment by transglutaminase—a conserved early innate immune mechanism. PLoS Pathog. 6, e1000763.
Waterhouse, R.M., Kriventseva, E.V., Meister, S., Xi, Z., Alvarez, K.S., Bartholomay, L.C., Barillas-Mury, C., Bian, G., Blandin, S., Christensen, B.M., Dong, Y., Jiang, H., Kanost, M.R., Koutsos, A.C., Levashina, E.A., Li, J., Ligoxygakis, P., MacCallum, R.M., Mayhew, G.F., Mendes, A., Michel, K., Osta, M.A., Paskewitz, S., Shin, S.W., Vlachou, D., Wang, L., Wei, W., Zheng, L., Zou, Z., Severson, D.W., Raikhel, A.S., Kafatos, F.C., Dimopoulos, G., Zdobnov, E.M., Christophides, G.K., 2007. Evolutionary dynamics of immune-related genes and pathways in disease-vector mosquitoes. Science 316, 1738–1743.
Wertheim, B., Kraaijeveld, A.R., Schuster, E., Blanc, E., Hopkins, M., Pletcher, S.D., Strand, M.R., Partridge, L., Godfray, H.C.J., 2005. Genome-wide gene expression in response to parasitoid attack in *Drosophila*. Genome Biol. 6, R94.
Wyatt, G.R., 1961. The biochemistry of insect hemolymph. Annu. Rev. Entomol. 6, 75–102.
Xia, X., Yu, L., Xue, M., Yu, X., Vasseur, L., Gurr, G.M., Baxter, S.W., Lin, H., Lin, J., You, M., 2015. Genome-wide characterization and expression profiling of immune genes in the diamondback moth, *Plutella xylostella* (L.). Sci. Rep. 5, 9877. http://dx.doi.org/10.1038/srep09877.
You, H., Lee, W.J., Lee, W.J., 2014. Homeostasis between gut-associated microorganisms and the immune system in *Drosophila*. Curr. Opin. Immunol. 30, 48–53.
Zasloff, M., 2002. Antimicrobial peptides of multicellular organisms. Nature 415, 389–395.
Zhang, Z., Zhu, S., 2012. Comparative genomics analysis of five families of antimicrobial peptide-like genes in seven ant species. Dev. Comp. Immunol. 38, 262–274.
Zou, Z., Evans, J.D., Lu, Z., Zhao, P., Williams, M., Sumathipala, N., Hetru, C., Hultmark, D., Jiang, H., 2007. Comparative genomic analysis of the *Tribolium* immune system. Genome Biol. 8, R177.

CHAPTER TWO

Phagocytosis in Insect Immunity

Ashley E. Nazario-Toole, Louisa P. Wu
Institute for Bioscience and Biotechnology Research, University of Maryland, College Park, MD, United States

Contents

1. Introduction 35
2. Insect Blood Cells 36
3. Phagocytic Receptors in Insects 37
   3.1 Scavenger Receptors 43
   3.2 Nimrod Receptor Superfamily 46
   3.3 Peptidoglycan-Recognition Receptors Important for Phagocytosis 52
   3.4 Integrins 57
   3.5 Down Syndrome Adhesion Molecule 1 (Dscam 1) 60
   3.6 Opsonins in Insect Phagocytosis 62
4. Regulation of Signalling During Phagocytosis 65
5. Phagosome Maturation 68
6. Conclusion 73
References 73

Abstract

The cellular immune response of insects is mediated by specialized blood cells that employ germline encoded receptors to recognize, engulf, and eliminate microbes through the process of phagocytosis. Phagocytosis is a vital component of innate immunity and research over the past 3 decades in model systems, such as the fruit fly *Drosophila melanogaster* and mosquitoes, has led to the identification of an array of processes that invertebrates rely on to respond to invading microorganisms. Valuable insights have been gleaned from studies of insect phagocytes and, in this review, we highlight research related to the roles of receptors and intracellular signalling molecules involved in phagocytosis of bacterial and parasitic microbes.

1. INTRODUCTION

Cells of the innate immune system serve as the initial line of defence against invading microbes. The importance of innate immunity is underscored by the fact that innate immune responses are found in nearly all

Advances in Insect Physiology, Volume 52
ISSN 0065-2806
http://dx.doi.org/10.1016/bs.aiip.2016.12.001

animals, while adaptive immune responses are restricted to the jawed vertebrate group and some jawless fishes (Lemaitre and Hoffmann, 2007). Phagocytosis, the cornerstone of a robust and powerful innate immune response, is a cell-mediated process that was first described over 100 years ago by Èlie Metchnikoff (Kaufmann, 2008). Phagocytic cells express germline-encoded pattern-recognition receptors (PRRs) that recognize common pathogen-associated molecular patterns (PAMPs) such as bacterial-derived lipopolysaccharide (LPS), lipoteichoic acid (LTA), peptidoglycan, double-stranded RNA, unmethylated CpG DNA, and β-glucan of fungi (Flannagan et al., 2012). Once bound to their cognate ligands, these receptors initiate signalling events that lead to the clearance of pathogens. The most extensively studied mammalian phagocytic receptors are the Fc receptor (FcR) and the complement receptor, CR3 (Griffin et al., 1975; Odin et al., 1991). These receptors are regarded as a general model for the cellular and molecular events that take place during phagocytosis. Ex vivo and in vitro cell biology and microscopy techniques are frequently used to examine the cellular mechanisms underlying phagocytosis in mammalian cells (Stuart and Ezekowitz, 2005). While these reductionist approaches have increased our understanding of the complex cell biology of phagocytosis, they do not address the relative importance of the cellular immune response in intact organisms. Over the last 30 years, analyses of insect immune responses have significantly enhanced our understanding of the roles of phagocytic cells in vivo. Here we review the current knowledge on the cellular immune response in insects, with specific emphasis on PRRs, particle internalization, and phagosome maturation in the fruit fly, *Drosophila melanogaster*, and mosquitoes (Diptera: Culicidae). Comprehensive investigations of immune pathways and receptors have been carried out in the fruit fly, aided by availability of genetic tools and mutant lines, as well as the fully sequenced genomes of 12 *Drosophila* species (Drosophila 12 Genomes et al., 2007). The sequencing of 16 *Anopheles* genomes and the use of RNA interference (RNAi)-based gene silencing have facilitated the characterization of mosquito PRRs, which is crucial for the development of novel strategies to prevent disease transmission (Neafsey et al., 2015).

2. INSECT BLOOD CELLS

The defence reactions of mosquitoes and flies consist of potent humoral and cellular responses (Lemaitre and Hoffmann, 2007). Humoral immunity is characterized by the systemic production of antimicrobial peptides (AMPs) after immune cells of the blood and fat body detect bacteria or fungi in the hemolymph. The cellular immune response is specifically carried out by specialized blood cells known as hemocytes. There are three classes of hemocytes in the fly: plasmatocytes, crystal cells, and lamellocytes (Lanot et al., 2001). Similarly, three types of hemocytes have been described in adult *Anopheles gambiae* and *Aedes aegypti* mosquitoes: granulocytes, oenocytoids, and prohemocytes (Castillo et al., 2006). *Drosophila* plasmatocytes are similar to mosquito granulocytes and both cell types are the functional equivalent of mammalian macrophages (Castillo et al., 2006; Williams, 2007). Plasmatocytes adhere to the heart tissue in the dorsal vessel where they function as the primary effector of cellular immunity in adult fruit flies (Elrod-Erickson et al., 2000). Similarly, mosquito granulocytes aggregate in heart peristaltic regions, where they phagocytose and eliminate pathogens such as the Gram-positive bacteria, *Staphylococcus aureus*, *Staphylococcus epidermidis*, *Micrococcus luteus*, and Gram-negative bacteria, *Escherichia coli* (Sigle and Hillyer, 2016). Survival experiments carried out using flies where plasmatocytes were genetically ablated show that phagocytosis is critical for the clearance of the Gram-positive bacteria, *Enterococcus faecalis* and *S. aureus* (Charroux and Royet, 2009; Defaye et al., 2009; Nehme et al., 2011; Shia et al., 2009). In this review, we will focus on the functions of phagocytic hemocytes with an emphasis on hemocyte PRRs and their ligands. The remaining types of *Drosophila* and mosquito blood cells have been recently reviewed elsewhere (Hillyer and Strand, 2014; Honti et al., 2014). Finally, the relationship between blood cells and cancer-like tumours has been explored in *Drosophila* and recently reviewed (Wang et al., 2014).

3. PHAGOCYTIC RECEPTORS IN INSECTS

D. melanogaster has been successfully utilized to identify several PRRs via large-scale RNAi screens and smaller, classical genetic screens. In mosquitoes, phagocytic receptors have been characterized using RNAi-mediated silencing of candidate genes in cell culture and adults. Phagocytic receptors identified in both organisms are grouped according to receptor families in Table 1. The predicted mammalian orthologs for each receptor family is also given in Table 1 (Hu et al., 2011). Finally, the overlap of the ligand specificity of the receptors is illustrated in Fig. 1.

Table 1 Cell Surface Recognition Receptors and Opsonins

Receptor Family	Receptor		Ligands and References		Related Mammalian Molecules
	Drosophila	Mosquito	Drosophila	Mosquito	
Scavenger receptors	Croquemort		Apoptotic cells (Franc et al., 1999)		CD36
			Staphylococcus aureus (Stuart et al., 2005)		
	Peste		*Mycobacterium fortuitum* (Agaisse et al., 2005; Philips et al., 2005)		SCARBI
			Mycobacterium smegmatis (Philips et al., 2005)		
			Listeria monocytogenes (Agaisse et al., 2005)		
			NOT *Escherichia coli* or *S. aureus* (Philips et al., 2005)		
	SR-CI		*E. coli* (Ramet et al., 2001)		None found
			S. aureus (Ramet et al., 2001)		
			NOT *Candida salvatica* (Ramet et al., 2001)		
			Double-stranded RNA (Ulvila et al., 2006)		

Nimrod receptor superfamily	Eater		*E. coli*, *Serratia marcescens*, and *S. aureus* (Chung and Kocks, 2011; Kocks et al., 2005)		SREC: MEGF10; MEGF11; CD91; Stabilin 1 and Stabilin 2
			Enterococcus faecalis (Chung and Kocks, 2011; Nehme et al., 2011)		
			NOT *Micrococcus luteus* (Chung and Kocks, 2011; Nehme et al., 2011)		
			Double-stranded RNA (Ulvila et al., 2006)		
	NimC1		*E. coli* (Kurucz et al., 2007a)		
			S. aureus (Kurucz et al., 2007a)		
	Draper	AgNimB2	Apoptotic cells (Freeman et al., 2003; Manaka et al., 2004)	*S. aureus* (Midega et al., 2013)	
			Axon pruning (Awasaki et al., 2006; MacDonald et al., 2006)		
			S. aureus (Cuttell et al., 2008; Shiratsuchi et al., 2012)		
			E. coli (Cuttell et al., 2008)		

Continued

Table 1 Cell Surface Recognition Receptors and Opsonins—cont'd

Receptor Family	Receptor		Ligands and References		Related Mammalian Molecules
	Drosophila	Mosquito	Drosophila	Mosquito	
Peptidoglycan recognition proteins	PCRP-LC	AgPGRP-LC	*E. coli* (Bergeret et al., 2008; Ramet et al., 2002)	*E. coli* (Moita et al., 2005)	Mammalian PGRPs
			NOT *S. aureus* (Ramet et al., 2002)		
			NOT *E. coli* or *S. aureus* (Choe et al., 2002; Garver et al., 2006)		
	PGRP-SC1/ picky		*S. aureus* (Garver et al., 2006)		
			NOT *E. coli* or *Saccharomyces cerevisiae* (Garver et al., 2006)		
	PGRP-SA		*S. aureus* (Garver et al., 2006)		
			NOT *E. coli* (Garver et al., 2006)		
Integrins	αPS3/βν	Bint2	Apoptotic cells (Nagaosa et al., 2011; Nonaka et al., 2013)	*E. coli* (Moita et al., 2005, 2006)	ITCA4 (CD49D) ITGBI
			S. aureus (Nonaka et al., 2013; Shiratsuchi et al., 2012)	NOT *S. aureus* (Moita et al., 2005)	
Ig-like	Dscam1	AgDscam	*E. coli* (Watson et al., 2005)	*E. coli* (Dong et al., 2006b)	DSCAM
				S. aureus (Dong et al., 2006b)	

TEPs		AgTEPI		*E. coli* and *S. aureus* (Levashina et al., 2001; Moita et al., 2005)	Complement components
				Plasmodium (Blandin et al., 2004)	
	TEPII		*E. coli* (Stroschein-Stevenson et al., 2006)		
	TEPIII		*S. aureus* (Stroschein-Stevenson et al., 2006)		
	Mcr(TEPVI)		*Candida albicans* (Stroschein-Stevenson et al., 2006)		

Drosophila and mosquito phagocytic receptors are grouped according functional properties and common features. EGF-like repeat containing protein orthologs were identified in Kocks et al. (2005). All other mammalian orthologs were identified using the Drosophila RNAi Screening Center (DRSC) Integrative Ortholog Prediction Tool (DIOPST) (Hu et al., 2011). CD36, CD36 (thrombospondin receptor); SCARB1, scavenger receptor class B, member 1; SR-C1, scavenger receptor class C, type 1; αPS3, integrin alpha PS3 subunit (encoded by *scab*); βν, integrin beta subunit (encoded by *Itgbn*); Ig-like, Immunoglobulin-like; Dscam, Down syndrome cell adhesion molecule; TEP, Thioester-containing protein.

		Gram (–) bacteria		Gram (+) bacteria				Mycobacteria		Other		Yeast		Parasite
		S. marcescens	*E. coli*	*S. aureus*	*L. monocytogenes*	*E. faecalis*	*M. luteus*	*M. fortuitum*	*M. smegmatis*	Apoptotic cells	dsRNA	*C. albicans*	*S. cerevisiae*	*Plasmodium*
Drosophila melanogaster	Croquemort			✓						✓				
	Peste		o	o	✓			✓	✓					
	SR-CI		✓	✓							✓	o		
	Eater	✓	✓	✓		✓	o				✓			
	NimC1		✓	✓										
	Draper		✓	✓						✓				
	PGRP-LC		✓	✓										
	PGRP-SC1A		o	✓									o	
	PGRP-SA		o	✓										
	αPS3/βν			✓						✓				
	Dscam1		✓											
	TEPII		✓											
	TEPIII			✓										
	Mcr											✓		
Anopheles gambiae	AgNimB2			✓										
	AgPGRP-LC		✓											
	Bint2		✓	o										
	AgDscam		✓	✓										
	AgTEPI		✓	✓										✓

Fig. 1 Overlap of phagocytic PRR binding affinities in *Drosophila melanogaster* (top) and *Anopheles gambiae* (bottom). ✓ Indicates that the PRR mediates phagocytosis of the indicated microbe. o Indicates that the loss of the PRR did not affect phagocytosis of the microbe.

Phagocytosis is initiated when cell surface receptors recognize their target ligands and trigger engulfment of molecules into a nascent organelle, the phagosome. Phagocytic receptors can either directly bind to ligands expressed on the surface of target cells or recognize targets coated by opsonins, soluble host factors that bind to foreign bodies. Additionally, due to the inherent diversity of particles that are taken up by phagocytosis, multiple receptors are simultaneously engaged to ligands on the surface of target particles to facilitate uptake. The overlap and redundancy in receptor ligand specificities help in the formation of strong interactions between the target particle and the phagocyte. Receptor redundancy is also evolutionarily advantageous as it allows the host cell to combat pathogens that have developed mechanisms to evade detection by a particular receptor.

Phagocytic cells target two main classes of particles: apoptotic cells and microorganisms. Removal of apoptotic cells is key during embryogenesis and development and many receptors that recognize apoptotic cell ligands also recognize microbial ligands (Arandjelovic and Ravichandran, 2015). During the immune response, both professional phagocytic cells and non-professional phagocytic cells, such as endothelial and epidermal cells, are able to phagocytose invasive bacteria. Mammalian professional phagocytes, macrophages, neutrophils, and dendritic cells, respond to infection by migrating towards infected tissue. In *Drosophila* larvae, plasmatocytes are the primary phagocytic immune cells, and these cells also migrate to sites of infection (Babcock et al., 2008). In *Drosophila* adults, plasmatocytes do not freely circulate, but mostly adhere to tissues such as the dorsal vessel (Elrod-Erickson et al., 2000; Lanot et al., 2001). Bacterial infections trigger mitosis in a subset of heart-associated plasmatocytes (Ghosh et al., 2015). In adult mosquitoes some hemocytes circulate freely while others are sessile, heart-associated hemocytes. Bacterial infections induce the migration of circulating hemocytes to regions surrounding the heart ostia (Castillo et al., 2006; Sigle and Hillyer, 2016). In mosquitoes and *D. melanogaster*, all hemolymph flows through the heart and hemocytes aggregated along the heart rapidly phagocytose microbes swept by the hemolymph.

Detection of PAMPs on the surface of microorganisms is the first step in phagocytosis of commensal and pathogenic microbes. Receptors present on the extracellular side of the plasma membrane of phagocytes directly bind to the microbes or to opsonins that are deposited on the microbial surface. Some of the receptors that participate in phagocytosis in *Drosophila* and mosquitoes have mammalian orthologs with similar functions, while others are unique to insects.

3.1 Scavenger Receptors

Scavenger receptors are a group of structurally unrelated receptors with shared functional properties that bind multiple polyanionic ligands

(Canton et al., 2013). The receptors have heterogeneous structures and are subdivided into nine classes (Classes A–I) based on shared domain architecture. A common feature of scavenger receptors is that they exhibit broad ligand specificity. For example, the mammalian Class B scavenger receptor CD36 recognizes altered self-ligands, acetylated, and/or oxidized low-density lipoprotein (LDL) and phosphatidylserine, as well as conserved microbial PAMPs from Gram-negative and Gram-positive bacteria.

During insect embryogenesis, hemocytes differentiate into macrophages that phagocytose apoptotic cells (Tepass et al., 1994). Based on this observation, Ezekowitz and group (Franc et al., 1996) identified Croquemort, the *Drosophila* paralog of mammalian CD36. Using immunohistochemistry, they found Croquemort was expressed on embryonic hemocytes after these cells developed the ability to phagocytose apoptotic cell corpses. Importantly, nonphagocytic mammalian COS-7 cells expressing Croquemort bound to apoptotic cells in vitro. Genetic follow-up studies by the Ezekowitz group (Franc et al., 1999), using *croquemort* null flies, revealed that Croquemort is essential for phagocytosis of apoptotic cells in vivo.

Croquemort also participates in the cellular immune response to bacteria. Moore and colleagues found that Croquemort is a receptor for *S. aureus*, but not *E. coli*, in a forward genetic screen using RNAi in *Drosophila* embryonic S2 cells (Stuart et al., 2005). This was the first study to demonstrate that CD36-related receptors act as receptors for bacteria.

One of the four mosquito orthologs of *Drosophila* Croquemort, SCRBQ2, interacts with the malaria parasite *Plasmodium berghei*. After bacterial challenge, *SCRBQ2* expression is upregulated in two immortalized *An. gambiae* cell lines (Dimopoulos et al., 2000, 2002). *SCRBQ2* is highly expressed in the mosquito midgut epithelium—which malaria parasites ookinetes pass through to carry on their life cycle (Dong et al., 2006a). In female adult *An. gambiae*, the expression of *SCRBQ2* is upregulated after noninfected and *P. berghei*-infected blood meals (Gonzalez-Lazaro et al., 2009). Interestingly, RNAi-mediated silencing of *SCRBQ2* led to ~60% reduction in the number of parasitic oocysts in the midgut of *An. gambiae*, suggesting that SCRBQ2 assists in parasite development. However, it is unknown if SCRBQ2 directly binds to *P. berghei*. Additionally, it is not known whether SCRBQ2 participates in the phagocytosis of bacteria. Thus, the precise function of SCRBQ2 during mosquito–parasite interactions has yet to be fully established. Further work is necessary to clarify the roles of *Croquemort*, SCRBQ2, and the remaining mosquito Croquemort orthologs in the phagocytosis of bacteria or other microbes.

Class C scavenger receptors are unique to insects. To date, the only extensively characterized member is Scavenger receptor class C, type I (SR-CI) from *D. melanogaster*. In embryos, SR-CI is expressed in macrophages and when expressed in mammalian CHO cells, SR-CI exhibited high binding affinity for LDL (Pearson et al., 1995). In 2001, the Ezekowitz group (Ramet et al., 2001) identified SR-CI as a receptor for Gram-negative and Gram-positive bacteria. This study was also notable because the authors established an in vitro insect cell model to study phagocytosis. They compared the phagocytic potential of ex vivo hemocytes to S2 cells, a macrophage-like cell culture derived from late-stage *Drosophila* embryos. They found that both cell types efficiently phagocytose bacteria and yeast. They also examined whether *E. coli* and *S. aureus* were recognized by the same or different S2 cell receptors by performing cross-competition experiments in which they coincubated cells with unlabelled and fluorescently labelled bacteria. Unlabelled *E. coli* was able to decrease the amount of phagocytosed fluorescently labelled *S. aureus*, and vice versa, suggesting the existence of common PRR(s) for *S. aureus* and *E. coli*. Interestingly, neither *E. coli* nor *S. aureus* inhibited the association of fluorescently labelled yeast, *Candida silvatica*, indicating that the receptors for *C. silvatica* do not overlap with those for *E. coli* or *S. aureus*. Acetylated LDL, LTA, and polyinosinic acid inhibited the binding of both *E. coli* and *S. aureus* in a dose-dependent manner.

There are four members of the Class C SR family in Drosophila, dSR-CI, CII, CIII, and CIV. SR-CI and CII are membrane-bound receptors while CIII and CIV are predicted to encode secreted proteins. RNAseq analysis shows that SR-CIII and SR-CIV are expressed at low levels in S2 cells (Graveley et al., 2011). Additionally, the tissue and temporal expression analysis showed that SR-CI is expressed in larval hemocytes and throughout the life of the fly whereas the other Class C SRs are only expressed in the early stages of development. The expression analysis, coupled with binding profiles for SR-CI indicated that it was a potential PRR candidate. Natural polymorphisms in *SR-CI* are associated with varying levels of resistance to the Gram-negative entomopathogen *Serratia marcescens*, indicating that SC-RI likely plays an important role in the immune response among wild fruit flies (Lazzaro, 2005; Lazzaro et al., 2004). Finally, in S2 cells, SR-CI, along with Eater, was shown to mediate more than 90% of the uptake of dsRNA (Ulvila et al., 2006).

Mycobacterium marinum causes a lethal infection in *Drosophila* and during the early stages of infection the bacteria grows in phagocytes (Dionne et al., 2003). To identify potential receptors for *Mycobacterium fortuitum*, a human pathogen, the Perrimon group conducted a genome-wide RNAi screen in S2 cells and identified another CD36 homolog, class B scavenger receptor Peste (Philips et al., 2005). The group treated S2 cells with dsRNA to deplete specific host genes, and then infected the cells with *M. fortuitum* that expressed GFP under the control of the *map24* promoter. *map24* is responsive to low pH that the internalized bacteria encounter in the lumen of the phagosome. Silencing of *peste* in S2 cells blocked invasion by *M. fortuitum*, but did not affect uptake of *S. aureus* or *E. coli*. Peste was also required for the uptake of the nonpathogenic *Mycobacterium smegmatis*, suggesting that Peste is a PRR for *Mycobacteria* species. Human embryonic kidney (HEK) 293 cells, which are normally refractory to infection by *M. fortuitum*, could be infected when *peste* was heterologously expressed. Furthermore, heterologous expression of Peste in HEK293 cells caused a small increase in the uptake of *S. aureus* and *E. coli*. This result was not seen in experiments of silencing Peste in S2 cells, most likely due to the presence of multiple receptors for *S. aureus* and *E. coli* on the surface of S2 cells. Finally, RNAi of *Peste* in S2 cells also causes a decreased uptake of another intracellular bacteria, *Listeria monocytogenes* (Agaisse et al., 2005; Philips et al., 2005). Both *M. fortuitum* and *L. monocytogenes* grow in *Drosophila* hemocytes (Dionne et al., 2003; Mansfield et al., 2003). Survival experiments in *Peste* mutant or RNAi flies are necessary to assess the importance of Peste-mediated phagocytosis of intracellular bacteria by hemocytes in vivo.

3.2 Nimrod Receptor Superfamily

The Nimrod superfamily is a diverse class of proteins characterized by the presence of epidermal growth factor (EGF)-like repeats called NIM repeats (Kurucz et al., 2007a). EGF repeats have roles in extracellular adhesion, coagulation, and receptor–ligand interactions. The NIM repeat, also known as an EGF-like repeat, is a special type of the EGF domain that is shifted one cysteine unit compared to the typical EGF repeat. Based on the shared structural characteristics, the family is divided into three types: (1) Draper-type genes (*Drosophila nimrod A* and *Draper* and Mosquito *draper*) as well as proteins containing many NIM domains (poly-NIM proteins), (2) Nimrod B-types (*Drosophila nimrod B 1–5* and Mosquito *NimB2*), and (3) Nimrod C-types (*Drosophila nimrod C 1–4* and *eater* and Mosquito *eater*) (Estevez-Lao and Hillyer, 2014; Somogyi et al., 2008). The Draper-type proteins have an EMI domain at the N-terminus and one copy of the

NIM motif followed by several EGF domains. The *Nimrod B* genes lack transmembrane domains and are most likely secreted proteins (Kurucz et al., 2007a). *Nimrod C* family genes such as *eater* are transmembrane proteins with a variable number of NIM repeats. Phylogenetic analysis of *Drosophila* Draper and Nimrods A, B, and C indicate that Draper-type molecules evolved first—likely from a protein with a poly-EGF run, giving rise to Nimrod-family proteins (Somogyi et al., 2008). Draper-type proteins show wide taxonomic distribution and are important for phagocytosis in several organisms, including *C. elegans*, humans, mosquitoes, and *Drosophila*. Poly-NIM proteins are to unique to insects where they evolved from a Draper-type ancestor. The Nimrod C-type emerged first and, as will be discussed later, members of this family function as phagocytic receptors. Loss of the transmembrane domain of a Nimrod C family protein may have led to the formation of the secreted Nimrod B-type proteins—members of which recognize pathogens in mosquitoes and the large beetle, *Holotrichia domphalia* (Ju et al., 2006; Midega et al., 2013).

The Ezekowitz group performed microarrays of S2 cells and identified 46 genes with signal sequences and transmembrane domains that were downregulated over twofold after RNAi of a vital hemocyte GATA transcription factor, *serpent* (Kocks et al., 2005; Ramet et al., 2002). They tested for effects on phagocytosis by silencing these candidates using RNAi in S2 cells and looking for binding of *S. aureus* and *E. coli* to the cells. One gene, which the researchers named *eater*, encodes a predicted cell surface receptor and showed strong reduction in *S. aureus* and *E. coli* phagocytosis. The Eater protein contains an N-terminal signal peptide, 32 EGF-like domains, a transmembrane domain, and an intracellular C-terminal domain with a predicted tyrosine phosphorylation motif. The Eater protein has a low level homology (25% amino acid identity overall) to *C. elegans* CED-1, a receptor for apoptotic cells. The extracellular domain of CED-1 is homologous to a human scavenger receptor on endothelial cells, SREC. Analysis of *eater* expression showed that it is restricted to the plasmatocyte lineage.

The first four EGF-like repeats of Eater have a high level of amino acid diversity, suggesting that the N-terminal part of the protein may be important for binding to ligands. Amino acids 1–199 strongly bind to the Gram-negative bacteria *S. marcescens*, an entomopathogen, and the Gram-positive bacteria *S. aureus*. Larval hemocytes from *eater* null flies showed significantly impaired phagocytosis of both *S. marcescens* and *S. aureus*. Adult *eater* null hemocytes were significantly impaired for phagocytosis of both *E. coli* and *S. aureus*. To assess the effects of phagocytic defects in a natural infection model the researchers tested survival phenotypes of *eater* null flies that had been fed *S. marcescens*. *eater* null flies were more susceptible to *S. marcescens*, and this increased susceptibility was accompanied by 10,000-fold higher levels of *S. marcescens* in the fly hemolymph. The AMPs *Drosomycin* and *Diptericin* were induced normally in *eater* null mutants indicating that the susceptibility to *S. marcescens* was not attributed to defects in the humoral immune response but instead were likely caused by the impaired cellular immune response.

A soluble Fc-tagged construct of Eater (Eater-Fc—199 amino acids of the N-terminus of Eater), bound to live or inactivated Gram-positive bacteria, *S. aureus* and *E. faecalis* (Chung and Kocks, 2011). In contrast, Eater-Fc was unable to bind live or heat-killed Gram-negative bacteria *E. coli*, *S. marcescens*, and *Pseudomonas aeruginosa*. Interestingly, membrane-disrupting treatments, such as treating bacteria with the cationic AMP, Cecropin A, unmasked Eater-Fc ligands on Gram-negative bacteria. To assess the relative importance of Eater recognition and phagocytosis of Gram-positive pathogens, the group conducted in vitro binding assays, in vivo phagocytosis assays, and survival assays for three types of Gram-positive bacterial pathogens. Eater is important for the phagocytosis of *S. aureus* and *E. faecalis* in vitro, but it is not required for phagocytosis of *M. luteus* (Chung and Kocks, 2011; Nehme et al., 2011). Consistent with the phagocytic characteristics of Eater, *eater* null flies were susceptible to *S. aureus* and *E. faecalis* but showed little to no susceptibility after *M. luteus* infection (Nehme et al., 2011). Additionally, augmenting the host response by systemically overexpressing the AMP Defensin or the Toll signalling pathway protected *eater* null flies against *E. faecalis* but not against *S. aureus*. Thus, Eater-mediated phagocytosis is required for host defence against some Gram-positive bacteria (*S. aureus*) but is dispensable for defence against others (*E. faecalis*). These experiments indicate that the cellular and humoral responses, as well as the interactions between them, are tailored to each pathogen the fly encounters, highlighting the flexibility of the innate immune response in *Drosophila*.

To identify hemocyte-specific molecules, the Hultmark and Ando groups generated a set of monoclonal antibodies against hemocytes (Kurucz et al., 2003). The research group used the antibodies to identify a plasmatocyte-specific EGF domain containing transmembrane protein, Nimrod C1 (NimC1) (Kurucz et al., 2007b). Immunofluorescence studies revealed that two monoclonal antibodies, P1 and P2, that recognized different epitopes on NimC1 bound to the majority of larval hemocytes (with plasmatocyte morphology) but were absent on lamellocytes and crystal cells. Immunoprecipitation followed by MALDI-TOF mass spectrometry identified the P1 target as a 90–100 kDa single-pass transmembrane protein with 10 NIM repeats, which the authors named Nimrod C1. Interestingly, the P1 antibodies did not recognize any antigens on S2 cells, indicating that *nimC1* is not expressed in this cell line, perhaps explaining why this receptor was not identified in the whole genome RNAi screens conducted by the Ezekowitz group. NimC1 localizes to the plasma membrane of larval hemocytes. RNAi-mediated silencing of *nimC1* in larval hemocytes decreased *S. aureus* uptake to one-third of the controls but had no effect on *E. coli* phagocytosis. However, overexpression of NimC1 in S2 cells stimulated uptake of *S. aureus* and *E. coli* by 2.5- and 2-fold, respectively. Thus, similar to Eater, NimC1 is important for *S. aureus* phagocytosis in plasmatocytes, and may play a redundant role for *E. coli* phagocytosis. Interestingly, NimC1 overexpression did not change the amount bacteria that bound to S2 cells, but did lead to increased uptake. Based on this result, it is unlikely that NimC1 directly binds to the microbe, but instead it may act as a coreceptor, perhaps with Eater. Alternatively, NimC1 could be important for a later stage of the phagocytic process, such as particle engulfment.

The importance of Eater and NimC1 for *E. coli* phagocytosis was recently assessed in *eater* or *nimC1* RNAi flies (Horn et al., 2014). The researchers counted the number of bacteria in adult hemocytes by injecting fluorescently labelled *E. coli* and imaging individual hemocytes in the dissected dorsal vessel. Downregulation of *eater* and *nimC1* caused a modest but significant ($p < 0.05$) reduction in *E. coli* phagocytosis in adult hemocytes. In contrast, the Hultmark group's ex vivo larval phagocytosis assay showed that *E. coli* phagocytosis was not significantly affected in *nimC1*-silenced hemocytes. The discrepancy between the function of NimC1 in larval hemocytes vs adult hemocytes could also be attributed to the nature of the experiments themselves or that NimC1 may need to interact with host factors that are only present in the hemolymph. In the future, in vivo larval phagocytosis assays could help to clarify the function of NimC1 in the animal, since in vivo experiments more closely reflect physiological conditions.

The final *Drosophila* Nimrod family member characterized as a phagocytic receptor is Draper, which is expressed in two types of phagocytes, glial cells and hemocytes (Freeman et al., 2003). *Draper* is the homolog of *ced-1*, a gene that encodes a receptor for apoptotic cells in *C. elegans* (Zhou et al., 2001). Nakanishi and group showed that Draper is an important receptor involved in the phagocytosis of apoptotic cells by glia and hemocytes

(Manaka et al., 2004). This paper reported that, in S2 cells, Draper does not recognize the common 'eat me' signal, phosphatidylserine (PS) (Manaka et al., 2004). The same group later determined that the endosomal protein Pretaporter is exposed on the surface of apoptotic cells and serves as a ligand for Draper (Kuraishi et al., 2009). Additionally, their genetic studies found that the Rho-GTPases *Rac1* and *Rac2* are involved in the Draper/Ced-6 pathway to engulf apoptotic cells. Finally, they reexamined the possibility Draper binds PS using biochemical and genetic techniques and found full-length Draper efficiently binds to PS while truncated Draper proteins did not (Tung et al., 2013).

Franc and group confirmed the role for Draper and Ced-6 in phagocytosis of apoptotic cells and further examined the possibility of a role for Draper in the phagocytosis of bacteria (Cuttell et al., 2008). The heat-killed bioparticles used by the authors were conjugated to the pH-sensitive dye pHrodo, which fluoresces at low pH (~4.5). After being engulfed by the cell, pHrodo-labelled particles will only fluoresce in the acidic environment of the phagolysosome. *draper* RNAi-treated S2 cells showed significantly less fluorescence of pHrodo-labelled *E. coli* and *S. aureus*. The lack of fluorescence could be indicative of a decreased uptake and/or impaired maturation of the phagosome. Additionally, *draper* and *ced-6* mutant flies injected with pHrodo-*S. aureus* and *E. coli* showed reduced fluorescence in hemocytes near the dorsal vessel. Interestingly, a study by Shiratsuchi and colleagues found that Draper recognizes LTA in the cell wall of *S. aureus* (Hashimoto et al., 2009).

Several intracellular signalling molecules and scaffold proteins that participate in Draper-mediated phagocytosis have been described. In *C. elegans*, the Draper homolog, CED-1, acts upstream CED-6, an adaptor protein that serves as a molecular scaffold for signalling complexes at the phagocytic cup (Zhou et al., 2001). CED-1 contains an intracellular NPxY motif that is a binding site for proteins containing a phosphotyrosine binding (PTB) domain that is a potential binding site for proteins containing Src-homology-2 domains (SH2). In *C. elegans*, the PTB domain adaptor protein, CED-6 binds to the CED-1 NPxY motif to promote phagocytosis of cell corpses (Liu and Hengartner, 1998; Su et al., 2002) *Drosophila* Ced-6 is an adaptor protein with an SH2 domain and a Pleckstrin homology (PH) domain. The SH2 domain of Ced-6 binds to a phosphorylated tyrosine in the NPxY motif in the intracellular region of activated Draper, and the PDZ domain recruits downstream factors important for apoptotic cell clearance and phagocytosis of bacteria (Awasaki et al., 2006; Fujita, 2012;

MacDonald et al., 2006; Ziegenfuss et al., 2008). Furthermore, Draper-mediated phagocytosis of apoptotic corpses requires phosphorylation of the immune receptor tyrosine-based activation motif (ITAM) in the Draper intracellular domain. Shark, the *Drosophila* counterpart of Syk and Zap70, is a Src-family kinase that mediates Draper ITAM phosphorylation in glial cells, promoting apoptotic cell phagocytosis (Fujita, 2012; Ziegenfuss et al., 2008). It is unknown, however, if Shark, or some other *Drosophila* tyrosine kinase, phosphorylates the Draper ITAM in response to bacteria. Thus, activation of Draper and the assembly of downstream signalling cascades may be ligand dependent, a possibility that might add yet another layer of complexity to the function of Draper.

Together this data demonstrate Draper is a multifunctional receptor with wide-ranging ligand specificity, an important feature for a receptor found on the surface of invertebrate phagocytes. It recognizes LTA on the surface of Gram-positive bacteria, phosphatidylserine, and Pretaporter on the surface of apoptotic cells, and undetermined ligands on the surface of Gram-negative bacteria.

The same study that identified and characterized *Drosophila nimC1* also identified six putative Nimrod family members in the malaria vector *An. gambiae* (Kurucz et al., 2007a). Using a bioinformatics approach, the group identified *An. gambiae eater, draper,* one Nimrod A subclass gene, one Nimrod B subclass gene, and two Nimrod C subclass genes. However, a reannotation of the *An. gambiae* genome predicted just four Nimrod gene family members (Li et al., 2006). To address these discrepancies, and to elucidate the immune roles of Nimrod family members in *An. gambiae*, the Hillyer and Christophides groups conducted detailed bioinformatics, molecular, and functional analyses of the mosquito Nimrods (Estevez-Lao and Hillyer, 2014; Midega et al., 2013). The Christophides group carried out an extensive bioinformatics comparison of the *An. gambiae* and *D. melanogaster* genomes that honed in on two NIM repeat containing proteins in the mosquito, *AgNimB2* and *AgEater* (Midega et al., 2013). The AgEater protein is predicted to be a plasma membrane-bound receptor with 21 NIM repeats and AgNimB2 is predicted to be secreted. Injection of *AgEater* dsRNA into adult mosquitoes, to specifically silence the expression of *AgEater*, did not affect phagocytosis of either *S. aureus* or *E. coli.* Likewise, a separate report by the Christophides group found that dsRNA-mediated silencing of *AgEater* in mosquito cell lines did not alter *E. coli* phagocytosis (Lombardo et al., 2013). Together, these findings suggest that AgEater is not required, or plays a redundant role, in bacterial phagocytosis in the mosquito.

Much less is known about the functional role of mosquito NimB2. Christophides and colleagues showed that AgNimB2 exhibits specificity for *S. aureus* (Midega et al., 2013). dsRNA-mediated silencing of *AgNimB2* reduced the fluorescence of pHrodo-labelled *S. aureus* by ~50% in adult mosquitoes. To determine if AgNimB2 or AgEater directly binds to the surface of microbes, the authors carried out in vitro binding assays with conditioned media containing each of the proteins. Interestingly, neither AgNimB2 nor AgEater directly bound to *S. aureus*, *E. coli*, or *Plasmodium* ookinetes, suggesting an indirect mechanism of action for each protein during phagocytosis.

The Hillyer group also investigated whether the Nimrod gene family participates in the immune response of *An. gambiae* (Estevez-Lao and Hillyer, 2014). The group did not examine the phagocytic roles. Instead, they measured survival and bacterial loads after dsRNA-mediated silencing of one or multiple genes. Adult mosquitoes were injected with *E. coli* following the knockdown of *AgEater*, *AgNimB2*, or *AgDraper*. Loss of *AgEater* led to an increased number of viable *E. coli* 24 h post infection, while knockdown of the other two Nimrod genes had no effect. However, this increased microbial load was not associated with an increased susceptibility to *E. coli*, with about 90% of control and knockdown mosquitoes succumbing to the infection within 10 days. Subsequent survival experiments revealed that: (1) *AgEater* and *AgDraper* are important for mosquito survival following infection with the Gram-positive bacteria *S. epidermidis* and (2) knockdown of individual Nimrod family members does not affect the mosquito's survival following *E. coli*, *M. luteus*, or *S. aureus* infection. Together, studies from the Hillyer and Chrisophides groups demonstrate that *AgEater*, *AdNimB2*, or *AgDraper* regulate the mosquito antibacterial response. In order to definitively determine if the mosquito Nimrod family members are genuine PRRs, additional experiments will be necessary. For example, an examination of the binding of exogenously expressed mosquito Nimrods to components of Gram-positive and Gram-negative bacterial cell walls should address the questions of specificity and redundancy in pathogen recognition.

3.3 Peptidoglycan-Recognition Receptors Important for Phagocytosis

The peptidoglycan recognition proteins (PGRPs) are important microbial receptors that were first identified in the hemolymph of silkworms, *Bombyx mori* (Yoshida et al., 1996). PGRPs were subsequently found in moth (*Trichoplusia ni*) larvae challenged with the Gram-positive bacteria *Enterobacter cloacae*

(Kang et al., 1998). Moth PGRPs bind to peptidoglycan (PGN), a complex polymer consisting of sugars and amino acids that is restricted to the cell wall of Gram-positive and Gram-negative bacteria (Yoshida et al., 1996). Peptidoglycan is made up of alternating *N*-acetylglucosamine (GlcNAc) and *N*-acetylmuramic acid (MurNAc) residues that are cross linked to each other by short peptide bridges of three to five amino acids. PGN of most Gram-positive bacteria contain a lysine residue as the third amino acid in the peptide chain and is known as Lys-PGN. Gram-positive bacilli and Gram-negative bacteria have mesodiaminopimelic acid as the third amino acid (DAP-type PGN). Another feature that is unique to DAP-type PGN is the presence of a monomer, known as tracheal cytotoxin (TCT), on the terminal PGN unit. Finally, Gram-positive and Gram-negative bacteria PGN differ in their location in the cell wall. DAP-type PGN forms a single layer that is hidden in the periplasmic space beneath the outer membrane and LPS layer of the cell wall of Gram-negative bacteria. In Gram-positive bacteria, PGN is highly abundant and can account for half of the mass of the cell wall. Gram-positive bacteria PGN form a multilayer structure that is exposed on the surface (Royet and Dziarski, 2007).

The Steiner group performed an elegant study to characterize the structure and relatedness of insect and mammalian PGRPs. They cloned PGRP from moth, mouse, and human samples and found that transcripts corresponding to PGRP were highly expressed in organs of the immune system. Comparison of the predicted amino acid sequences of PGRPs revealed that murine and human PGRPs share 43% sequence identity with *T. ni* PGRP. Additionally, mammalian PGRPs function in a manner analogous to insect PGRPs, as demonstrated by an experiment where recombinant murine PGRP bound to PGN in a manner similar to *T. ni* PGRP. Further examination of the structure of *T. ni* PGRP revealed that the protein shared 28% identity and 50% similarity with bacteriophage T3 lysozyme, a zinc-dependent *N*-acetylmuramoyl-L-alanine amidase (Kang et al., 1998). *N*-acetylmuramoyl-L-alanine amidases cleave peptidoglycan at the lactylaminde bond, removing the peptidic bridge from the sugar backbone. Interestingly, recombinant *T. ni* PGRP showed no amidase activity on *E. coli* cell walls. This observed lack of amidase activity could be explained by the fact that *T. ni* PGRP lacks the zinc-binding residues present in the phage enzyme and suggests the primary function of the *T. ni* PGRP was recognition and binding of PGN (and not cleavage).

There are 13 PGRP genes in *Drosophila* and studies in the fruit fly model system have provided the most comprehensive data on PGRPs. *Drosophila*

PGRPs recognize microbial ligands upstream of the Toll and IMD signalling pathways, the major signalling cascades regulating the humoral innate immune response. Briefly, the Toll signalling pathway is activated after infection with fungi, Gram-positive bacteria, or Drosophila X virus, while the IMD pathway is activated by Gram-negative bacteria. Activation of the Toll and IMD pathways leads to the production of systemic AMPs and other immune responsive effectors.

Six *Drosophila* PGRP genes code for long (L) forms, four of which are transmembrane proteins localized at the plasma membrane. The remaining seven PGRP genes are short (S) forms that are predicted to be secreted (Werner et al., 2000). *Drosophila* PGRPs can also be divided based on their recognition and/or catalytic properties. Members of the noncatalytic group (PGRP-SA, SD, LA, LC, LD, LE, and LF) serve as microbial sensors and PRRs. These PGRPs lack the critical cysteine residue in the enzymatic pocket of the PGRP domain and are unable to degrade PGN (Mellroth et al., 2003). The second group, catalytic PGRPs have either been experimentally verified (PGRP-SC1, LB, and SB1) or predicted (SC2 and SB2) to possess amidase activity needed to degrade PGN (Bischoff et al., 2006; Mellroth and Steiner, 2006; Mellroth et al., 2003; Zaidman-Remy et al., 2006, 2011).

Finally, recognition of PGN plays a critical role in host defence in *Drosophila*. Evidence supporting the importance of recognition of PGN by phagocytes recently came from adult and larval phagocytosis studies using an *S. aureus* strain with a temperature-sensitive mutation in UDP-*N*-acetylenolpyruvylglucosamine reductase (*murB*). This reductase is a key enzyme in bacterial peptidoglycan synthesis, and *murB* mutant *S. aureus* contain significantly less peptidoglycan polymers in their cell wall than their wild-type counterparts. *Drosophila* hemocytes phagocytosed *murB* bacteria 50% less efficiently than wild-type *S. aureus*, and this phenotype could be rescued by complementation of the mutant strain with the wild-type gene (Shiratsuchi et al., 2012).

The Wu laboratory reported that PGRP-SC1a is a receptor for the Gram-positive bacteria, *S. aureus*, and not for Gram-negative bacteria (Garver et al., 2006). Using an adult, in vivo phagocytosis assay, the group screened a collection of ethylmethane sulfonate (EMS) mutated flies (Koundakjian et al., 2004). One mutant, *picky eater* (*picky*), was defective for *S. aureus* phagocytosis—only 25% of tested flies took up the fluorescein-labelled particles. However, *picky* flies were able to efficiently phagocytose *E. coli* and *Saccharomyces cerevisiae* zymosan particles as well as live, GFP-expressing *Bacillus subtilis* (a Gram-positive bacteria possessing DAP-type PGN). The *picky* mutant was also impaired for survival after *S. aureus* infection. The *picky* mutation mapped to the catalytic PGRP gene, *PGRP-SC1a.* Both the impaired recognition of *S. aureus* and survival of *picky* mutants were rescued by transgenic expression of *PGRP-SC1a.* The catalytic activity of PGRP-SC1a was required for phagocytosis and clearance of *S. aureus* since a noncatalytic PGRP-SC1a (in which the critical cysteine residue was replaced by a serine) was not sufficient to rescue phagocytosis or survival after *S. aureus* infection. This data provide strong evidence for the role of PGRP-SC1a as a PRR in the fruit fly.

PGRP-SA is another PRR with dual roles in *Drosophila* humoral and cellular immunity. A screen to identify mutations that impair the production of the Toll pathway AMP, Drosomycin, identified the mutation *semmelweis* (*seml*) (Michel et al., 2001). The *seml* mutation is caused by an amino acid change in the PGRP domain, Cysteine-80 to Tyrosine, and this change effectively inactivated the PGRP-SA protein. After infection with Gram-positive bacteria, but not fungi, *seml* flies are unable to produce Drosomycin due to impaired Toll activation (Michel et al., 2001). In addition to its role in activating the Toll pathway, PGRP-SA may also be important for phagocytosis of Gram-positive bacteria. Garver et al. tested *seml* mutants using the adult in vivo phagocytosis assay and found that 94% of *seml* mutant flies efficiently phagocytosed *E. coli*, while only 25% were able to phagocytose *S. aureus* (Garver et al., 2006). In contrast, a separate study looking at in vivo phagocytosis of *S. aureus* in *seml* mutants failed to observe an effect on *S. aureus* phagocytosis in adult flies (Nehme et al., 2011). The reason for the discrepancy observed between the two papers is not clear. In particular, the experiment should be performed in triplicate in order to draw reliable conclusions about the phagocytosis phenotype of the *seml* mutant. We recently conducted additional, independent, in vivo *S. aureus* phagocytosis experiments using *seml* mutants. We found that *seml* flies, in our hands, indeed show defective *S. aureus* phagocytosis (A. Nazario-Toole, unpublished) as previously reported.

An RNAi screen in S2 cells conducted by the Ezekowitz group identified the noncatalytic, membrane-bound PGRP-LC as important for phagocytosis of Gram-negative (*E. coli*) but not Gram-positive (*S. aureus*) bacteria (Ramet et al., 2002). Decreased expression of *PGRP-LC* reduced *E. coli* phagocytosis by 30% and also affected binding of the bacteria to the cell surface. This modest decrease in phagocytosis was likely due to the fact that other receptors, such as Eater or SR-CI, participate in recognition and uptake of Gram-negative bacteria. The group assessed the humoral immune role of PGRP-LC by using oligonucleotide microarrays to measure the induction of genes 6 h after *E. coli* exposure. RNAi of *PGRP-LC* in S2 cells dramatically reduced the expression of genes regulated by the IMD pathway, such as the AMP *Attacin*. To test the function of PGRP-LC in vivo, the group generated *PGRP-LC* mutants: *Δ5* is a null allele and *N18* is a hypomorphic allele. Both *Δ5* and *N18* flies were more susceptible to *E. coli* and this susceptibility was accompanied by a reduced expression of IMD-regulated AMPs (Ramet et al., 2002). Around the same time, two additional reports, one by the Anderson group and another by the Royet group, identified separate mutations in PRGP-LC that disrupted IMD pathway induction (Choe et al., 2002; Gottar et al., 2002). Together, the work of these groups clearly established that PGRP-LC is the major receptor upstream of the IMD pathway in vivo and in vitro.

Although the data supporting a role for PGRP-LC during the humoral response to Gram-negative bacteria was in agreement, the importance of PGRP-LC as a phagocytic receptor in the fruit fly was less clear. The Ezekowitz group showed that silencing *PGRP-LC* led to a modest decrease in *E. coli* phagocytosis in vitro, but the two other groups reported that PGRP-LC mutant blood cells efficiently phagocytose Gram-positive and Gram-negative bacteria (Choe et al., 2002; Garver et al., 2006; Ramet et al., 2002). This discrepancy may be attributed to the fact that the Ezekowitz group silenced all three *PGRP-LC* isoforms, *PGRP-LCa*, *-LCx*, and *-LCy* (Werner et al., 2000). PGRP-LC protein variants share the same intracellular signalling domains but have unique extracellular domains. PGRP-LCx recognizes PGN purified from *E. coli* as well as the TCT fragment of PGN. In contrast PGRP-LCa specifically recognizes TCT (Kaneko et al., 2004). The *PGRP-LC ird7* mutation caused an amino acid change in the PGRP domain of only PGRP-LCx (Choe et al., 2002). In contrast, RNAi treatment decreased the expression of all isoforms. It is possible that the unaltered expression of *PGRP-LCa* in the *ird7* mutant is sufficient to allow for PGRP-LC-mediated uptake of *E. coli* in vivo. Additionally, other *E. coli* PRRs may participate in phagocytosis in vivo.

Through studying the role of an evolutionarily conserved family of extracellular proteins, the nonaspanins, the Fauvarque group corroborated findings that PGRP-LC is important for phagocytosis of Gram-negative bacteria (Bergeret et al., 2008; Perrin et al., 2015). Nonaspanins (TM9 protein family) are characterized by the presence of a large extracellular N-terminal domain and nine transmembrane domains and, members of this protein family mediate adhesion and phagocytosis in *Dictoyostelium* amoebae (Cornillon et al., 2006). The Fauvarque group initiated a genetic and phenotypic analysis to characterize the function of *TM9SF4* in *Drosophila* (Bergeret et al., 2008). The *TM9SF4* deletion mutant was more susceptible to pathogenic Gram-negative infection (*Klebsiella pneumoniae* and *E. cloacae*) but showed normal resistance to nonpathogenic Gram-negative bacteria (*E. coli*) and Gram-positive bacteria. Additionally, the Toll and IMD responsive genes were not affected in *TM9SF4* mutants. *TM9SF4* mutant adults did show impaired phagocytosis of GFP-labelled *K. pneumoniae* and the hemolymph of mutant flies was carrying a higher bacterial load. An ex vivo phagocytosis assay showed that *TM9SF4* mutant larval hemocytes phagocytosed *E. coli* two times less efficiently than wild-type hemocytes. The group explored the possibility that PGRP-LC and TM9SF4 interact in vivo due to the fact that *TM9SF4* mutants and *PGRP-LC* RNAi of S2 cells both exhibit specific defects in the phagocytosis of Gram-negative bacteria (Perrin et al., 2015). In S2 cells, GFP-tagged TM9SF4 and V5 epitope-tagged PGRP-LC coimmunoprecipitate indicating that the proteins do indeed interact. Furthermore, in the fat body, the functional equivalent of the mammalian liver, GFP-tagged TM9SF4 and FLAG-tagged PGRP-LC colocalized at the plasma membrane. Importantly, in S2 cells, TM9SF4 is required for PGRP-LC localization to the plasma membrane and the observed reduced phagocytosis of Gram-negative bacteria in *TM9SF4* null flies may be due to a loss of PGRP-LC at the plasma membrane. Finally, a study in adult *An. gambiae* mosquitoes provides evidence that PGRP-LC is a phagocytic receptor in other insects. This study, by Moita and colleagues, showed that downregulation of *PGRP-LC* by injecting dsRNA specifically impaired phagocytosis of *E. coli* in adult mosquitoes (Moita et al., 2005).

3.4 Integrins

In addition to identifying Draper as a phagocytic receptor, the Nakanishi group found that the integrin heterodimer, αPS3 and βν, is a receptor for *S. aureus* and apoptotic cells in Drosophila (Nagaosa et al., 2011; Nonaka et al., 2013; Shiratsuchi et al., 2012). Using a procedure described in Kurucz et al., the group raised monoclonal antibodies against *Drosophila* hemocytes by immunizing mice with larval hemocytes (Nagaosa et al., 2011). They then added each antibody to culture dishes containing the larval phagocytic blood cell line, l(2)mbn, and then coincubated those cells with chemically killed S2 cells to look for effects on phagocytosis. Treatment with an antibody that recognized an extracellular Perlecan-like protein, Trol, led to decreased uptake of the dead S2 cells. The extracellular region of the Trol protein has three RGD domains, a motif recognized by a subset of integrins. The binding of integrin to RGD ligands induces the phosphorylation of the tyrosine kinase, Focal adhesion kinase (FAK) (Shattil et al., 2010). l(2)mbn cells treated with recombinant Trol protein showed a 1.5-fold increase in the levels of phosphorylated FAK, indicating that Trol binds to integrin on the surface of the cells.

Integrin functions as a heterodimer of two transmembrane subunits, α and β integrin. In the *Drosophila* genome, five genes code for the α subunit and two genes code for the β subunit (Brown et al., 2000). The Nakanishi group initially focused on the examining the two β subunits. To do so, they examined phagocytosis of dead S2 cells by Croquemort-positive hemocytes from mutant flies lacking either integrin β subunit gene. Cells from *Integrin beta nu* (*Itgbn*) mutant embryos displayed apoptotic cell phagocytosis defects. Antibody staining confirmed that the βν protein is normally found on the surface of embryonic hemocytes. To determine the relationship of integrin βν with the known apoptotic cell phagocytic receptor, Draper, the authors analysed phagocytosis of S2 dead cells by embryonic hemocytes derived from *Intbn* and *drpr* single and double mutants. Simultaneous loss of both receptors decreased phagocytosis of dead S2 cells to half that of single mutants, indicating that Draper and integrin βν are independent receptors for apoptotic cells (Nagaosa et al., 2011).

Subsequent studies examined a role for integrin βν in the phagocytosis of *S. aureus*. Similar to the results with apoptotic cells, *Intbn* and *drpr* double-mutant flies phagocytosed *S. aureus* less efficiently than single-mutant flies, indicating that Draper and integrin βν also act independently as receptors during *S. aureus* phagocytosis. Importantly, integrin βν-deficient adult and larval hemocytes phagocytose *S. aureus* less efficiently than control flies, but are able to phagocytose *E. coli* and the DAP-type PGN-containing Gram-positive bacteria *B. subtilis* with the same efficiency as wild-type flies. Adult flies lacking the integrin βν subunit were more susceptible to septic *S. aureus* infection. And, these flies carried a higher bacterial load indicating that integrin βν-mediated phagocytosis of *S. aureus* limits bacterial growth within the fly. Thus, integrin βν is a receptor for *S. aureus* (and possibly other Gram-positive bacteria) that plays a critical role in the host cellular immune response.

The Nakanishi group then sought to determine which component of the cell wall of *S. aureus* is recognized by integrin βν by carrying out binding assays with *murB* mutated *S. aureus*. The authors incubated either *murB* mutant or parental *S. aureus* bacteria with a GST-fused recombinant integrin βν protein. Western blotting of cell lysates, using anti-GST antibodies, revealed that wild-type *S. aureus* bound to integrin βν more efficiently than the *murB* mutant strain. Finally, GST-fused integrin βν bound to culture dishes was able to adhere to a solid phase preparation of *S. aureus* peptidoglycan in a dose-dependent manner, but this was not the case with GST alone. Together, these binding assays show that integrin βν binds to *S. aureus* peptidoglycan and this physical association may be critical for integrin βν's role as a phagocytic receptor.

Drosophila integrin functions as a heterodimer of β and α subunits. The identity of the α subunit that forms a complex with integrin βν was determined in a recent paper from the Nakanishi lab (Nonaka et al., 2013). Nonaka and colleagues utilized RNA interference to silence the expression of each of the five *Drosophila* α subunit genes specifically in hemocytes. They assessed ability of the α integrin-depleted larval hemocytes to phagocytose apoptotic cells and identified one subunit, αPS3, whose loss impaired phagocytosis. αPS3 integrin protein is encoded by *scab* (*scb*). *scb* deficiency flies (mutant flies with a deletion in the chromosome region that includes *scb*) and flies with a P-element insertion that disrupts the coding region of *scb*, both show a reduction in the level of apoptotic cell clearance. Forced expression of a transgenic wild-type αPS3 in the *scb* deficiency mutant was sufficient to restore the phagocytosis of apoptotic cells, indicating that αPS3 is the *Drosophila* α integrin subunit required for the recognition and uptake of dead cells in vivo. To assess a functional interaction between βν and αPS3, the authors used RNAi to silence βν (*Itgbn*) or αPS3 (*scb*) in hemocytes—singly or in combination. In all three types of embryos, only about 20% of the hemocytes were able to phagocytose apoptotic cells. Because phagocytosis of apoptotic cells occurred almost equally in the three fly lines, βν and αPS3 function in the same pathway during phagocytosis in embryonic hemocytes. A similar approach was taken to assess a functional interaction between αPS3 and βν during phagocytosis of *S. aureus*. Embryonic hemocytes from flies with *Itgbn* and *scb* silenced together, or alone, showed equal levels of *S. aureus* phagocytosis. These results indicate that αPS3 and βν form a heterodimer that serves as a phagocytic receptor for *S. aureus*. Importantly, the physical association of αPS3 and βν was confirmed through immunoprecipitation and Western blotting of l(2) mbn cell lysates. In conjunction, the genetic and biochemical analyses carried out by the Nakanishi group clearly establish that the αPS3/βν integrin heterodimer is a phagocytic receptor for apoptotic cells and *S. aureus* in *Drosophila* embryos.

In *An. gambiae*, a member of the β-integrin family, *β-integrin 2* (*BINT2*), was shown to be important for phagocytosis of *E. coli*, but not *S. aureus*, in adult mosquitoes injected with *BINT2*-dsRNA (Moita et al., 2005). Moita and colleagues cloned *BINT2* from an abdominal cDNA library and phylogenetic analyses revealed that BINT2 is a 1:1 ortholog to *Drosophila* βν integrin (Moita et al., 2006). Treating phagocytic 5.1* mosquito cells with *BINT2*-dsRNA reduced *E. coli* phagocytosis by 40%. The authors note that 5.1* cells phagocytose *S. aureus* poorly. Thus, they only examined how knockdown of *BINT2* affected *E. coli* phagocytosis. A bioinformatics examination of the *An. gambiae* genome identified two β-integrin genes, *BINT1* and *BINT2*, but only BINT2 played a role in *E. coli* phagocytosis. The group also identified four α-integrin genes. Two of these genes, *AINT3A* and *AINT3B*, form an orthologous group with *Drosophila αPS3*, *αPS4*, and *αPS5*. We know that *Drosophila* αPS3 and βν form a heterodimer that serves as a receptor for *S. aureus*. Perhaps heterodimers of BINT2/AINT3A or BINT2/AINT3B act as phagocytic receptors in mosquito blood cells, a possibility that could be investigated in vivo using dsRNA to simultaneously silence α- and β-integrin genes.

3.5 Down Syndrome Adhesion Molecule 1 (Dscam 1)

The PRRs of innate immunity are effective in recognizing a wide array of PAMPs. However, innate immune responses are constrained to structures that are common to pathogens and conserved during evolution. In contrast, receptors of the adaptive immune response are able to recognize an almost infinite diversity of antigens through somatic rearrangement of genes. Members of the immunoglobulin superfamily (IgSF), such as antibodies and the antigen receptors found on the surface of B and T lymphocytes, are an essential part of mammalian adaptive immune responses.

One IgSF member found in *Drosophila* is *Down syndrome adhesion molecule 1* (*Dscam1*). The *Drosophila* genome contains four Dscam-like genes and the most extensively characterized is *Dscam1* (Armitage et al., 2012; Vogel et al., 2003). The *Dscam1* gene is arranged into clusters of variable exons (exons 4, 6, 9, and 17) that are flanked by constant exons. Mutually exclusive alternative splicing of the variable exons generates a large protein isoform repertoire that has the potential to recognize and bind diverse ligands (Schmucker et al., 2000). Dscam1 is critical for nervous system development and is essential for axon guidance and the formation of neural connections in *Drosophila* (Wojtowicz et al., 2004; Zhan et al., 2004).

To explore a potential role for the hypervariable Dscam1 receptor in the immune response to bacteria, the Schmucker group conducted a functional analysis of *Dscam1* expression in immune competent tissues of *Drosophila* (Watson et al., 2005). In situ hybridization of larval tissue revealed that *Dscam1* is expressed in neural tissue, hemocytes, and fat body tissue. cDNAs derived from each of the three tissues were hybridized to microarrays containing 50-mer oligos for all alternatively spliced exons. Based on the number of alternatively spliced exons detected, an estimated 18,000 Dscam1 receptor isoforms are expressed in hemocytes and the fat body. The Dscam1 protein is expressed in immune tissues; antibodies against the common Dscam cytoplasmic region recognized Dscam in S2 cells, larval hemocytes, and larval fat bodies. Western blots using the anti-Dscam1 antibody revealed the presence of a soluble Dscam1 protein in S2 cell-conditioned media and larval hemolymph raising the possibility that secreted Dscam1 proteins act as opsonins or receptors that recognize microbes present in the hemolymph of the fly.

GFP-positive hemocytes were purified from wild-type and *Dscam* mutant larvae (with a transallelic combination of a hypomorphic and an amorphic *Dscam1*) and the amount of fluorescently labelled *E. coli* phagocytosed by these hemocytes was determined using flow cytometry. Fifty-five percent of mutant hemocytes phagocytosed the bacteria while 90% of wild-type hemocytes took up the bacteria. Silencing *Dscam1* expression using a hemocyte-specific promoter led to a 60% reduction in the number of hemocytes that phagocytosed *E. coli*. Additionally, treating S2 cells with anti-Dscam1 antibody (to block Dscam1 function by binding to the extracellular domain) also reduced the number of cells that could phagocytose *E. coli*. Both loss of Dscam1 expression and blocking Dscam1 function with antibodies caused significant phagocytosis defects indicating that Dscam1 functions as a receptor for *E. coli*. To determine if Dscam1 directly binds to *E. coli*, the authors tested the binding of certain Dscam1 isoforms to live DH5α *E. coli*. Two Dscam1 isoforms, the full extracellular domain and the N-terminal Ig-like domain with the first Fibronectin III domain, were able to bind to the bacteria. However, an alternatively spliced Dscam1 isoform, containing the complete extracellular domain, was unable to bind to *E. coli*. The distinct binding properties of the tested isoforms hinted at the possibility that Dscam1 isoforms bind distinct microbial ligands, thus increasing the number of possible ligands recognized by this receptor.

Studies using the mosquito *An. gambiae* clearly demonstrated that Dscam isoforms show specificity for different microbes (Dong et al., 2006b). The immune competent mosquito cell line, Sua5B, was challenged with Gram-positive bacteria, Gram-negative bacteria, and the malaria parasite *P. berghei*. Quantitative RT-PCR analysis of transcripts from challenged cells revealed rapid and robust changes in *AgDscam* exon usage. Each microbe-induced distinctive splice isoform repertoires that would result in the production of AgDscam molecules with diverse binding properties. The splice isoforms elicited after a specific immune challenge showed higher binding affinity to the specific microorganism used during the challenge, suggesting that alternative splicing of *AgDscam* plays a role in the mosquito's immune receptor diversity and specificity. Importantly, the occurrence of pathogen-induced *AgDscam* alternative splicing was also observed in vivo. Injecting bacteria into adult mosquitoes triggered pathogen-specific alternative splicing of *AgDscam*. Furthermore, dsRNA-mediated silencing of *AgDscam* led to increased susceptibly to *S. aureus* and *E. coli* infection compared to controls treated with *GFP* dsRNA.

Studies of the immune function of the Dscam homolog in the crayfish, *Pacifastacus leniusculus*, also showed that bacterial infection induced the alternative splicing of isoforms with specific affinity to the bacteria used to infect the animal. Furthermore, as with the fruit fly and the mosquito, crayfish Dscam was shown to mediate bacteria clearance and phagocytosis (Watthanasurorot et al., 2011). Therefore, Dscam is a phagocytic receptor for bacteria and this function is conserved in invertebrates. It will be interesting to determine if Dscam pathogen-specific isoforms persist in the animal after an infection is cleared and if they serve to prime the immune response to subsequent infections—which may be an example of convergent evolution of adaptive immune responses.

3.6 Opsonins in Insect Phagocytosis

Opsonization is the process by which soluble host molecules bind to and alter the surface of a pathogen or particle so that it can be ingested more efficiently by phagocytes. In mammals, antibodies and complement factors act as opsonins. Insect thioester-containing proteins (TEPs) share sequence similarities with the vertebrate complement factors C3/C4/C4 and the α2-macroglobulin family of serine proteases. In vertebrate immunity, activated complement proteins, such as iC3b, form covalent bonds with molecules on the surfaces of pathogens or altered self. Complement attachment to the surface of target particles marks these cells for opsonization.

In *Drosophila*, the TEP family is made up of six genes, *TEPI–TEPVI*, of which one, *TEPV*, does not seem to be expressed (Lagueux et al., 2000). All genes of this family possess a signal peptide, indicating that they are secreted proteins. TEPI–TEPIV are most closely related to complement factors, as they share a common CGEQ amino acid motif critical for the formation of thioester bonds with target surfaces. *TEPVI*, also known as *macroglobulin complement related* (*Mcr*), differs from the other TEPs in that it lacks the critical cysteine residue in the thioester-binding site (Stroschein-Stevenson et al., 2006). Phylogenetic analysis reveals that TEP proteins are found in nematodes, insects, mollusks, fish, birds, and mammals (Nonaka, 2000). A population genetic analysis of the TEPI–TEPIV proteins in *Drosophila* showed that *TEPI* is under positive selection and is one of the most rapidly evolving genes in the *Drosophila* genome (Jiggins and Kim, 2006). Thus, it is possible that *TEPI* is evolving to adapt to new pathogens encountered in the wild. The authors also found evidence of less intense positive selection acting on *TEPII*. In contrast, there was no evidence that *TEPIII* or *TEPIV* are evolving under positive selection.

In *Drosophila* larvae, *TEPI*, *TEPII*, *TEPIII*, *TEPIV*, and *Mcr* are expressed in plasmatocytes and *TEPI*, *TEPII*, and *TEPIV* are expressed in the fat body, consistent with a role for these genes in innate immunity. Several studies have shown that *TEPI–IV* and *Mcr* expression is upregulated in larval hemocytes, larval fat body, and whole adult flies after bacterial infection (Bou Aoun et al., 2011; Dionne et al., 2006; Irving et al., 2005; Lagueux et al., 2000). Additionally, transcriptome and QPCR analyses of gene expression in fly larvae following parasitoid wasp infection showed that *TEPI* is massively upregulated and may be important for the encapsulation and melanization of wasp eggs (Salazar-Jaramillo et al., 2014; Wertheim et al., 2005).

A large-scale RNAi screen in S2 cells found that Mcr is required for phagocytosis of the fungus *Candida albicans* (Stroschein-Stevenson et al., 2006). Interestingly, *Mcr* RNAi treatment of S2 cells did not affect phagocytosis of *E. coli* or *S. aureus*. Despite the lack of an active thioester motif, Mcr specifically binds to the surface of *C. albicans*. It does not, however, bind to another fungal pathogen, *S. cerevisiae*, indicating that Mcr recognizes some feature unique to the *C. albicans* cell wall. The addition of conditioned media from untreated S2 cells to *Mcr* RNAi-treated S2 cells is sufficient to rescue the *C. albicans* phagocytosis defect caused by the loss of *Mcr*. Hence, secreted Mcr is required to facilitate *C. albicans* phagocytosis in S2 cells. The same study also looked at the role of TEPII and TEPIII in phagocytosis. *TEPII* RNAi led to a modest (about 25%) decrease in phagocytosis of *E. coli*. *TEPIII* RNAi had a similar effect on *S. aureus* phagocytosis.

More recently, an in vivo study was carried out to investigate the function of TEPs in the defence against pathogens. *TEPI*-RNAi and *TEPII–TEPIV* null mutants were challenged with septic infections of Gram-positive bacteria (*S. pyogenes*, *S. aureus*, *E. faecalis*, and *L. monocytogenes*), Gram-negative bacteria (*E. coli* and *E. cloacae*), *M. marinum*, or a fungal pathogen (*B. bassiana*) (Bou Aoun et al., 2011). Surprisingly, the *TEPI*, *TEPII*, *TEPIII*, and *TEPIV* deficient flies were not more susceptible to the bacterial or fungal infections than wild-type flies. A similar phenotype was observed in *TEPII/TEPIII* null double mutants and *TEPII/TEPIII/TEPIV* mutants (obtained by crossing the *TEPII/TEPIII* null double to a P-element insertion mutant). Finally, adult hemocytes from triple *TEPII–IV* mutants were able to phagocytose *E. coli* pHrodo-labelled bioparticles (Bou Aoun et al., 2011). At this point, whether the TEPs play a role during opsonization of bacteria is still unknown. The immune function of TEPs may be difficult to decipher in vivo if the effects of loss-of-function mutations are masked by other opsonins in *Drosophila* and further experiments in TEP-null mutants may be necessary.

The best evidence that TEPs act as opsonins is from *An. gambiae* studies. A family of 19 TEP genes has been identified in the genome of *An. gambiae* (Christophides et al., 2002). The most extensively studied TEP in mosquitoes is *An. gambiae TEP1* (*AgTEPI*). In *An. gambiae*, *P. berghei* ookinetes that traversed the midgut epithelium were bound by AgTEP1, resulting in parasite killing and clearance (Blandin et al., 2004). Additionally, during bacterial infections, expression of *AgTEPI* increased after septic infection with a mixture of *E. coli* and *M. luteus* (Levashina et al., 2001). AgTEPI can be detected at high levels in mosquito hemolymph and in conditioned media from a mosquito hemocyte cell line, 5.1*. In vivo, secreted AgTEPI originates from mosquito hemocytes as immunofluorescence studies found that AgTEPI is selectively expressed in hemocytes throughout the body cavity. To assess the binding of AgTEPI to bacteria, *E. coli* or *S. aureus* were incubated with 5.1* conditioned media, precipitated, and probed with the anti-AgTEP1 antibody. AgTEPI bound to both bacteria in a thioester-dependent manner. Phagocytosis of *E. coli* by mosquito 5.1* cells was dramatically enhanced after the addition of conditioned media. This effect was lost when the added media was either pretreated to chemically inactivate TEPs or when it was obtained from 5.1* cells treated with dsRNA against *AgTEPI*. Similar effects were observed with two additional Gram-negative bacteria (*S. marcescens* and *Salmonella typhimurium*), but not with the phagocytosis of Gram-positive bacteria (*B. subtilis*, *M. luteus*, and

S. aureus). Thus, secreted AgTEPI opsonizes and enhances the phagocytosis of Gram-negative bacteria by mosquito 5.1* cells. Levashina and group confirmed Moita's findings and noted that 5.1* cells showed only low levels of phagocytosis of Gram-positive bacteria. It is possible that 5.1* cells may simply be less capable of phagocytosis of Gram-positive bacteria.

The antibacterial role of *AgTEPI* in mosquitoes was confirmed in a small-scale in vivo dsRNA screen (Moita et al., 2005). Downregulation of *AgTEPI* in adult mosquitoes led to a 60% reduction in *E. coli* phagocytosis and a nearly 40% decrease in phagocytosis of *S. aureus*. Furthermore, *AgTEPIII* downregulation decreased *E. coli* phagocytosis by about 50%. *AgTEPIV* dsRNA treatment had the most dramatic effect on phagocytosis, with a 60% reduction in phagocytosis of both pathogens. These findings underscore the importance of in vivo experiments in testing the function of immune genes and indicate that opsonization of microbes is a crucial component of the mosquito immune response.

An opsonin-like role has also been described for Dscam1, as soluble Dscam1 is present in S2 cell-conditioned media and larval hemolymph. Thus, in addition to TEPs, Dscam1 may function to opsonize microbes present in the hemolymph of the insect. However, the identity of receptors for specific isoforms of Dscam1 or TEPs on the surface of phagocytes has yet to be determined and could be an area for future study.

Finally, the Christophides group predicted an interaction between two opsonin-like PRRs, AgNimB2 and AgTEP1, in mosquitoes (Midega et al., 2013). AgNimB2 was shown to mediate phagocytosis of *S. aureus* in vivo, but biochemical analysis revealed that AgNimB2 does not directly bind to the surface of *S. aureus*. The authors suggest that AgNimB2 binds to opsonins-like AgTEP1 on the surface of bacteria and mediates the phagocytosis of opsonized bacteria by then binding to membrane-bound receptors on phagocytes.

4. REGULATION OF SIGNALLING DURING PHAGOCYTOSIS

Humoral immune signalling cascades have been extensively studied in *Drosophila* (reviewed in Cherry and Silverman, 2006; Ganesan et al., 2011; Lemaitre and Hoffmann, 2007). In comparison, relatively less is known about the regulation of signalling during phagocytosis in insect blood cells. Signalling from bound phagocytic receptors triggers coordinated rearrangements of the actin cytoskeleton. Small GTPases of the Ras superfamily, such as the Rho-GTPases Cdc42, Rac1, and Rac2 are recruited to the plasma membrane, where they associate with membrane phospholipids and proteins. Rho-GTPases function as molecular switches that alternate between active (GTP-bound) and inactive (GDP-bound) states. They are activated by guanine nucleotide exchange factors (GEFs), which facilitate the binding of GTP, and are inhibited by the hydrolysis of GTP, which is carried out by guanine nucleotide dissociation inhibitors. The *Drosophila* gene *Ziziman-related* (*Zir*) is a Rho-GEF that interacts genetically with Cdc42 and Rac2 to mediate larval hemocyte phagocytosis of *E. coli* and *S. aureus* (Sampson et al., 2012). The primary function of Rho-GTPases during phagocytosis is the regulation and activation of cytoskeletal remodeling enzymes. Rac1 activates WAVE, a member of the Wiskott–Aldrich syndrome protein (WASP) family. WAVE then activates the Arp 2/3 complex, which stimulates actin nucleation, the initial step required for the formation of new actin filament structures. Cdc42 activates WAS(p), the founding member of the WASP family, which in turn activates the Arp 2/3 complex. Cofilin and cofilin-like proteins control the debranching and disassembly of actin filaments to facilitate recycling of actin monomers and structural changes necessary for cytoskeletal reorganization (Chan et al., 2009). Cdc42, Rac1, Rac2, and the Arp 2/3 complex were all identified in RNAi screens and genetic studies to find factors that mediate phagocytosis in S2 cells (Agaisse et al., 2005; Philips et al., 2005; Stroschein-Stevenson et al., 2006; Stuart et al., 2005).

To understand how Rho-GTPases control phagocytosis of bacterial pathogens in *Drosophila* hemocytes, the Fauvarque group generated transgenic *Drosophila* mutants that expressed the Gram-negative pathogen *P. aeruginosa* exotoxin, ExoS, specifically in hemocytes (Avet-Rochex et al., 2005). ExoS contains an N-terminal GTPase activating (GAP) domain that inactivates Rho-GTPases and Rho-dependent signalling. Expressing ExoSGAP in blood cells led to significantly reduced *E. coli* uptake by both adult and larval hemocytes. Mutant flies lacking individual Rho-GTPases (Rho1, Rac1, Rac2, or Cdc42) were challenged with *P. aeruginosa* and only *Rac2* mutants showed significant susceptibility to the bacteria (Avet-Rochex et al., 2007). Furthermore, *Rac2* mutants were also more susceptible to infection with other Gram-negative (such as *E. coli*) and Gram-positive (*E. faecalis* and *S. aureus*) bacterial pathogens. Larval hemocytes from *Rac2* mutants showed a 35% decrease in uptake of *E. coli* and a 55% decrease in *S. aureus* phagocytosis.

Studies of Draper-mediated phagocytosis have provided the most complete picture of intracellular signalling cascades that take place during *Drosophila* phagocytosis. During *S. aureus* phagocytosis, Draper signals through Rho-GTPases, Rac1 or Rac2 (Hashimoto et al., 2009). Larval hemocytes from flies with loss of one copy of *draper*, *Rac1*, or *Rac2* show no phagocytosis defects. However, hemocytes from flies with simultaneous heterozygous loss of *draper* and *Rac1* or *Rac2* were dramatically impaired for bacterial phagocytosis. Thus, after Draper binds *S. aureus*, Rac1 and/or Rac2 are required for the engulfment of the microbe.

The cytoplasmic signalling complex that controls Draper-mediated phagocytosis of apoptotic cells has been examined using classical genetic approaches (Ziegenfuss et al., 2008). In *Drosophila* glial cells, Draper physically interacts with Shark, an SH2 domain containing nonreceptor tyrosine kinase similar to mammalian Zap-70, and this interaction is dependent upon the Src-tyrosine kinase, Src42A. Based on the genetic and biochemical studies, Ziegenfuss and colleagues proposed the following model of signalling: Draper binds to target ligands on cell corpses, Src42A phosphorylates tyrosines located in the intracellular ITAM motif of Draper, the SH2 domain of Shark associates with the Draper's phosphorylated ITAM domain, and Shark activates further downstream signalling events required for apoptotic cell uptake. To determine if Shark plays a role in Draper-mediated uptake of *S. aureus*, Hashimoto and colleagues tested *S. aureus* phagocytosis in flies carrying one copy of mutated alleles for both genes (Hashimoto et al., 2009). Larval hemocytes from single- and double-heterozygous flies showed no difference in the uptake of *S. aureus*. This finding may indicate that Shark does not act downstream of Draper to mediate uptake of *S. aureus*. However, studies using RNAi to target both genes or with flies carrying homozygous mutant alleles may help clarify the role of Shark during Draper-mediated phagocytosis of *S. aureus*.

A number of genetic screens and RNAi screens have been conducted to identify proteins that regulate actin cytoskeleton reorganization during phagocytosis in the fruit fly. A forward genetic screen identified the *Drosophila* homolog of WAVE, D-SCAR, as an important regulator of *E. coli* and *S. aureus* phagocytosis in *Drosophila* larval hemocytes (Pearson et al., 2003). The study also analysed the role of the *Drosophila* WAS(p) (D-WAS(p)) homolog. *D-WAS(p)* RNAi impaired *S. aureus* uptake by S2 cells. The differences observed after loss of D-SCAR or D-WAS(p) may indicate that these proteins function in independent pathways, perhaps downstream of receptors with distinct ligand specificity. Increased phagocytosis of *E. coli* and *S. aureus* was observed in larval hemocytes obtained from a line with a P-element insertion in *chickadee*, the gene encoding the *Drosophila* homolog of profilin. Profilin sequesters free actin, and loss of profilin in *chic* mutants may lead to increased phagocytosis due to the higher availability of free actin and increased spontaneous actin nucleation. Loss of profilin also leads to decreased phagocytosis of *M. fortuitum*, demonstrating that profilin is a host factor required for general phagocytosis in *Drosophila* (Philips et al., 2005).

5. PHAGOSOME MATURATION

The process of particle internalization culminates in the formation of a membrane-bound vesicle, the phagosome, which contains the microbe or cell corpse. Phagosome formation is followed by rapid series of biochemical and cellular changes that convert the nascent phagosome into a potent microbicidal and acidic organelle (Desjardins et al., 1994). Almost immediately, newly formed phagosomes undergo a series of highly ordered fusion and fission events with components of the endosomal pathway. This process, termed phagosome maturation, produces a highly acidic and hydrolytic phagolysosome designed to destroy the cargo (Kinchen and Ravichandran, 2008). The maturation of phagosomes involves interactions with other cellular organelles, including early endosomes, recycling endosomes, late endosomes, and lysosomes (Vieira et al., 2002). Our understanding of the complexity of this organelle has been enhanced by large-scale proteomic analyses of latex bead containing phagosomes in humans and S2 cells (Garin et al., 2001; Stuart et al., 2007). A general overview of the topic is discussed later, with specific details regarding *Drosophila* homologs of the following components: Rab GTPases, phosphatidylinositol 3-kinase, Vacuolar H+-ATPase, the Endosomal sorting complex required for transport (ESCRT) complex, and the Vacuolar protein sorting-C (VPS-C) complex.

Phagosomes formed by receptor-mediated particle internalization quickly fuse with early endosomal vesicles (Mayorga et al., 1991). The small Rab GTPase, Rab5, coordinates early endosomal targeting, tethering, and fusion with the nascent phagosome (Bucci et al., 1992). Rab5 is recruited to newly formed phagosomes by the GTPase Dynamin (Kinchen et al., 2008). Overexpression of Rab5 in *Drosophila* hemocytes leads to an accumulation of *E. coli*-containing vesicles in larval hemocytes (Horn et al., 2014). Rab5 recruits multiple effectors to the early endosomal/phagosomal membrane, including the early endosome antigen 1 (EEA1), SNARE proteins (which are required for membrane fusion), and Vps34 and its regulatory subunit, Vps15 (also known as p150).

Vps15 is a serine–threonine kinase that recruits Vps34 to the early phagosome. Vps34 is a class III phosphatidylinositol-3 kinase (PI3-kinase) that generates phosphatidylinositol-3-phosphate (PI(3)P) on the early phagosomal membrane (Vieira et al., 2001). PI(3)P interacts with proteins containing FYVE (for conserved in **F**ab1, **YO**TB, **V**ac1, and **E**EA1) domains. The *Drosophila* homolog of mammalian Vps34, phosphatidylinositol-3 kinase 59F (Pi3K59F) functions during the cellular immune response to bacterial and fungal pathogens (Qin et al., 2008, 2011). Similar to its counterpart in mammals, *Drosophila*'s homolog of EEA1, Rabenosyn-5, is a FYVE domain containing protein that binds to PI(3)P and Rab5 on the surface of the phagosome, where it is required for fusion of endocytic vesicles and early endosomes (Morrison et al., 2008; Simonsen et al., 1998). The generation of PI(3)P is essential for the progression of phagosome maturation. In mouse fibroblasts, PI(3)P stabilizes the interaction of EEA1 on the early phagosome. Loss of PI3-kinase activity leads to decreased association of EEA1 and blockage of phagosome maturation (Vieira et al., 2001).

In eukaryotes, Vps15 is known for its role in endocytosis and phagocytosis as part of the PI3-kinase complex with Vps34. In an effort to identify genes that regulate activation of the *Imd* pathway after *E. coli* infection, the Wu group conducted a forward genetic screen of EMS mutant flies. This screen identified a *Drosophila* Vps15 mutant, *ird1*, as important for IMD pathway activation (Wu et al., 2007). *ird1* mutants were shown to be more susceptible to infection with *E. coli* or *M. luteus*, and had impaired AMP synthesis in the mutant, which may account for this effect. It is also possible that *ird1* mutants are defective for phagosome maturation of bacteria, but this has not been experimentally verified.

The vacuolar H+-ATPase (V-ATPase) complex is found on the phagosome membrane at very early stages and is required to acidify the phagosomal lumen during phagosome maturation (Beyenbach and Wieczorek, 2006). In *Drosophila*, the V-ATPase complex is made up of multiple subunits. The Perrimon group identified three components of the V-ATPase in a genome-wide RNAi screen looking for genes that altered the expression of GFP from the *map24* promoter of *M. fortuitum* (Philips et al., 2005). The *map24* promoter is responsive at low pH and silencing of V-ATPase components increases the pH of the lumen of the phagosome, thereby decreasing GFP expression under the *map24* promoter. Additionally, eight V-ATPase subunits were identified in a genome-wide S2 cell RNAi screen for genes that are important for the pathogenesis of the facultative intracellular Gram-positive bacteria, *L. monocytogenes* (Cheng et al., 2005).

During the transition from the early to the late phagosome stage, multivesicular bodies (MVBs) begin to appear within the phagosome. MVBs are luminal vesicles that arise from inward budding and scission of portions of the limiting membrane of endosomes and phagosomes. In the endosomal pathway, transmembrane proteins that are destined for degradation are ubiquitinated and then sorted into MVBs (Lee et al., 2005). Work from the Perrimon group illustrated a role for the ESCRT complex in restricting the intracellular growth of *Mycobacterium* species in the fruit fly. Double-stranded RNAs targeting ESCRT factors *Vps28*, *CG8055*, *Tsg101*, and *Vps4* led to a decreased GFP expression under the pH responsive *map24* promoter in *M. fortuitum*. dsRNAs targeting ESCRT impaired the formation of MVBs and effectively halted phagosome maturation at a stage before the induction of *map24* (Philips et al., 2005, 2008). The group then examined how silencing of ESCRT factors affects bacterial growth of the nonpathogenic *M. smegmatis* (Philips et al., 2008). S2 cells normally restrict the growth of *M. smegmatis*, but silencing of the ESCRT factors led to an increase in bacterial growth. These results indicate that the knockdown created a permissive phagosome environment allowing *M. smegmatis* to survive, grow, and disseminate to other cells. ESCRT-mediated sorting of ubiquitinated proteins is nearly absent when S2 cells are treated with dsRNA targeting ESCRT components, as is evidenced by an accumulation of ubiquitin in the vesicular compartments in treated cells. In ESCRT depleted S2 cells, *M. smegmatis* colocalized with vesicles containing excess ubiquitin, revealing that the ESCRT complex normally functions within phagosomes that contain bacteria. The ESCRT machinery works in an analogous manner in mammalian cells. RNAi depletion of *Tg101* and *Vps28* in the mammalian macrophage cell line, RAW267.7, led to significantly higher *M. smegmatis* growth and increased ubiquitin in bacteria containing phagosomes.

After the formation of MVBs, the phagosome transitions to the late stage which is characterized by a more acidic luminal pH. The late phagosome is characterized by the presence of several molecules including lysosomal-associated membrane proteins (LAMPs) and hydrolases. In mammalian cells, LAMPs are required for the last stage of phagosome maturation, the fusion of the phagosome with the lysosome (Huynh et al., 2007).

A recent study in *Drosophila* to identify host factors that are required for phagocytosis and intracellular maintenance of the protozoan parasite *Leishmania donovani* found that *L. donavani* amastigotes colocalize with vesicles that are positive for *Drosophila* Lamp1 (DmLamp1 formerly known as CG3305) within S2 cells (Peltan et al., 2012). As proof-of-concept that *L. donovani* can infect S2 cells, the authors carried out immunofluorescence studies of infected S2 cells expressing a DmLamp1-GFP fusion protein. Late-stage *L. donovani*-containing phagosomes are positive for DmLamp1 indicating that phagocytosed parasites undergo phagosome maturation within S2 cells. The relative importance of DmLamp1 during the cellular immune response to bacterial pathogens should be assessed in flies expressing *DmLamp1*-RNAi in hemocytes to more fully characterize the importance of Lamp1 during phagosome maturation in vivo.

Additional V-ATPases are acquired by late phagosomes, and the vesicles also acquire the small Rab GTPase, Rab7, a characteristic marker of late phagosomes (Desjardins et al., 1994). Rab7 is key to phagosome trafficking between late endosomes and lysosomes. It recruits effectors such as Rab-interacting lysosomal protein, which tethers the vesicle to the dynein–dynactin motor, facilitating the movement of the phagosome towards the centre of the cell (Harrison et al., 2003; Jordens et al., 2001). VPS-C complexes interact with SNAREs and Rabs during the phagosome maturation process. There are two VPS-C complexes: CORVET (class C core vacuole/endosome tethering) and HOPS (homotypic fusion and vacuole protein sorting) (reviewed in Balderhaar and Ungermann, 2013; Solinger and Spang, 2013). The CORVET complex interacts with Rab5-GTP and promotes early endosome/phagosome fusion. The HOPS complex interacts with Rab7-GTP on late endosomes/MVBs to facilitate their fusion with lysosomes. Additionally, live cell imaging studies in a human cell line indicate that the HOPS complex exchanges Rab5 for Rab7 to facilitate the transition from early to late phagosomes (Rink et al., 2005). Much of the early work defining the composition of each complex was carried out in yeast. Both complexes are heterohexameric: they are composed of four shared class C subunits (Vps11, Vps16, Vps18, and Vps33) and two Rab-specific subunits. In *Drosophila*, both Vps33 and Vps16 have two homologs (*car* and *Vps33B*, *Vps16A*, and *Vps16B*, respectively) (Li and Blissard, 2015; Pulipparacharuvil et al., 2005). Vps16A and Vps16B are predicted to associate with different HOPS complexes, and this association may dictate the function of the complex (Pulipparacharuvil et al., 2005). *Vps16A* mutant

Drosophila larvae are unable to clear autophagosomes following starvation-induced autophagy, indicating that the Vps16A subunit of HOPS is essential for the fusion of autophagosomes with lysosomes (Takats et al., 2015). *Vps16B* mutants, *full of bacteria* (*fob*), are highly susceptible to *E. coli*, which can be rescued by specifically expressing wild-type *fob* in hemocytes (Akbar et al., 2011). *fob* hemocytes are able to engulf *E. coli* but show defects in phagosome acidification. These hemocytes show no defects in the acquisition of early endosome markers such as Rab5 and Rbsn-5, but have significantly higher numbers of Rab-7-positive phagosomes, suggesting that phagosome maturation is stalled at this stage. To test this hypothesis, wild-type and *fob* mutant hemocytes were treated with Alexa-488 labelled dextrans, which, when internalized by fluid-phase endocytosis, label lysosomes. The hemocytes were then challenged with fluorescein-labelled *E. coli* and colocalization of dextran and bacteria was examined using immunofluorescence. Approximately 30% of bacteria-positive phagosomes colocalized with dextran-labelled lysosomes in wild-type hemocytes. Fewer, only 9%, of bacteria containing phagosomes colocalize with lysosomes in *fob* mutant hemocytes. These results confirm that Vps16B mediates phagosome to lysosome fusion in *Drosophila*.

The final step in the maturation process is the formation of the phagolysosome (pH ~4.5). Phagolysosomes are highly effective microbicidal organelles that are equipped with host factors that impede microbial growth while simultaneously attacking and degrading the pathogen. Key cofactors of bacterial housekeeping enzymes (such as free iron or divalent metal ions (Fe^{2+}, Zn^{2+}, and Mn^{2+})) are removed from the phagosomal lumen to prevent bacterial growth. Free iron is sequestered by lactoferrin, a glycoprotein found in the phagosome lumen while divalent metal ions are actively removed from the phagosome by NRAMP, an integral membrane protein. Reactive oxygen (ROS) and nitrogen (RNS) species attack bacterial DNA, proteins, and lipids to destroy the pathogen. ROS are generated through the action of the membrane-bound NOX2 NADPH oxidase, which transfers electrons from cytosolic NADPH to molecular oxygen (O_2^-) and releases the O_2^- into the phagosomal lumen. Superoxide dismutase catalyses the dismutation of O_2^- into H_2O_2, which in turn can be converted into additional toxic ROS species (hypochlorous acid and chloramines) that kill microoganisms. RNS are also important antimicrobial factors. The enzyme inducible nitric oxide synthase, iNOS, catalyses the formation of nitric oxide on the cytoplasmic side of the phagosome. Nitric oxide diffuses across the bilayer into the phagosome, where it encounters

ROS and converts it to various RNS that are highly toxic to the microorganism. Phagolysosomes are also equipped with an assortment of bactericidal elements: AMPs, peptidases, lipases, and hydrolases.

6. CONCLUSION

In this review, we have highlighted the current knowledge of phagocytic PRRs and the signalling pathways that facilitate particle internalization and phagosome maturation in insect hemocytes. Over the last 3 decades, classical genetics, molecular genetics, and systems biology approaches using insect models have advanced our understanding of the complexity and specificity of cellular immune responses. *D. melanogaster* is a relatively straightforward and genetically tractable system that has provided valuable insights into the complex cell biology of phagocytes. Pivotal findings in fruit fly immunity have also facilitated comparative research and the identification of novel regulators in other organisms, such as the important disease vectors *An. gambiae* and *Ae. agypti.*

Comparative genomic and evolutionary studies of insect immunity have revealed that similar and divergent components mediate the immune response across insect orders (Viljakainen, 2015). Future studies examining phagocytosis and hemocyte-mediated responses in more insect models will be required to characterize functional similarities among species. Finally, it will be interesting to explore how insect immune responses are tailored for specific pathogens.

REFERENCES

Agaisse, H., Burrack, L.S., Philips, J.A., Rubin, E.J., Perrimon, N., Higgins, D.E., 2005. Genome-wide RNAi screen for host factors required for intracellular bacterial infection. Science 309, 1248–1251.

Akbar, M.A., Tracy, C., Kahr, W.H., Kramer, H., 2011. The full-of-bacteria gene is required for phagosome maturation during immune defense in Drosophila. J. Cell Biol. 192, 383–390.

Arandjelovic, S., Ravichandran, K.S., 2015. Phagocytosis of apoptotic cells in homeostasis. Nat. Immunol. 16, 907–917.

Armitage, S.A., Freiburg, R.Y., Kurtz, J., Bravo, I.G., 2012. The evolution of Dscam genes across the arthropods. BMC Evol. Biol. 12, 53.

Avet-Rochex, A., Bergeret, E., Attree, I., Meister, M., Fauvarque, M.O., 2005. Suppression of Drosophila cellular immunity by directed expression of the ExoS toxin GAP domain of *Pseudomonas aeruginosa*. Cell. Microbiol. 7, 799–810.

Avet-Rochex, A., Perrin, J., Bergeret, E., Fauvarque, M.O., 2007. Rac2 is a major actor of Drosophila resistance to Pseudomonas aeruginosa acting in phagocytic cells. Genes Cells 12, 1193–1204.

Awasaki, T., Tatsumi, R., Takahashi, K., Arai, K., Nakanishi, Y., Ueda, R., Ito, K., 2006. Essential role of the apoptotic cell engulfment genes draper and ced-6 in programmed axon pruning during Drosophila metamorphosis. Neuron 50, 855–867.

Babcock, D.T., Brock, A.R., Fish, G.S., Wang, Y., Perrin, L., Krasnow, M.A., Galko, M.J., 2008. Circulating blood cells function as a surveillance system for damaged tissue in Drosophila larvae. Proc. Natl. Acad. Sci. U.S.A. 105, 10017–10022.

Balderhaar, H.J., Ungermann, C., 2013. CORVET and HOPS tethering complexes—coordinators of endosome and lysosome fusion. J. Cell Sci. 126, 1307–1316.

Bergeret, E., Perrin, J., Williams, M., Grunwald, D., Engel, E., Thevenon, D., Taillebourg, E., Bruckert, F., Cosson, P., Fauvarque, M.O., 2008. TM9SF4 is required for Drosophila cellular immunity via cell adhesion and phagocytosis. J. Cell Sci. 121, 3325–3334.

Beyenbach, K.W., Wieczorek, H., 2006. The V-type H+ ATPase: molecular structure and function, physiological roles and regulation. J. Exp. Biol. 209, 577–589.

Bischoff, V., Vignal, C., Duvic, B., Boneca, I.G., Hoffmann, J.A., Royet, J., 2006. Downregulation of the Drosophila immune response by peptidoglycan-recognition proteins SC1 and SC2. PLoS Pathog. 2, e14.

Blandin, S., Shiao, S.H., Moita, L.F., Janse, C.J., Waters, A.P., Kafatos, F.C., Levashina, E.A., 2004. Complement-like protein TEP1 is a determinant of vectorial capacity in the malaria vector *Anopheles gambiae*. Cell 116, 661–670.

Bou Aoun, R., Hetru, C., Troxler, L., Doucet, D., Ferrandon, D., Matt, N., 2011. Analysis of thioester-containing proteins during the innate immune response of *Drosophila melanogaster*. J. Innate Immun. 3, 52–64.

Brown, N.H., Gregory, S.L., Martin-Bermudo, M.D., 2000. Integrins as mediators of morphogenesis in Drosophila. Dev. Biol. 223, 1–16.

Bucci, C., Parton, R.G., Mather, I.H., Stunnenberg, H., Simons, K., Hoflack, B., Zerial, M., 1992. The small GTPase rab5 functions as a regulatory factor in the early endocytic pathway. Cell 70, 715–728.

Canton, J., Neculai, D., Grinstein, S., 2013. Scavenger receptors in homeostasis and immunity. Nat. Rev. Immunol. 13, 621–634.

Castillo, J.C., Robertson, A.E., Strand, M.R., 2006. Characterization of hemocytes from the mosquitoes *Anopheles gambiae* and *Aedes aegypti*. Insect Biochem. Mol. Biol. 36, 891–903.

Chan, C., Beltzner, C.C., Pollard, T.D., 2009. Cofilin dissociates Arp2/3 complex and branches from actin filaments. Curr. Biol. 19, 537–545.

Charroux, B., Royet, J., 2009. Elimination of plasmatocytes by targeted apoptosis reveals their role in multiple aspects of the Drosophila immune response. Proc. Natl. Acad. Sci. U.S.A. 106, 9797–9802.

Cheng, L.W., Viala, J.P., Stuurman, N., Wiedemann, U., Vale, R.D., Portnoy, D.A., 2005. Use of RNA interference in Drosophila S2 cells to identify host pathways controlling compartmentalization of an intracellular pathogen. Proc. Natl. Acad. Sci. U.S.A. 102, 13646–13651.

Cherry, S., Silverman, N., 2006. Host–pathogen interactions in Drosophila: new tricks from an old friend. Nat. Immunol. 7, 911–917.

Choe, K.M., Werner, T., Stoven, S., Hultmark, D., Anderson, K.V., 2002. Requirement for a peptidoglycan recognition protein (PGRP) in Relish activation and antibacterial immune responses in Drosophila. Science 296, 359–362.

Christophides, G.K., Zdobnov, E., Barillas-Mury, C., Birney, E., Blandin, S., Blass, C., Brey, P.T., Collins, F.H., Danielli, A., Dimopoulos, G., Hetru, C., Hoa, N.T., Hoffmann, J.A., Kanzok, S.M., Letunic, I., Levashina, E.A., Loukeris, T.G., Lycett, G., Meister, S., Michel, K., Moita, L.F., Muller, H.M., Osta, M.A., Paskewitz, S.M., Reichhart, J.M., Rzhetsky, A., Troxler, L., Vernick, K.D.,

Vlachou, D., Volz, J., Von Mering, C., Xu, J., Zheng, L., Bork, P., Kafatos, F.C., 2002. Immunity-related genes and gene families in Anopheles gambiae. Science 298, 159–165.
Chung, Y.S., Kocks, C., 2011. Recognition of pathogenic microbes by the Drosophila phagocytic pattern recognition receptor Eater. J. Biol. Chem. 286, 26524–26532.
Cornillon, S., Gebbie, L., Benghezal, M., Nair, P., Keller, S., Wehrle-Haller, B., Charette, S.J., Bruckert, F., Letourneur, F., Cosson, P., 2006. An adhesion molecule in free-living Dictyostelium amoebae with integrin beta features. EMBO Rep. 7, 617–621.
Cuttell, L., Vaughan, A., Silva, E., Escaron, C.J., Lavine, M., Van Goethem, E., Eid, J.P., Quirin, M., Franc, N.C., 2008. Undertaker, a Drosophila Junctophilin, links Draper-mediated phagocytosis and calcium homeostasis. Cell 135, 524–534.
Defaye, A., Evans, I., Crozatier, M., Wood, W., Lemaitre, B., Leulier, F., 2009. Genetic ablation of Drosophila phagocytes reveals their contribution to both development and resistance to bacterial infection. J. Innate Immun. 1, 322–334.
Desjardins, M., Huber, L.A., Parton, R.G., Griffiths, G., 1994. Biogenesis of phagolysosomes proceeds through a sequential series of interactions with the endocytic apparatus. J. Cell Biol. 124, 677–688.
Dimopoulos, G., Casavant, T.L., Chang, S., Scheetz, T., Roberts, C., Donohue, M., Schultz, J., Benes, V., Bork, P., Ansorge, W., Soares, M.B., Kafatos, F.C., 2000. *Anopheles gambiae* pilot gene discovery project: identification of mosquito innate immunity genes from expressed sequence tags generated from immune-competent cell lines. Proc. Natl. Acad. Sci. U.S.A. 97, 6619–6624.
Dimopoulos, G., Christophides, G.K., Meister, S., Schultz, J., White, K.P., Barillas-Mury, C., Kafatos, F.C., 2002. Genome expression analysis of *Anopheles gambiae*: responses to injury, bacterial challenge, and malaria infection. Proc. Natl. Acad. Sci. U.S.A. 99, 8814–8819.
Dionne, M.S., Ghori, N., Schneider, D.S., 2003. *Drosophila melanogaster* is a genetically tractable model host for *Mycobacterium marinum*. Infect. Immun. 71, 3540–3550.
Dionne, M.S., Pham, L.N., Shirasu-Hiza, M., Schneider, D.S., 2006. Akt and FOXO dysregulation contribute to infection-induced wasting in Drosophila. Curr. Biol. 16, 1977–1985.
Dong, Y., Aguilar, R., Xi, Z., Warr, E., Mongin, E., Dimopoulos, G., 2006a. *Anopheles gambiae* immune responses to human and rodent Plasmodium parasite species. PLoS Pathog. 2, e52.
Dong, Y., Taylor, H.E., Dimopoulos, G., 2006b. AgDscam, a hypervariable immunoglobulin domain-containing receptor of the *Anopheles gambiae* innate immune system. PLoS Biol. 4, e229.
Drosophila 12 Genomes, C., Clark, A.G., Eisen, M.B., Smith, D.R., Bergman, C.M., Oliver, B., Markow, T.A., Kaufman, T.C., Kellis, M., Gelbart, W., Iyer, V.N., Pollard, D.A., Sackton, T.B., Larracuente, A.M., Singh, N.D., Abad, J.P., Abt, D.N., Adryan, B., Aguade, M., Akashi, H., Anderson, W.W., Aquadro, C.F., Ardell, D.H., Arguello, R., Artieri, C.G., Barbash, D.A., Barker, D., Barsanti, P., Batterham, P., Batzoglou, S., Begun, D., Bhutkar, A., Blanco, E., Bosak, S.A., Bradley, R.K., Brand, A.D., Brent, M.R., Brooks, A.N., Brown, R.H., Butlin, R.K., Caggese, C., Calvi, B.R., Bernardo De Carvalho, A., Caspi, A., Castrezana, S., Celniker, S.E., Chang, J.L., Chapple, C., Chatterji, S., Chinwalla, A., Civetta, A., Clifton, S.W., Comeron, J.M., Costello, J.C., Coyne, J.A., Daub, J., David, R.G., Delcher, A.L., Delehaunty, K., Do, C.B., Ebling, H., Edwards, K., Eickbush, T., Evans, J.D., Filipski, A., Findeiss, S., Freyhult, E., Fulton, L., Fulton, R., Garcia, A.C., Gardiner, A., Garfield, D.A., Garvin, B.E., Gibson, G., Gilbert, D., Gnerre, S., Godfrey, J., Good, R., Gotea, V., Gravely, B., Greenberg, A.J., Griffiths-Jones, S., Gross, S., Guigo, R., Gustafson, E.A., Haerty, W., Hahn, M.W., Halligan, D.L.,

Halpern, A.L., Halter, G.M., Han, M.V., Heger, A., Hillier, L., Hinrichs, A.S., Holmes, I., Hoskins, R.A., Hubisz, M.J., Hultmark, D., Huntley, M.A., Jaffe, D.B., et al., 2007. Evolution of genes and genomes on the Drosophila phylogeny. Nature 450, 203–218.

Elrod-Erickson, M., Mishra, S., Schneider, D.S., 2000. Interactions between the cellular and humoral immune responses in Drosophila. Curr. Biol. 10, 781–784.

Estevez-Lao, T.Y., Hillyer, J.F., 2014. Involvement of the *Anopheles gambiae* Nimrod gene family in mosquito immune responses. Insect Biochem. Mol. Biol. 44, 12–22.

Flannagan, R.S., Jaumouille, V., Grinstein, S., 2012. The cell biology of phagocytosis. Annu. Rev. Pathol. 7, 61–98.

Franc, N.C., Dimarcq, J.L., Lagueux, M., Hoffmann, J., Ezekowitz, R.A., 1996. Croquemort, a novel Drosophila hemocyte/macrophage receptor that recognizes apoptotic cells. Immunity 4, 431–443.

Franc, N.C., Heitzler, P., Ezekowitz, R.A., White, K., 1999. Requirement for croquemort in phagocytosis of apoptotic cells in Drosophila. Science 284, 1991–1994.

Freeman, M.R., Delrow, J., Kim, J., Johnson, E., Doe, C.Q., 2003. Unwrapping glial biology: Gcm target genes regulating glial development, diversification, and function. Neuron 38, 567–580.

Fujita, Y., 2012. Role of NPxY motif in Draper-mediated apoptotic cell clearance in Drosophila. Drug Discov. Ther. 6 (6), 291–297.

Ganesan, S., Aggarwal, K., Paquette, N., Silverman, N., 2011. NF-kappaB/Rel proteins and the humoral immune responses of *Drosophila melanogaster*. Curr. Top. Microbiol. Immunol. 349, 25–60.

Garin, J., Diez, R., Kieffer, S., Dermine, J.F., Duclos, S., Gagnon, E., Sadoul, R., Rondeau, C., Desjardins, M., 2001. The phagosome proteome: insight into phagosome functions. J. Cell Biol. 152, 165–180.

Garver, L.S., Wu, J., Wu, L.P., 2006. The peptidoglycan recognition protein PGRP-SC1a is essential for Toll signaling and phagocytosis of *Staphylococcus aureus* in Drosophila. Proc. Natl. Acad. Sci. U.S.A. 103, 660–665.

Ghosh, S., Singh, A., Mandal, S., Mandal, L., 2015. Active hematopoietic hubs in Drosophila adults generate hemocytes and contribute to immune response. Dev. Cell 33, 478–488.

Gonzalez-Lazaro, M., Dinglasan, R.R., Hernandez-Hernandez, L., Rodriguez, M.H., Laclaustra, M., Jacobs-Lorena, M., Flores-Romo, L., 2009. *Anopheles gambiae* Croquemort SCRBQ2, expression profile in the mosquito and its potential interaction with the malaria parasite *Plasmodium berghei*. Insect Biochem. Mol. Biol. 39, 395–402.

Gottar, M., Gobert, V., Michel, T., Belvin, M., Duyk, G., Hoffmann, J.A., Ferrandon, D., Royet, J., 2002. The Drosophila immune response against Gram-negative bacteria is mediated by a peptidoglycan recognition protein. Nature 416, 640–644.

Graveley, B.R., Brooks, A.N., Carlson, J.W., Duff, M.O., Landolin, J.M., Yang, L., Artieri, C.G., Van Baren, M.J., Boley, N., Booth, B.W., Brown, J.B., Cherbas, L., Davis, C.A., Dobin, A., Li, R., Lin, W., Malone, J.H., Mattiuzzo, N.R., Miller, D., Sturgill, D., Tuch, B.B., Zaleski, C., Zhang, D., Blanchette, M., Dudoit, S., Eads, B., Green, R.E., Hammonds, A., Jiang, L., Kapranov, P., Langton, L., Perrimon, N., Sandler, J.E., Wan, K.H., Willingham, A., Zhang, Y., Zou, Y., Andrews, J., Bickel, P.J., Brenner, S.E., Brent, M.R., Cherbas, P., Gingeras, T.R., Hoskins, R.A., Kaufman, T.C., Oliver, B., Celniker, S.E., 2011. The developmental transcriptome of *Drosophila melanogaster*. Nature 471, 473–479.

Griffin, F.M., Griffin, J.A., Leider, J.E., Silverstein, S.C., 1975. Studies on the mechanism of phagocytosis. I. Requirements for circumferential attachment of particle-bound ligands to specific receptors on the macrophage plasma membrane. J. Exp. Med. 142, 1263–1282.

Harrison, R.E., Bucci, C., Vieira, O.V., Schroer, T.A., Grinstein, S., 2003. Phagosomes fuse with late endosomes and/or lysosomes by extension of membrane protrusions along microtubules: role of Rab7 and RILP. Mol. Cell. Biol. 23, 6494–6506.
Hashimoto, Y., Tabuchi, Y., Sakurai, K., Kutsuna, M., Kurokawa, K., Awasaki, T., Sekimizu, K., Nakanishi, Y., Shiratsuchi, A., 2009. Identification of lipoteichoic acid as a ligand for draper in the phagocytosis of *Staphylococcus aureus* by Drosophila hemocytes. J. Immunol. 183, 7451–7460.
Hillyer, J.F., Strand, M.R., 2014. Mosquito hemocyte-mediated immune responses. Curr. Opin. Insect Sci. 3, 14–21.
Honti, V., Csordas, G., Kurucz, E., Markus, R., Ando, I., 2014. The cell-mediated immunity of *Drosophila melanogaster*: hemocyte lineages, immune compartments, microanatomy and regulation. Dev. Comp. Immunol. 42, 47–56.
Horn, L., Leips, J., Starz-Gaiano, M., 2014. Phagocytic ability declines with age in adult Drosophila hemocytes. Aging Cell 13, 719–728.
Hu, Y., Flockhart, I., Vinayagam, A., Bergwitz, C., Berger, B., Perrimon, N., Mohr, S.E., 2011. An integrative approach to ortholog prediction for disease-focused and other functional studies. BMC Bioinformatics 12, 357.
Huynh, K.K., Eskelinen, E.L., Scott, C.C., Malevanets, A., Saftig, P., Grinstein, S., 2007. LAMP proteins are required for fusion of lysosomes with phagosomes. EMBO J. 26, 313–324.
Irving, P., Ubeda, J.M., Doucet, D., Troxler, L., Lagueux, M., Zachary, D., Hoffmann, J.A., Hetru, C., Meister, M., 2005. New insights into Drosophila larval haemocyte functions through genome-wide analysis. Cell. Microbiol. 7, 335–350.
Jiggins, F.M., Kim, K.W., 2006. Contrasting evolutionary patterns in Drosophila immune receptors. J. Mol. Evol. 63, 769–780.
Jordens, I., Fernandez-Borja, M., Marsman, M., Dusseljee, S., Janssen, L., Calafat, J., Janssen, H., Wubbolts, R., Neefjes, J., 2001. The Rab7 effector protein RILP controls lysosomal transport by inducing the recruitment of dynein–dynactin motors. Curr. Biol. 11, 1680–1685.
Ju, J.S., Cho, M.H., Brade, L., Kim, J.H., Park, J.W., Ha, N.C., Soderhall, I., Soderhall, K., Brade, H., Lee, B.L., 2006. A novel 40-kDa protein containing six repeats of an epidermal growth factor-like domain functions as a pattern recognition protein for lipopolysaccharide. J. Immunol. 177, 1838–1845.
Kaneko, T., Goldman, W.E., Mellroth, P., Steiner, H., Fukase, K., Kusumoto, S., Harley, W., Fox, A., Golenbock, D., Silverman, N., 2004. Monomeric and polymeric Gram-negative peptidoglycan but not purified LPS stimulate the Drosophila IMD pathway. Immunity 20, 637–649.
Kang, D., Liu, G., Lundstrom, A., Gelius, E., Steiner, H., 1998. A peptidoglycan recognition protein in innate immunity conserved from insects to humans. Proc. Natl. Acad. Sci. U.S.A. 95, 10078–10082.
Kaufmann, S.H., 2008. Immunology's foundation: the 100-year anniversary of the Nobel Prize to Paul Ehrlich and Elie Metchnikoff. Nat. Immunol. 9, 705–712.
Kinchen, J.M., Ravichandran, K.S., 2008. Phagosome maturation: going through the acid test. Nat. Rev. Mol. Cell Biol. 9, 781–795.
Kinchen, J.M., Doukoumetzidis, K., Almendinger, J., Stergiou, L., Tosello-Trampont, A., Sifri, C.D., Hengartner, M.O., Ravichandran, K.S., 2008. A pathway for phagosome maturation during engulfment of apoptotic cells. Nat. Cell Biol. 10, 556–566.
Kocks, C., Cho, J.H., Nehme, N., Ulvila, J., Pearson, A.M., Meister, M., Strom, C., Conto, S.L., Hetru, C., Stuart, L.M., Stehle, T., Hoffmann, J.A., Reichhart, J.M., Ferrandon, D., Ramet, M., Ezekowitz, R.A., 2005. Eater, a transmembrane protein mediating phagocytosis of bacterial pathogens in Drosophila. Cell 123, 335–346.

Koundakjian, E.J., Cowan, D.M., Hardy, R.W., Becker, A.H., 2004. The Zuker collection: a resource for the analysis of autosomal gene function in *Drosophila melanogaster*. Genetics 167, 203–206.

Kuraishi, T., Nakagawa, Y., Nagaosa, K., Hashimoto, Y., Ishimoto, T., Moki, T., Fujita, Y., Nakayama, H., Dohmae, N., Shiratsuchi, A., Yamamoto, N., Ueda, K., Yamaguchi, M., Awasaki, T., Nakanishi, Y., 2009. Pretaporter, a Drosophila protein serving as a ligand for Draper in the phagocytosis of apoptotic cells. EMBO J. 28, 3868–3878.

Kurucz, E., Zettervall, C.J., Sinka, R., Vilmos, P., Pivarcsi, A., Ekengren, S., Hegedus, Z., Ando, I., Hultmark, D., 2003. Hemese, a hemocyte-specific transmembrane protein, affects the cellular immune response in Drosophila. Proc. Natl. Acad. Sci. U.S.A. 100, 2622–2627.

Kurucz, E., Markus, R., Zsamboki, J., Folkl-Medzihradszky, K., Darula, Z., Vilmos, P., Udvardy, A., Krausz, I., Lukacsovich, T., Gateff, E., Zettervall, C.J., Hultmark, D., Ando, I., 2007a. Nimrod, a putative phagocytosis receptor with EGF repeats in Drosophila plasmatocytes. Curr. Biol. 17, 649–654.

Kurucz, E., Vaczi, B., Markus, R., Laurinyecz, B., Vilmos, P., Zsamboki, J., Csorba, K., Gateff, E., Hultmark, D., Ando, I., 2007b. Definition of Drosophila hemocyte subsets by cell-type specific antigens. Acta Biol. Hung. 58 (Suppl.), 95–111.

Lagueux, M., Perrodou, E., Levashina, E.A., Capovilla, M., Hoffmann, J.A., 2000. Constitutive expression of a complement-like protein in toll and JAK gain-of-function mutants of Drosophila. Proc. Natl. Acad. Sci. U.S.A. 97, 11427–11432.

Lanot, R., Zachary, D., Holder, F., Meister, M., 2001. Postembryonic hematopoiesis in Drosophila. Dev. Biol. 230, 243–257.

Lazzaro, B.P., 2005. Elevated polymorphism and divergence in the class C scavenger receptors of *Drosophila melanogaster* and *D. simulans*. Genetics 169, 2023–2034.

Lazzaro, B.P., Sceurman, B.K., Clark, A.G., 2004. Genetic basis of natural variation in *D. melanogaster* antibacterial immunity. Science 303, 1873–1876.

Lee, W.L., Kim, M.K., Schreiber, A.D., Grinstein, S., 2005. Role of ubiquitin and proteasomes in phagosome maturation. Mol. Biol. Cell 16, 2077–2090.

Lemaitre, B., Hoffmann, J., 2007. The host defense of *Drosophila melanogaster*. Annu. Rev. Immunol. 25, 697–743.

Levashina, E.A., Moita, L.F., Blandin, S., Vriend, G., Lagueux, M., Kafatos, F.C., 2001. Conserved role of a complement-like protein in phagocytosis revealed by dsRNA knockout in cultured cells of the mosquito, *Anopheles gambiae*. Cell 104, 709–718.

Li, Z., Blissard, G., 2015. The vacuolar protein sorting genes in insects: a comparative genome view. Insect Biochem. Mol. Biol. 62, 211–225.

Li, J., Riehle, M.M., Zhang, Y., Xu, J., Oduol, F., Gomez, S.M., Eiglmeier, K., Ueberheide, B.M., Shabanowitz, J., Hunt, D.F., Ribeiro, J.M., Vernick, K.D., 2006. *Anopheles gambiae* genome reannotation through synthesis of ab initio and comparative gene prediction algorithms. Genome Biol. 7, R24.

Liu, Q.A., Hengartner, M.O., 1998. Candidate adaptor protein CED-6 promotes the engulfment of apoptotic cells in *C. elegans*. Cell 93, 961–972.

Lombardo, F., Ghani, Y., Kafatos, F.C., Christophides, G.K., 2013. Comprehensive genetic dissection of the hemocyte immune response in the malaria mosquito *Anopheles gambiae*. PLoS Pathog. 9, e1003145.

MacDonald, J.M., Beach, M.G., Porpiglia, E., Sheehan, A.E., Watts, R.J., Freeman, M.R., 2006. The Drosophila cell corpse engulfment receptor Draper mediates glial clearance of severed axons. Neuron 50, 869–881.

Manaka, J., Kuraishi, T., Shiratsuchi, A., Nakai, Y., Higashida, H., Henson, P., Nakanishi, Y., 2004. Draper-mediated and phosphatidylserine-independent phagocytosis of apoptotic cells by Drosophila hemocytes/macrophages. J. Biol. Chem. 279, 48466–48476.

Mansfield, B.E., Dionne, M.S., Schneider, D.S., Freitag, N.E., 2003. Exploration of host–pathogen interactions using Listeria monocytogenes and *Drosophila melanogaster*. Cell. Microbiol. 5, 901–911.
Mayorga, L.S., Bertini, F., Stahl, P.D., 1991. Fusion of newly formed phagosomes with endosomes in intact cells and in a cell-free system. J. Biol. Chem. 266, 6511–6517.
Mellroth, P., Steiner, H., 2006. PGRP-SB1: an N-acetylmuramoyl L-alanine amidase with antibacterial activity. Biochem. Biophys. Res. Commun. 350, 994–999.
Mellroth, P., Karlsson, J., Steiner, H., 2003. A scavenger function for a Drosophila peptidoglycan recognition protein. J. Biol. Chem. 278, 7059–7064.
Michel, T., Reichhart, J.M., Hoffmann, J.A., Royet, J., 2001. Drosophila Toll is activated by Gram-positive bacteria through a circulating peptidoglycan recognition protein. Nature 414, 756–759.
Midega, J., Blight, J., Lombardo, F., Povelones, M., Kafatos, F., Christophides, G.K., 2013. Discovery and characterization of two Nimrod superfamily members in *Anopheles gambiae*. Pathog. Glob. Health 107, 463–474.
Moita, L.F., Wang-Sattler, R., Michel, K., Zimmermann, T., Blandin, S., Levashina, E.A., Kafatos, F.C., 2005. In vivo identification of novel regulators and conserved pathways of phagocytosis in A. gambiae. Immunity 23, 65–73.
Moita, L.F., Vriend, G., Mahairaki, V., Louis, C., Kafatos, F.C., 2006. Integrins of *Anopheles gambiae* and a putative role of a new beta integrin, BINT2, in phagocytosis of E. coli. Insect Biochem. Mol. Biol. 36, 282–290.
Morrison, H.A., Dionne, H., Rusten, T.E., Brech, A., Fisher, W.W., Pfeiffer, B.D., Celniker, S.E., Stenmark, H., Bilder, D., 2008. Regulation of early endosomal entry by the Drosophila tumor suppressors Rabenosyn and Vps45. Mol. Biol. Cell 19, 4167–4176.
Nagaosa, K., Okada, R., Nonaka, S., Takeuchi, K., Fujita, Y., Miyasaka, T., Manaka, J., Ando, I., Nakanishi, Y., 2011. Integrin betanu-mediated phagocytosis of apoptotic cells in Drosophila embryos. J. Biol. Chem. 286, 25770–25777.
Neafsey, D.E., Waterhouse, R.M., Abai, M.R., Aganezov, S.S., Alekseyev, M.A., Allen, J.E., Amon, J., Arca, B., Arensburger, P., Artemov, G., Assour, L.A., Basseri, H., Berlin, A., Birren, B.W., Blandin, S.A., Brockman, A.I., Burkot, T.R., Burt, A., Chan, C.S., Chauve, C., Chiu, J.C., Christensen, M., Costantini, C., Davidson, V.L., Deligianni, E., Dottorini, T., Dritsou, V., Gabriel, S.B., Guelbeogo, W.M., Hall, A.B., Han, M.V., Hlaing, T., Hughes, D.S., Jenkins, A.M., Jiang, X., Jungreis, I., Kakani, E.G., Kamali, M., Kemppainen, P., Kennedy, R.C., Kirmitzoglou, I.K., Koekemoer, L.L., Laban, N., Langridge, N., Lawniczak, M.K., Lirakis, M., Lobo, N.F., Lowy, E., Maccallum, R.M., Mao, C., Maslen, G., Mbogo, C., McCarthy, J., Michel, K., Mitchell, S.N., Moore, W., Murphy, K.A., Naumenko, A.N., Nolan, T., Novoa, E.M., O'Loughlin, S., Oringanje, C., Oshaghi, M.A., Pakpour, N., Papathanos, P.A., Peery, A.N., Povelones, M., Prakash, A., Price, D.P., Rajaraman, A., Reimer, L.J., Rinker, D.C., Rokas, A., Russell, T.L., Sagnon, N., Sharakhova, M.V., Shea, T., Simao, F.A., Simard, F., Slotman, M.A., Somboon, P., Stegniy, V., Struchiner, C.J., Thomas, G.W., Tojo, M., Topalis, P., Tubio, J.M., Unger, M.F., Vontas, J., Walton, C., Wilding, C.S., Willis, J.H., Wu, Y.C., Yan, G., Zdobnov, E.M., Zhou, X., Catteruccia, F., Christophides, G.K., Collins, F.H., Cornman, R.S., et al., 2015. Mosquito genomics. Highly evolvable malaria vectors: the genomes of 16 Anopheles mosquitoes. Science 347, 1258522.
Nehme, N.T., Quintin, J., Cho, J.H., Lee, J., Lafarge, M.C., Kocks, C., Ferrandon, D., 2011. Relative roles of the cellular and humoral responses in the Drosophila host defense against three gram-positive bacterial infections. PLoS One 6, e14743.

Nonaka, M., 2000. Origin and evolution of the complement system. Curr. Top. Microbiol. Immunol. 248, 37–50.

Nonaka, S., Nagaosa, K., Mori, T., Shiratsuchi, A., Nakanishi, Y., 2013. Integrin alphaPS3/betanu-mediated phagocytosis of apoptotic cells and bacteria in Drosophila. J. Biol. Chem. 288, 10374–10380.

Odin, J.A., Edberg, P.C., Painter, C.J., Kimberly, R.P., Unkeless, J.C., 1991. Regulation of phagocytosis and [Ca2+]i flux by distinct regions of an Fc receptor. Science 254, 1785–1788.

Pearson, A., Lux, A., Krieger, M., 1995. Expression cloning of dSR-CI, a class C macrophage-specific scavenger receptor from *Drosophila melanogaster*. Proc. Natl. Acad. Sci. U.S.A. 92, 4056–4060.

Pearson, A.M., Baksa, K., Rämet, M., Protas, M., McKee, M., Brown, D., Ezekowitz, R.A.B., 2003. Identification of cytoskeletal regulatory proteins required for efficient phagocytosis in Drosophila. Microbes Infect. 5, 815–824.

Peltan, A., Briggs, L., Matthews, G., Sweeney, S.T., Smith, D.F., 2012. Identification of Drosophila gene products required for phagocytosis of *Leishmania donovani*. PLoS One 7, e51831.

Perrin, J., Mortier, M., Jacomin, A.C., Viargues, P., Thevenon, D., Fauvarque, M.O., 2015. The nonaspanins TM9SF2 and TM9SF4 regulate the plasma membrane localization and signalling activity of the peptidoglycan recognition protein PGRP-LC in Drosophila. J. Innate Immun. 7, 37–46.

Philips, J.A., Rubin, E.J., Perrimon, N., 2005. Drosophila RNAi screen reveals CD36 family member required for mycobacterial infection. Science 309, 1251–1253.

Philips, J.A., Porto, M.C., Wang, H., Rubin, E.J., Perrimon, N., 2008. ESCRT factors restrict mycobacterial growth. Proc. Natl. Acad. Sci. U.S.A. 105, 3070–3075.

Pulipparacharuvil, S., Akbar, M.A., Ray, S., Sevrioukov, E.A., Haberman, A.S., Rohrer, J., Kramer, H., 2005. Drosophila Vps16A is required for trafficking to lysosomes and biogenesis of pigment granules. J. Cell Sci. 118, 3663–3673.

Qin, Q.M., Pei, J., Ancona, V., Shaw, B.D., Ficht, T.A., De Figueiredo, P., 2008. RNAi screen of endoplasmic reticulum-associated host factors reveals a role for IRE1alpha in supporting Brucella replication. PLoS Pathog. 4, e1000110.

Qin, Q.M., Luo, J., Lin, X., Pei, J., Li, L., Ficht, T.A., De Figueiredo, P., 2011. Functional analysis of host factors that mediate the intracellular lifestyle of *Cryptococcus neoformans*. PLoS Pathog. 7, e1002078.

Ramet, M., Pearson, A.M., Manfruelli, P., Li, X., Koziel, H., Gobel, V., 2001. Drosophila scavenger receptor CI is a pattern recognition receptor for bacteria. Immunity 15, 1027–1038.

Ramet, M., Manfruelli, P., Pearson, A.M., Mathey-Prevot, B., Ezekowitz, R.A., 2002. Functional genomic analysis of phagocytosis and identification of a Drosophila receptor for *E. coli*. Nature 416, 644–648.

Rink, J., Ghigo, E., Kalaidzidis, Y., Zerial, M., 2005. Rab conversion as a mechanism of progression from early to late endosomes. Cell 122, 735–749.

Royet, J., Dziarski, R., 2007. Peptidoglycan recognition proteins: pleiotropic sensors and effectors of antimicrobial defences. Nat. Rev. Microbiol. 5, 264–277.

Salazar-Jaramillo, L., Paspati, A., Van De Zande, L., Vermeulen, C.J., Schwander, T., Wertheim, B., 2014. Evolution of a cellular immune response in Drosophila: a phenotypic and genomic comparative analysis. Genome Biol. Evol. 6, 273–289.

Sampson, C.J., Valanne, S., Fauvarque, M.O., Hultmark, D., Ramet, M., Williams, M.J., 2012. The RhoGEF Zizimin-related acts in the Drosophila cellular immune response via the Rho GTPases Rac2 and Cdc42. Dev. Comp. Immunol. 38, 160–168.

Schmucker, D., Clemens, J.C., Shu, H., Worby, C.A., Xiao, J., Muda, M., Dixon, J.E., Zipursky, S.L., 2000. Drosophila Dscam is an axon guidance receptor exhibiting extraordinary molecular diversity. Cell 10, 671–684.

Shattil, S.J., Kim, C., Ginsberg, M.H., 2010. The final steps of integrin activation: the end game. Nat. Rev. Mol. Cell Biol. 11, 288–300.
Shia, A.K., Glittenberg, M., Thompson, G., Weber, A.N., Reichhart, J.M., Ligoxygakis, P., 2009. Toll-dependent antimicrobial responses in Drosophila larval fat body require Spatzle secreted by haemocytes. J. Cell Sci. 122, 4505–4515.
Shiratsuchi, A., Mori, T., Sakurai, K., Nagaosa, K., Sekimizu, K., Lee, B.L., Nakanishi, Y., 2012. Independent recognition of *Staphylococcus aureus* by two receptors for phagocytosis in Drosophila. J. Biol. Chem. 287, 21663–21672.
Sigle, L.T., Hillyer, J.F., 2016. Mosquito hemocytes preferentially aggregate and phagocytose pathogens in the periostial regions of the heart that experience the most hemolymph flow. Dev. Comp. Immunol. 55, 90–101.
Simonsen, A., Lippe, R., Christoforidis, S., Gaullier, J.M., Brech, A., Callaghan, J., Toh, B.H., Murphy, C., Zerial, M., Stenmark, H., 1998. EEA1 links PI(3)K function to Rab5 regulation of endosome fusion. Nature 394, 494–498.
Solinger, J.A., Spang, A., 2013. Tethering complexes in the endocytic pathway: CORVET and HOPS. FEBS J. 280, 2743–2757.
Somogyi, K., Sipos, B., Penzes, Z., Kurucz, E., Zsamboki, J., Hultmark, D., Ando, I., 2008. Evolution of genes and repeats in the Nimrod superfamily. Mol. Biol. Evol. 25, 2337–2347.
Stroschein-Stevenson, S.L., Foley, E., O'Farrell, P.H., Johnson, A.D., 2006. Identification of Drosophila gene products required for phagocytosis of *Candida albicans*. PLoS Biol. 4, e4.
Stuart, L.M., Ezekowitz, R.A., 2005. Phagocytosis: elegant complexity. Immunity 22, 539–550.
Stuart, L.M., Deng, J., Silver, J.M., Takahashi, K., Tseng, A.A., Hennessy, E.J., Ezekowitz, R.A., Moore, K.J., 2005. Response to *Staphylococcus aureus* requires CD36-mediated phagocytosis triggered by the COOH-terminal cytoplasmic domain. J. Cell Biol. 170, 477–485.
Stuart, L.M., Boulais, J., Charriere, G.M., Hennessy, E.J., Brunet, S., Jutras, I., Goyette, G., Rondeau, C., Letarte, S., Huang, H., Ye, P., Morales, F., Kocks, C., Bader, J.S., Desjardins, M., Ezekowitz, R.A., 2007. A systems biology analysis of the Drosophila phagosome. Nature 445, 95–101.
Su, H.P., Nakada-Tsukui, K., Tosello-Trampont, A.C., Li, Y., Bu, G., Henson, P.M., Ravichandran, K.S., 2002. Interaction of CED-6/GULP, an adapter protein involved in engulfment of apoptotic cells with CED-1 and CD91/low density lipoprotein receptor-related protein (LRP). J. Biol. Chem. 277, 11772–11779.
Takats, S., Varga, A., Pircs, K., Juhasz, G., 2015. Loss of Drosophila Vps16A enhances autophagosome formation through reduced Tor activity. Autophagy 11, 1209–1215.
Tepass, U., Fessler, L.I., Aziz, A., Hartenstein, V., 1994. Embryonic origin of hemocytes and their relationship to cell death in Drosophila. Development 120, 1829–1837.
Tung, T.T., Nagaosa, K., Fujita, Y., Kita, A., Mori, H., Okada, R., Nonaka, S., Nakanishi, Y., 2013. Phosphatidylserine recognition and induction of apoptotic cell clearance by Drosophila engulfment receptor Draper. J. Biochem. 153, 483–491.
Ulvila, J., Parikka, M., Kleino, A., Sormunen, R., Ezekowitz, R.A., Kocks, C., Ramet, M., 2006. Double-stranded RNA is internalized by scavenger receptor-mediated endocytosis in Drosophila S2 cells. J. Biol. Chem. 281, 14370–14375.
Vieira, O.V., Botelho, R.J., Rameh, L., Brachmann, S.M., Matsuo, T., Davidson, H.W., Schreiber, A., Backer, J.M., Cantley, L.C., Grinstein, S., 2001. Distinct roles of class I and class III phosphatidylinositol 3-kinases in phagosome formation and maturation. J. Cell Biol. 155, 19–25.
Vieira, O.V., Botelho, R.J., Grinstein, S., 2002. Phagosome maturation: aging gracefully. Biochem. J. 366, 689–704.
Viljakainen, L., 2015. Evolutionary genetics of insect innate immunity. Brief. Funct. Genomics 14, 407–412.

Vogel, C., Teichmann, S.A., Chothia, C., 2003. The immunoglobulin superfamily in *Drosophila melanogaster* and *Caenorhabditis elegans* and the evolution of complexity. Development 130, 6317–6328.

Wang, L., Kounatidis, I., Ligoxygakis, P., 2014. Drosophila as a model to study the role of blood cells in inflammation, innate immunity and cancer. Front. Cell. Infect. Microbiol. 3, 113.

Watson, F.L., Puttmann-Holgado, R., Thomas, F., Lamar, D.L., Hughes, M., Kondo, M., Rebel, V.I., Schmucker, D., 2005. Extensive diversity of Ig-superfamily proteins in the immune system of insects. Science 309, 1874–1878.

Watthanasurorot, A., Jiravanichpaisal, P., Liu, H., Soderhall, I., Soderhall, K., 2011. Bacteria-induced Dscam isoforms of the crustacean, *Pacifastacus leniusculus*. PLoS Pathog. 7, e1002062.

Werner, T., Liu, G., Kang, D., Ekengren, S., Steiner, H., Hultmark, D., 2000. A family of peptidoglycan recognition proteins in the fruit fly *Drosophila melanogaster*. Proc. Natl. Acad. Sci. U.S.A. 97, 13772–13777.

Wertheim, B., Kraaijeveld, A.R., Schuster, E., Blanc, E., Hopkins, M., Pletcher, S.D., Strand, M.R., Partridge, L., Godfray, H.C., 2005. Genome-wide gene expression in response to parasitoid attack in Drosophila. Genome Biol. 6, R94.

Williams, M.J., 2007. Drosophila hemopoiesis and cellular immunity. J. Immunol. 178, 4711–4716.

Wojtowicz, W.M., Flanagan, J.J., Millard, S.S., Zipursky, S.L., Clemens, J.C., 2004. Alternative splicing of Drosophila Dscam generates axon guidance receptors that exhibit isoform-specific homophilic binding. Cell 118, 619–633.

Wu, J., Randle, K.E., Wu, L.P., 2007. ird1 is a Vps15 homologue important for antibacterial immune responses in Drosophila. Cell. Microbiol. 9, 1073–1085.

Yoshida, H., Kinoshita, K., Ashida, M., 1996. Purification of a peptidoglycan recognition protein from hemolymph of the silkworm, Bombyx mori. J. Biol. Chem. 271, 13854–13860.

Zaidman-Remy, A., Herve, M., Poidevin, M., Pili-Floury, S., Kim, M.S., Blanot, D., Oh, B.H., Ueda, R., Mengin-Lecreulx, D., Lemaitre, B., 2006. The Drosophila amidase PGRP-LB modulates the immune response to bacterial infection. Immunity 24, 463–473.

Zaidman-Remy, A., Poidevin, M., Herve, M., Welchman, D.P., Paredes, J.C., Fahlander, C., Steiner, H., Mengin-Lecreulx, D., Lemaitre, B., 2011. Drosophila immunity: analysis of PGRP-SB1 expression, enzymatic activity and function. PLoS One 6, e17231.

Zhan, X.L., Clemens, J.C., Neves, G., Hattori, D., Flanagan, J.J., Hummel, T., Vasconcelos, M.L., Chess, A., Zipursky, S.L., 2004. Analysis of Dscam diversity in regulating axon guidance in Drosophila mushroom bodies. Neuron 43, 673–686.

Zhou, Z., Hartwieg, E., Horvitz, H.R., 2001. CED-1 is a transmembrane receptor that mediates cell corpse engulfment in *C. elegans*. Cell 104, 43–56.

Ziegenfuss, J.S., Biswas, R., Avery, M.A., Hong, K., Sheehan, A.E., Yeung, Y.G., Stanley, E.R., Freeman, M.R., 2008. Draper-dependent glial phagocytic activity is mediated by Src and Syk family kinase signalling. Nature 453, 935–939.

CHAPTER THREE

The Melanization Response in Insect Immunity

Johnny Nakhleh[1], Layla El Moussawi[1], Mike A. Osta
American University of Beirut, Beirut, Lebanon

Contents

1. Introduction 84
2. Melanin Biosynthesis Pathways in Insects 86
3. PPO Activation Pathways in Model Insects 87
   3.1 *Drosophila melanogaster* 87
   3.2 *Manduca sexta* 88
   3.3 *Tenebrio molitor* 90
   3.4 *Anopheles gambiae* 92
   3.5 *Aedes aegypti* 93
4. The Contribution of the Melanization Response to Insect Immune Defence 93
5. Crosstalk Between Melanization and Other Immune Pathways 96
   5.1 Melanization and AMP Synthesis 96
   5.2 Melanization and Complement Pathway 100
6. Concluding Remarks 101
References 102

Abstract

Melanization plays important roles in diverse physiological processes in insects including wound healing, tanning of the cuticle and immunity. Upon infection, pattern recognition receptors activate downstream serine protease cascades that culminate in the activation of prophenoloxidase (PPO), the rate-limiting enzyme in the process of melanogenesis. During the last two decades, diverse genetic and biochemical approaches have been adopted to characterize this process, and a wealth of information has been generated concerning the molecular events that control PPO activation. Importantly, the melanization reaction was shown to be toxic to parasites, bacteria, fungi and recently viruses. Several studies pointed also to the existence of significant crosstalk between melanization and other immune responses possibly to coordinate immune attack against invaders. Here, we provide a critical review of the role of melanization in insect immunity, highlighting the important discoveries but also the gaps that remain to be explored in future studies.

[1] Contributed equally to this work.

Advances in Insect Physiology, Volume 52
ISSN 0065-2806
http://dx.doi.org/10.1016/bs.aiip.2016.11.002

1. INTRODUCTION

Insects, unlike vertebrates, depend solely on their innate immune system for defence against infectious agents. Insect immune responses are initiated when soluble or membrane-bound pattern recognition receptors (PRRs) recognize pathogen-associated molecular patterns triggering eventually several effector responses such as phagocytosis, lysis or melanization that result in the elimination of the invader. Melanization is an immune effector response that is triggered locally in response to cuticle injury or systemically following microbial invasion of the hemocoel. It is characterized by the synthesis of melanin and its cross-linking with molecules on microbial surfaces or in injured areas resulting in the killing of the invader and hardening of the wound clot. Despite that clotting occurs in insects independent of the melanization response, recent studies in several insects inform the existence of a potential link between the coagulation system and melanization, whereby the former initiates the clotting process and the latter contributes to the hardening of the clots (reviewed in Eleftherianos and Revenis, 2011). In addition to its role in immunity, melanization is essential for cuticle sclerotization or tanning that leads to the hardening of the insect exoskeleton by cross-linking the cuticular proteins by quinones generated during that process (Andersen, 2010). A key enzyme in melanin biosynthesis is phenoloxidase (PO) which mediates the oxidation of tyrosine to dihydroxyphenylalanine and the oxidation of dihydroxyphenylalanine and dopamine to their respective quinones which are precursors of melanin formation (reviewed in Vavricka et al., 2010). PO is produced as prophenoloxidase (PPO) zymogen which is converted to active PO by a clip domain serine proteinase (CLIP). CLIPs are specific to invertebrates and act in cascades to modulate several immune responses including coagulation, melanization and the synthesis of antimicrobial peptides (AMPs) through Toll pathway activation. Not all CLIPs are catalytic; those that lack one or more of the three residues (His, Asp, Ser) that form the catalytic triad are noncatalytic [also known as clip-domain containing serine proteinase homologs (cSPHs)], while the rest are catalytic and will be referred to henceforth as cSP (clip domain containing serine proteinases).

Several methods have been utilized to classify CLIPs. Initially, the amino acid sequence length between cysteines 3 and 4 of the clip domain was utilized to group CLIPs into two groups, 1 and 2 (Jiang and Kanost, 2000). Later, this classification was further refined by adding an additional criteria related to the type of the amino acid residue before the cutting site of the zymogen resulting in the subdivision of group 1 into 1a and 1b (Ross et al., 2003). Phylogenetic analysis of mosquito CLIPs based on whole sequence alignment lead to their classification into five groups A–E, whereby groups A and E include noncatalytic CLIPs, while groups B, C and D are catalytic (Waterhouse et al., 2007). More recent phylogenetic analysis of *Manduca sexta* clip domains and their comparative sequence alignment with clip domains from other insects resulted in a similar classification: A are cSPHs containing 1 or more group 3 clip domains; B are cSP with 1 or 2 group-2 clip domains; CLIPC, D1 and D2 are cSP containing one clip domain belonging to groups 1a, 1b and 1c, respectively (Cao et al., 2015). A structure–function analysis of *Drosophila* grass cSP and comparative analysis with other cSPs of known function allowed the classification of cSPs into two functional groups: Those which contain a 75-loop protruding from the calcium-binding 70-loop in close proximity to the activation site are considered terminal proteinases that are directly involved in the processing of PPO triggering the melanization response, or of proSpätzle leading to Toll pathway activation. The remaining cSPs that lack the 75-loop are penultimate proteinases that are likely to act upstream in the cascade (Kellenberger et al., 2011). In all characterized PPO activation cascades, the most upstream proteinase that is likely to interact with and relay information from a PRR is a modular serine proteinase (ModSp) that lacks a clip domain but contains other domains that mediate protein interactions (Buchon et al., 2009; Ji et al., 2004; Roh et al., 2009; Takahashi et al., 2015). ModSps are generally autoactivated resulting in the proteolytic cleavage and activation of a CLIPC which in turn activates a CLIPB. The PPO-activating proteinase (PAP) is always a CLIPB (reviewed in Kanost and Jiang, 2015). CLIP cascades that control PPO activation are tightly regulated by serine proteinase inhibitors, known as serpins. Serpins are the largest family of serine proteinase inhibitors which regulate several physiological processes in insects including embryonic development, wound clotting and host defence (reviewed in Gulley et al., 2013; Silverman et al., 2010). Serpins contain a reactive center loop that binds specifically to the active site of the target proteinase in a similar way as the substrate. Upon cleaving of the serpin scissile bond, the serpin forms a covalent complex with the proteinase that is eventually eliminated from the hemolymph (Dunstone and Whisstock, 2011; Huntington et al., 2000; Olson and Gettins, 2011).

While the biochemical pathways of melanogenesis downstream of PO are well characterized, our knowledge of the molecular events leading to

PPO activation remains fragmentary in most insects. Here, we review our current understanding of the melanization immune response in model insects, focusing on the molecular pathways leading to PPO activation and the crosstalk between melanization and other immune responses. The phylogenetic and structural analysis of CLIPs will not be covered herein; readers interested in these topics are referred to the following excellent reviews (Kanost and Jiang, 2015; Veillard et al., 2016). We will also highlight the future challenges that remain to be addressed in order to provide a comprehensive picture of this fascinating immune response in insects.

2. MELANIN BIOSYNTHESIS PATHWAYS IN INSECTS

Melanogenesis in insects is initiated by the hydroxylation of phenylalanine by phenylalanine 4-monooxygenase (PAH), to form the rate-limiting substrate tyrosine (Futahashi and Fujiwara, 2005; Gorman et al., 2007a). In addition to its role in cuticular sclerotization (Futahashi and Fujiwara, 2005, 2007), genetic analysis highlighted also a key role for PAH in the melanization immune response. Indeed, *PAH* transcript levels were significantly increased in mosquitoes upon challenge with *Dirofilaria immitis* microfilariae (Johnson et al., 2003) and bacteria (Oduol et al., 2000). Also, *PAH* knockdown (kd) significantly reduced the melatonic encapsulation of microfilariae worms in the mosquitoes *Aedes aegypti* and *Armigeres subalbatus* (Infanger et al., 2004) as well as of *Plasmodium berghei* ookinetes in *Anopheles gambiae* (Fuchs et al., 2014). The effect of *PAH* RNAi was also observed at the level of mosquito fertility leading to an impairment in the melanization of the chorion and reduced egg laying (Fuchs et al., 2014). Following tyrosine synthesis, tyrosinase-like POs then catalyse the subsequent oxidation of tyrosine into dihydroxyphenylalanine (Dopa) and the oxidation of Dopa into the intermediate molecule dopaquinone. In the presence of thiol compounds, dopaquinone is converted to cysteinyl and glutathionyl conjugates that mediate the synthesis of the cutaneous reddish pigment pheomelanin. Otherwise, Dopaquinone undergoes a spontaneous cyclization into dopachrome, which in turn is decarboxylated by dopachrome conversion enzyme to generate 5,6-dihyroxyindole (DHI). Following PO-mediated DHI oxidation, indole quinones polymerize and give rise to the heteropolymer eumelanin (reviewed in Vavricka et al., 2010). DHI-eumelanin can be also derived from dopamine produced early on upon the decarboxylation of dopa by dopa decarboxylase (DDC). This pathway seems to be particularly important for cuticular sclerotization because several insect cuticles were found to contain domapine melanin (Futahashi and Fujiwara, 2005; Hiruma and Riddiford, 1985; Hiruma et al., 1985; Koch et al., 2000). On the other hand, few studies pointed also to a potential role for DDC in immunity. In *Drosophila*, DDC expression was upregulated throughout the epidermis following infections with Gram-positive or Gram-negative bacteria but not after a sterile injury (Davis et al., 2008). DDC upregulation was also observed in lipopolysaccharide (LPS)-challenged *Tribolium castaneum* (Altincicek et al., 2013). In the medfly *Ceratitis capitata*, DDC activity was required for efficient phagocytosis, melanization and nodulation responses after *Escherichia coli* infections (Sideri et al., 2008).

The immunoprotective functions of melanization are attributed in part to the oxidoreductive properties of melanogenic precursors that engage in various redox reactions to create a biochemically hostile environment to invaders. Eumelanin and quinoide intermediates deposit as cross-linking complexes on foreign nucleophilic surfaces to effectively encapsulate, immobilize and deprive circulating pathogens from nutrients (Nappi and Christensen, 2005; Zhao et al., 2007).

3. PPO ACTIVATION PATHWAYS IN MODEL INSECTS

3.1 *Drosophila melanogaster*

The infection-induced melanization response in *Drosophila* requires two CLIPs termed MP1 and MP2 whereby MP2 is thought to act upstream of MP1 in the cascade. These proteinases were identified in a genetic screen for the suppression of tissue melanization induced in *Spn27A*$^{-/-}$ flies (Tang et al., 2006). Spn27A is a serpin whose loss of function triggers spontaenous tissue melanization and semilethality (De Gregorio et al., 2002; Ligoxygakis et al., 2002). Genetic studies combined with enzymatic assays for PO activity suggest that MP1 is involved in the melanization response against fungal and bacterial infections whereby MP2 is more specific for fungal infections (Tang et al., 2006). Although MP1 is suggested to act as a PPO-activating enzyme, direct cleavage of PPO has been shown only for MP2 (An et al., 2013). MP2 also formed SDS-stable complexes with Spn27A in vitro and in vivo. In addition to Spn27A, Spn77Ba was also shown to regulate melanization in the epithelium of the respiratory tract by targeting MP1 and MP2. *Spn77Ba* RNAi flies showed excessive melanization of the tracheal system, followed by death of almost all the larvae before reaching the pupal stage. Silencing either *MP1* or *MP2* in *Spn77Ba* RNAi larvae completely suppressed tracheal melanization. Altogether, these results suggest that the proteinase cascade involving MP1 and MP2 is required for systemic melanization in the hemolymph as well as for local epithelial melanization, possibly to block microbial entry into the hemolymph (Tang et al., 2008). However, it remains unclear what PRR acts upstream of the MP2–MP1 cascade in *Drosophila*. Despite the fact that PGRP-LE (Takehana et al., 2002) and GNBP3 (Matskevich et al., 2010) were shown to be involved in the melanization response to infections with Gram-negative bacteria and fungi, respectively, none of these PRRs was linked to the MP2–MP1 module. Additionally, the ModSp acting upstream of this cascade remains unidentified. In addition to MP2 and MP1, another CLIP called Hayan was identified as a key activator of PPO in the systemic wound response. *Hayan* RNAi flies exhibited reduced survival rate in response to sterile injury, whereas neither *MP1* nor *MP2* RNAi flies exhibited a similar phenotype (Nam et al., 2012). Active recombinant Hayan directly cleaved *Drosophila* PPO1 suggesting that it acts as a PAP in the wound-induced melanization response. Interestingly, *Hayan* kd completely suppressed melanization induced in several serpin kd including *Spn27A* and *Spn28D* which control hemolymph PO (De Gregorio et al., 2002; Ligoxygakis et al., 2002; Scherfer et al., 2008), and Spn77Ba involved in tracheal melanization (Tang et al., 2008). Despite the fact that Hayan is required for both, wound-induced and microbe-induced hemolymph PO activity (Nam et al., 2012), there is no evidence yet that it might be part of the MP2–MP1 module. Hence, the current data infer the presence of two distinct pathways for melanization in *Drosophila*: one controls wounding and the second responds to infection.

3.2 *Manduca sexta*

Manduca has been among the best insect models for the biochemical characterization of PPO activation cascades. The significant amounts of hemolymph extracted from the large size larvae (in comparison to flies and mosquitoes) render this insect more amenable to in vitro biochemical assays for monitoring PPO cleavage and activity, determining the hierarchical organization of CLIP cascades and the identification of CLIP–serpin complexes. In *Manduca*, two β-glucan recognition proteins (βGRP1 and βGRP2) bind to the surfaces of Gram-positive bacteria and fungi triggering PPO activation (Jiang et al., 2004; Ma and Kanost, 2000). Binding of βGRP2 to glucans recruits initially the ModSp HP14 which autoactivates by internal processing into two polypeptide chains that remain linked by an interchain disulphide bond (Wang and Jiang, 2006). The low-density lipoprotein receptor class A domain in the N-terminus of HP14 is required for its interaction with βGRP2 (Takahashi et al., 2015). Efficient activation of the PPO cascade seems to require the self-aggregation of βGRP2 on microbial sugars. The functional significance of βGRP2 oligomers remains unclear; however, it is proposed that they may form protein platforms required for the efficient recruitment of downstream factors in the pathway (Takahashi et al., 2014). Active HP14 cleaves the cSP proHP21 into active HP21 which in turn cleaves the PPO-activating proteinase-2 zymogen (PAP-2) into active PAP-2, the terminal cSP in the cascade that processes PPO into PO (Wang and Jiang, 2007). In addition to PAP-2, HP21 was also shown to cleave PAP-3 (Gorman et al., 2007b). It is worth noting here that in *Manduca*, three PAPs (PAP-1, 2 and 3) have so far been shown to be direct activators of PPO (Jiang et al., 1998, 2003a,b); while PAP-2 and 3 contain each two N-terminal clip domains, PAP-1 has only one. Despite that PAP-1 is an activator of PPO it is not regulated by the HP14–HP21 pathway but rather by a different pathway requiring HP6. HP6 is an apparent orthologue of *Drosophila* persephone which is cleaved upon treatment of *Manduca* plasma with bacteria or curdlan (An et al., 2009). Both HP6 and HP21 have been initially identified by immunoaffinity chromatography using antibodies against serpins 4 and 5 (Tong et al., 2005). In addition to controlling PPO cleavage, HP6 also controls Toll pathway activation by cleaving the spätzle-processing enzyme (SPE) HP8 (An et al., 2009). PPO activation in *Manduca* requires also the two cSPHs, SPH1 and SPH2 (Gupta et al., 2005; Yu et al., 2003). These cSPHs seem to be required as cofactors for the efficient processing and activation of PPO. Interestingly, the precursor forms of SPH1 and SPH2 cannot activate PPO (Lu and Jiang, 2008; Yu et al., 2003) but rather require processing by PAP-3 and PAP-1 (to a lesser extent) to become active (Wang and Jiang, 2008; Wang et al., 2014).

An interesting finding that emerged from the biochemical analysis of the PPO cascade in *Manduca* is that this cascade is subject to a positive feedback mechanism through PAPs. Indeed, PAP-1 was shown to activate HP6 zymogen through a yet unknown mechanism, hence triggering more activation of PAP-1 zymogen (Wang and Jiang, 2008). Also, recombinant active PAP-3 added to *Manduca* naive plasma cleaved PPO as well as the precursor forms of SPH1, SPH2 and of PAP-3 itself leading to a positive feedback loop that significantly enhanced PO activity (Wang et al., 2014). The strong impact of active PAP-3 on PO activation likely explains why it is targeted by several serpins including serpin 1J (Jiang et al., 2003b), serpin-3 (Christen et al., 2012), serpin-6 (Wang and Jiang, 2004b) and serpin-7 (Suwanchaichinda et al., 2013). Two other serpins, serpin-4 and -5, that were induced by bacterial infection were also shown to regulate the PPO cascade upstream of PAPs (Tong and Kanost, 2005). Affinity immunoprecipitation using antibodies against serpins-4 and -5 followed by mass spectrometry analysis revealed that serpin-4 is an inhibitor of HP21, HP6 and HP1 (Tong et al., 2005), and serpin-5 is an inhibitor of HP6 and HP1 (An and Kanost, 2010). This target redundancy of serpins that is observed at different levels and in different insects reveals the importance of keeping melanization under tight regulation, since an uncontrolled or exaggerated melanization response may have deleterious impact on the host (Ligoxygakis et al., 2002; Michel et al., 2005; Scherfer et al., 2008). While the roles of HP6 and HP21 in the melanization response are confirmed, a recent study suggests that HP1 may be involved in the PPO cascade; however, its in vivo targets and substrates remain to be determined (Yang et al., 2016).

3.3 *Tenebrio molitor*

In *Tenebrio molitor* incubation of crude hemolymph extracts with microbial cell wall components revealed that DAP-type peptidoglycan (DAP-PGN), lysine-type PGN (Lys-PGN), and β-1,3-glucan but not LPS were able to activate the PPO cascade (Park et al., 2006). Affinity purification with PGN fragments identified *Tenebrio* PGRP-SA as a PRR of the PPO cascade. Using as bait a recombinant PGRP-SA/Lys-PGN complex, two other components involved in initiating the PPO cascade were identified from PGRP-SA-deficient hemolymph; one is the PRR GNBP1 and the second is a ModSp (Park et al., 2006). It was also shown that PGRP-SA clustering on Lys-PGN is required to induce the robust activation of the melanization response (Park et al., 2007). Following the recruitment of ModSp by the PGRP-SA/GNBP1 complex, ModSp autoactivates and cleaves a downstream cSP called SAE (spätzle-processing enzyme-activating enzyme) which in turn activates SPE (Kim et al., 2008). In addition to processing spätzle leading to Toll pathway activation, SPE was also shown to process PPO and a precursor cSPH1 into their active forms triggering the melanization reaction (Kan et al., 2008). cSPH1 was previously shown to undergo limited proteolytic cleavage following exposure of hemolymph to β-1,3-glucan, and this processed form is required for the activation of *Tenebrio*

PPO in in vitro reconstitution assays (Lee et al., 2002). The processed active form of cSPH1 seems to form a complex with active PO on microbial surfaces called the melanization complex. The fact that cSPH1 (Zhang et al., 2003), like other cSPHs (Lee and Soderhall, 2001; Yu et al., 2003), can bind microbial cell wall components, the proposed role for such an association, over and above enhancing PPO activation, is probably to localize PO to microbial surfaces inhibiting its diffusion to the hemolymph. PPO itself was found to act as a negative regulator of this melanization complex by sequestering active cSPH1 (Kan et al., 2008).

Though noncatalytic, cSPHs have been shown to play essential roles in melanization in several insect species. These mainly act as cofactors for the efficient and correct cleavage of PPO by cSP as described earlier for *Manduca* and *Tenebrio*. The same was also observed for PPAFII, a cSPH from the beetle *Holotrichia diomphalia* which is also indispensable for PO activation. PPAFII is initially cleaved by a catalytic cSP called PPAFIII. Cleaved PPAFII acts as a cofactor for the catalytic cSP called PPAFI to cleave PPO into active PO (Kim et al., 2002). Structural studies revealed that following cleavage by PPAFIII, PPAFII self-oligomerizes to form a large complex in which each monomer recruits one molecule of active PO (PO76) through a cleft in the clip-domain, in a 1:1 ratio (Piao et al., 2005). PPOs cleaved by their specific cSPs in the absence of cSPH cofactors showed no activity in vitro even when the cSPH was added later to the activated PO (Gupta et al., 2005; Kan et al., 2008; Wang and Jiang, 2004a), indicating that cSPHs are required for the correct cleavage of PPOs.

Three serpins have been so far identified as regulators of the *Tenebrio* PPO activation cascade using immunoprecipitations with antibodies against the proteases ModSp, SAE and SPE. Serpin 40 was found in a complex with MSP; Spn55 and Spn48 were complexed with SAE and SPE, respectively. In vitro assays revealed that each of these serpins alone reduced significantly the proteolytic activity of SPE while adding all three serpins together completely abolished SPE activity (Jiang et al., 2009). In addition to serpins, a melanization-inhibiting protein (MIP) with no sequence identity to any known protein was identified in the hemolymph of *T. molitor*. Although the molecular mechanism by which MIP negatively regulates the melanization response is not completely understood, it is proposed that MIP does not act on components of the PPO cascade per se but may rather function more downstream as scavenger of reactive quinone products (Zhao et al., 2005).

3.4 *Anopheles gambiae*

In the malaria vector *A. gambiae* the melanization response is tightly controlled by the complement-like thioester-containing protein 1 (TEP1). RNAi-mediated *TEP1* kd abolished the melanization of *P. berghei* ookinetes present in the basal labyrinth of the midgut epithelium in different mosquito genetic backgrounds (Blandin et al., 2004; Povelones et al., 2011). Additionally, *TEP1* kd abolished PO activity in the hemolymph in response to *E. coli* systemic infections (Povelones et al., 2013). A systematic functional genetic screen of cSPs in the mosquito identified CLIPA8 as an essential component of the mosquito melanization response to *P. berghei* parasites (Volz et al., 2006). CLIPA8 is a cSPH which is cleaved soon after the injection of bacteria into the mosquito hemolymph (Schnitger et al., 2007). *TEP1* kd abolished CLIPA8 cleavage suggesting that it acts upstream of CLIPA8 in the melanization response. CLIPA2 is another cSPH that was identified in the screen as a negative regulator of the melanization response to *P. berghei*; silencing CLIPA2 significantly increased the numbers of melanized ookinetes in mosquito midguts (Volz et al., 2006). Subsequent studies revealed that CLIPA2 regulates melanization indirectly by controlling TEP1 activity during systemic infections. *CLIPA2* kd enhanced TEP1 activity leading to an exaggerated PO activity in the hemolymph following *E. coli* infections (Kamareddine et al., 2016; Yassine et al., 2014). SPCLIP1 is another cSPH that exhibits indirect control over melanization by acting as a positive regulator of TEP1. *SPCLIP1* kd, similar to that of TEP1, inhibited the cleavage of CLIPA8 and abolished PO activity in response to bacterial infections (Povelones et al., 2013). In addition to cSPHs, several cSPs have been shown to be required for the melanization of *P. berghei* ookinetes to different extents including, CLIPB17, CLIPB8, CLIPB3 and CLIPB4 (Volz et al., 2006). However, the hierarchical organization of these CLIPs and their relation to CLIPA8 await the biochemical characterization of mosquito cSP cascades, which is not a trivial task owing to the very small amounts of hemolymph present in mosquitoes compared to larger insects, such as *Manduca* and *Tenebrio*.

The PPO cascade in *A. gambiae* can be induced spontaneously in tissues forming melanotic pseudotumours following the kd of serpin-2 by RNAi. *Serpin-2* kd also increased the number of melanized *Plasmodium* ookinetes suggesting that it regulates more than one melanization pathway by targeting several cSPs (Michel et al., 2005). One of its target cSPs was identified later to be CLIPB9. *CLIPB9* kd partially reversed serpin-2 RNAi phenotype in vivo, and serpin-2 directly inhibited CLIPB9 proteolytic activity in vitro. Furthermore, CLIPB9 was shown to cleave directly *Manduca* PPO suggesting that it acts as a PAP in *A. gambiae* (An et al., 2011). *CLIPB8* kd was also shown to reverse partially the serpin-2 RNAi phenotype; however, it was not directly inhibited by serpin-2 in vitro, and its hierarchical position with respect to CLIPB9 remains to be determined (Zhang et al., 2016).

3.5 *Aedes aegypti*

Two melanization pathways were identified in *Ae. aegypti* using functional genetic analysis. Tissue melanization requires two cSPs, IMP1 (immune protease 1) and CLIPB8, and is negatively regulated by serpin-2 (Zou et al., 2010). Whereas hemolymph melanization induced by malaria parasites requires the two cSPs, IMP1 and IMP2, and is negatively regulated by serpin-1 (Zou et al., 2010). However, the hierarchical organization of these cSPs in *Aedes* remains to be determined. In addition to serpins, a modular serine protease called CLSP2 was shown to negatively regulate hemolymph PPO activity. CLSP2 seems to mediate its effect at the transcriptional level by suppressing the expression of several immunity genes involved in melanization (PPOs, serpins and cSP) and Toll pathway (Wang et al., 2015).

4. THE CONTRIBUTION OF THE MELANIZATION RESPONSE TO INSECT IMMUNE DEFENCE

Evidence supporting a substantial role of melanization in insect defence came from several studies. Probably one of the most convincing lines of evidence came from polydnaviruses carried by female parasitoid wasps (Lu et al., 2008) and from *Photorhabdus* bacteria pathogenic to *M. sexta* (Eleftherianos et al., 2007), which evolved independent specific strategies to counteract the host melanization response. Polydnaviruses integrate in the genome of the parasitoid wasp and are vertically transmitted to the offspring. During oviposition, the female wasp injects several viruses into the larval stage of the moth where they infect several tissues, yet they do not replicate but rather express a set of proteins that disable the host melanization response allowing the wasp offspring to survive (reviewed in Strand and Burke, 2013). Two of these proteins, Egf1.0 (Beck and Strand, 2007; Lu et al., 2008) and Egf1.5 (Lu et al., 2010) were shown to inhibit the processing and enzymatic activities of PAPs in hemolymph samples of several insect species. *Photorhabdus* is another virulent pathogen that overcomes its insect host immune defences. These Gram-negative bacteria infect a broad range of insects and are also symbionts of entomopathogenic nematodes (Forst et al., 1997). Using *M. sexta* as a model host, it was shown that a small molecule antibiotic (E)-1,3-dihydroxy-2-(isopropyl)-5-(2-phenylethenyl)benzene produced by *Photorhabdus luminescens* inhibits PO activity in vitro and in vivo (Eleftherianos et al., 2007). Screening of *P. luminescens* and *Photorhabdus asymbiotica* cosmid libraries in an *E. coli* host against *M. sexta* previously activated PO, identified a cosmid that disabled PO activity in vitro and in vivo (Eleftherianos et al., 2009). Insertional mutagenesis in that specific cosmid revealed that the inhibitory activity is associated with a locus (*mal*) of three ORFs encoding a maltodextrin phosphorylase (*malP*), amylomaltase (*malQ*) and a regulator of the *mal* operon (*malT*). How these proteins function to inhibit PO remain however unknown.

Genetic studies in the model dipteran *Drosophila* revealed initially conflicting data on the significance of melanization in the fly immune defence. Using a *Drosophila* mutant for PO-activating enzyme (PAE1) that fails to activate hemolymph PO following microbial infections, Leclerc et al. showed that PAE1 mutants exhibited survival rates similar to those of wild-type flies in response to infections with fungi, Gram-negative and Gram-positive bacteria (Leclerc et al., 2006). PAE1 mutants injected with *E. coli* harboured similar colony forming units as wildtypes, suggesting that PO activity does not contribute to the fly's resistance (i.e. microbial persistence in tissues) to infection. A similar conclusion was reached by another group using mutants for the MP1 and MP2 (same as PAE1) cSPs required for the activation of the melanization reaction in response to bacterial and fungal infections (Tang et al., 2006). However, in the latter study the authors proposed that melanization seems to enhance the effectiveness of the Toll and Imd pathways by comparing the survival rates of *Dif* and *dFadd* mutants with *Dif*/*MP1*(or *MP2*) and *dFadd*/*MP1*(or *MP2*), respectively. On the other hand, using a broad panel of bacterial to challenge *Drosophila*, it was shown that a single mutation in gene CG3066 (encoding PAE1) showed variations in the resistance and tolerance (i.e. ability to support the consequences of an infection on the host) properties of the flies with varying microbial challenge (Ayres and Schneider, 2008), providing support for a significant role of melanization in immune defence. More recently, the contribution of PO activity to antimicrobial defence has been addressed using flies carrying deletions in *PPO1* and *PPO2* genes (Binggeli et al., 2014). Survival assays revealed that single *PPO* mutants exhibited similar susceptibilities as wild-type flies in response to infections with fungi, Gram-negative and Gram-positive bacteria, whereas *PPO1*/*PPO2* double mutants ($PPO1^{\Delta}/PPO2^{\Delta}$) showed marked susceptibility to Gram-positive bacteria and fungal infections, indicating that melanization plays a nonredundant role in immune defence against these classes of pathogens. Interestingly, while $PPO1^{\Delta}/PPO2^{\Delta}$ *flies* showed increased susceptibility (or reduced tolerance) to infections with *Listeria monocytogenes*, *Enterococcus faecalis* and *Candida albicans*, the load of these microbes was similar to that in wild-type flies suggesting that PO activity might contribute to tolerance rather than resistance to these microorganisms. Only for *Staphylococcus aureus* infections, $PPO1^{\Delta}/PPO2^{\Delta}$ flies exhibited both reduced tolerance and resistance. Challenging melanization-defective flies with Gram-negative bacteria resulted in three different phenotypes: *Salmonella typhimurium* compromised the fly's survival rate and resistance (Ayres and Schneider, 2008); *Erwinia carotovora* only reduced the survival rate (Binggeli et al., 2014); whereas *Burkholderia cepacia* reduced the flies' resistance without affecting survival (Ayres and Schneider, 2008). These complex phenotypes observed in the absence of PO activity are most likely the result of several factors including the virulence of the bacterial strain, efficiency of bacterial neutralization by other effector mechanisms such AMPs and phagocytosis, and the extent of tissue damage inflicted on the host.

Abolishing hemolymph PO activity in the malaria vector *A. gambiae* by silencing CLIPA8 did not affect mosquito survival after infections with *E. coli* or *S. aureus* (Schnitger et al., 2007). Both bacterial species were cleared from CLIPA8-silenced mosquitoes as efficiently as from controls suggesting that melanization is not critical for antibacterial defence in the mosquito. However, these results must be interpreted cautiously as neither *E. coli* nor *S. aureus* are natural pathogens of the mosquito, hence the need of readdressing this question using a larger panel of bacteria including potentially virulent members of certain genera such as *Serratia*, *Pseudomonas*, *Enterobacter* and others that have been identified in metagenomic analysis of the mosquito microbiota (Boissiere et al., 2012; Dong et al., 2009; Gimonneau et al., 2014). In *A. gambiae*, the melanization response to *P. berghei* is also controlled by CLIPA8 (Volz et al., 2006), in addition to the complement-like protein TEP1 (Blandin et al., 2004) and two leucine-rich immune proteins, LRIM1 (Osta et al., 2004) and APL1C (Riehle et al., 2006, 2008). The latter two proteins form an obligate disulphide-linked heterodimer in the mosquito hemolymph that interacts with and stabilizes a cleaved form of TEP1 (Fraiture et al., 2009; Povelones et al., 2009). In addition to triggering ookinete lysis in the basal labyrinth of the midgut epithelium (Blandin et al., 2004; Osta et al., 2004;

Riehle et al., 2006), the TEP1/LRIM1/APL1C complex (henceforth TEP1 complex) is also required for the melanotic response to ookinetes in refractory mosquito genotypes (Blandin et al., 2004; Osta et al., 2004) as well as to bacteria injected directly into the hemolymph (unpublished data). Nevertheless, wild-type laboratory and field caught *A. gambiae* mosquitoes rarely melanize malaria parasites (Niare et al., 2002) indicating that this response is dispensable for anti-*Plasmodium* defence (Schnitger et al., 2007). On the other hand, *CLIPA8* kd *A. gambiae* mosquitoes exhibited compromised survival and resistance to natural infections with *Beauveria bassiana* indicating that melanization contributes significantly to antifungal defence in the mosquito (Yassine et al., 2012).

The role of melanization in immune defence seems to span viruses also. Despite the paucity of studies in this regard, it was clearly shown that Semliki Forest virus (SFV) triggered the PO activity in the conditioned medium of U4.4 cells of *Aedes albopictus* (Rodriguez-Andres et al., 2012). A recombinant SFV expressing the Egf1.0 PO inhibitor of the *Microplitis demolitor* bracovirus showed enhanced spreading through U4.4 cells. Importantly, adult *Ae. aegypti* mosquitoes that received a bloodmeal containing the recombinant SFV exhibited increased mortality and increased viral replication (Rodriguez-Andres et al., 2012). A similar role for melanization have been recently described in the crustacean *Penaeus monodon* (shrimp) whereby cosilencing the only two PPO genes by RNAi increased the shrimp mortality to infections with the white spot syndrome virus (Sutthangkul et al., 2015). While these studies clearly pinpoint a role for PO activity in antiviral defence, the nature of the receptors that recognize the virus and the mechanism by which melanization kills virus particles remain unknown. One possibility is that PO-generated reactive quinones such as 5,6-dihydroxyindole (DHI) or its oxidation products may directly damage viral proteins, especially that these molecules exhibit broad antimicrobial activities (Zhao et al., 2007).

5. CROSSTALK BETWEEN MELANIZATION AND OTHER IMMUNE PATHWAYS

5.1 Melanization and AMP Synthesis

The PO cascade does not seem to function as an independent branch of the humoral immune response of insects, but rather, there is substantial evidence for extensive crosstalk with different immune pathways, especially the Toll pathway (Fig. 1). In *Drosophila*, several reports revealed that gain-of-function

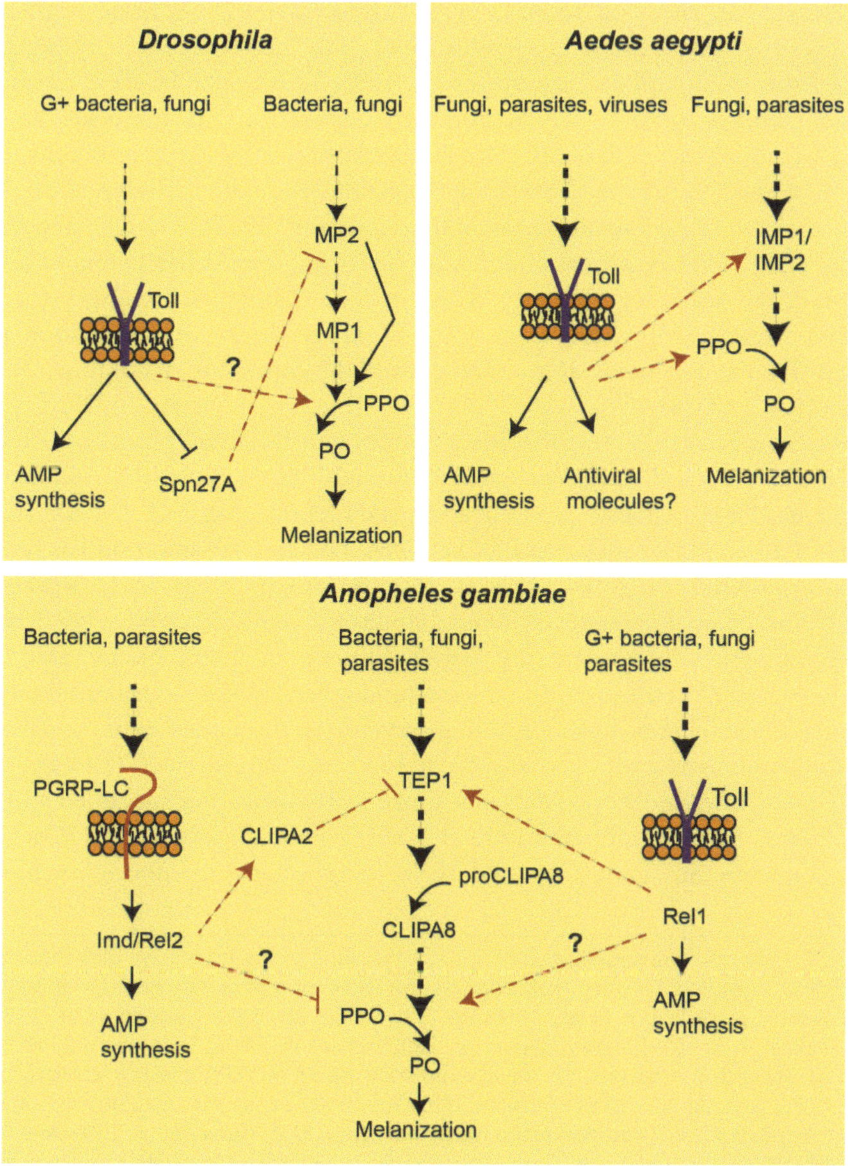

Fig. 1 Immune signalling pathways controlling AMP expression regulate the infection-induced melanization response in several insects. Schematic presentation of immune signalling pathways that crosstalk with PPO activation cascades in *Drosophila melanogaster*, *Anopheles gambiae* and *Aedes aegypti*. Regular *black dashed lines* indicate the presence of several known steps that were omitted for simplicity. *Thick dashed lines* correspond to steps that have not been thoroughly characterized. The *red dashed lines* indicate the functional connection between both pathways. In *Drosophila*, Toll controls *(Continued)*

mutations in the Toll receptor lead to spontaneous melanotic tumours suggesting the existence of functional link between both responses (Gerttula et al., 1988; Lemaitre et al., 1995; Qiu et al., 1998). This link was through Spn27A which controls melanization in *Drosophila* in a Toll-pathway dependent manner. Toll activation is required for the depletion of Spn27A from the hemolymph, possibly through expression of a degradative enzyme, relieving the inhibition over the PO cascade (De Gregorio et al., 2002; Ligoxygakis et al., 2002). Spn27A was later shown to bind to and inhibit the serine proteinase MP2 (An et al., 2013) which acts usptream of the proteinase MP1 to activate PPO (Tang et al., 2006). In addition to Spn27A, the Toll pathway may also control the expression of unknown factors that are required to cleave PPO to PO (Ligoxygakis et al., 2002). In this context, it is worth noting that PPO activation in *Drosophila* may also be triggered in a Toll-independent manner through the fungal receptor GNBP3 that assembles effector protein complexes containing PO (Matskevich et al., 2010). The functional link between Toll and melanization has been also observed in other insects. In the malaria vector *A. gambiae*, RNAi-mediated silencing of *Cactus*, inhibitor of the Toll pathway, rendered mosquitoes totally resistant to *P. berghei* (Frolet et al., 2006); No live oocysts were detected in the mosquito midguts and several ookinetes appeared melanized. The increased parasite killing and melanization phenotypes in *Cactus* kd mosquitoes is at least partially due to increased expression of *TEP1* by the Toll pathway. Indeed, several immunity genes were later identified to be regulated by the Toll pathway in *A. gambiae* including some cSPs (Garver et al., 2009). Similarly, *Cactus* kd in *Ae. aegypti* blocks the development of

Fig. 1—Cont'd the degradation of Spn27A that inhibits the proteinase MP2 required for PPO activation. Toll also likely controls the expression of certain genes in the PPO activation cascade (An et al., 2013; De Gregorio et al., 2002; Gerttula et al., 1988; Lemaitre et al., 1995; Ligoxygakis et al., 2002; Qiu et al., 1998; Tang et al., 2006). In *A. gambiae*, Toll regulates the basal expression levels of TEP1, the key activator of the PPO cascade, as well as of several other immunity genes that may be implicated in the mosquito melanization response (Frolet et al., 2006; Garver et al., 2009; Meister et al., 2005). On the other hand, the Imd/Rel2 pathway negatively regulates the infection-induced melanization response (Frolet et al., 2006; Meister et al., 2005), possibly by controlling the expression of *CLIPA2*, a negative regulator of TEP1 amplification (Kamareddine et al., 2016; Yassine et al., 2014). It is also possible that other yet unidentified mechanisms contribute to this regulation. Of note, TEP1 controls several effector responses such as lysis and phagocytosis which have been omitted from the figure for simplicity. In *Ae. aegypti*, the Toll pathway regulates the expression of *IMP1* and *IMP2* CLIPs as well as *PPO* genes that are required for the infection-induced melanization (Zou et al., 2008, 2010).

the avian malaria parasite *Plasmodium gallinaceum* and triggers ookinete melanization (Zou et al., 2008). *Ae. aegypti* Toll pathway seems to crosstalk with the immune melanization response by controlling the expression of the two catalytic cSPs *IMP-1* and *IMP-2*, and several *PPO* genes (Zou et al., 2010).

There is also evidence that PPO and Toll may be activated by a common factor that branches downstream to feed into both responses (Fig. 2). Using in vitro reconstitution assays with recombinant proteins it was shown that SPE of the beetle *T. molitor* cleaves spätzle (Kim et al., 2008) as well as PPO and cSPH1 zymogens (Kan et al., 2008), activating AMP synthesis and melanization, respectively. A similar biochemical approach in *M. sexta* revealed that HP6 activates on one hand proHP8, homologous to *Drosophila* and *T. molitor* SPE, leading to Toll pathway activation, and on the other proPAP1 leading to PPO activation (An et al., 2009). In the

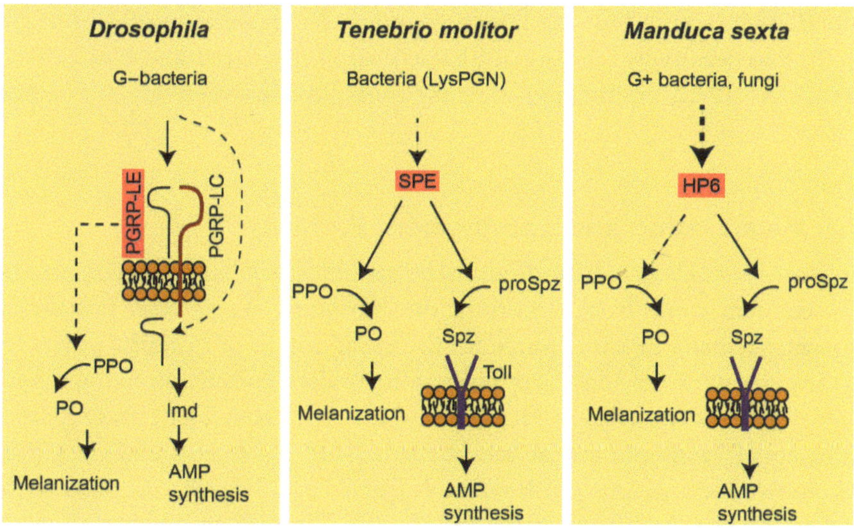

Fig. 2 The melanization response and the immune signalling pathways leading to AMP expression may be activated by common factors. Schematic presentation of immune signalling pathways and cascades leading to PPO activation in *Drosophila melanogaster*, *Tenebrio molitor* and *Manduca sexta*. In *pink* are the factors that can activate both the melanization response and the Toll or Imd pathways. Regular *black dashed lines* indicate the presence of several known steps that were omitted for simplicity. *Thick dashed lines* correspond to steps that have not been thoroughly characterized. In *Drosophila*, PGRP-LE activates the Imd pathway in concert with PGRP-LC or alone and may also activate the PPO-activating cascade (Takehana et al., 2002). In *Tenebrio*, SPE cleaves both proSpätzle (proSpz) and PPO into their active forms (Kan et al., 2008). In *Manduca*, HP6 directly cleaves proSpz and indirectly activates PPO by cleaving prophenoloxidase-activating proteinase 1 (proPAP1; not shown in figure) into active PAP1 (An et al., 2009).

silkworm *Bombyx mori*, serpin-5 was shown to regulate both Toll and PPO pathways (Li et al., 2016). Pull down assays identified the two proteinases BmHP6 and BmSP21 as potential targets for Serpin-5; however, it is not clear yet whether these proteinases are commonly activated upstream of both pathways. In addition to Toll, PGRP-LE which senses DAP-type PGN in *Drosophila* Imd pathway was also shown to integrate signals leading to both AMP synthesis and melanization. PGRP-LE overexpression in *Drosophila* larvae triggered constitutive expression of AMPs and spontaneous PPO activation (Takehana et al., 2002, 2004). However, whether Imd signalling is required for PGRP-LE induced melanization remains controversial (Takehana et al., 2002, 2004). In the mosquito, however, the Imd/Rel2 pathway seems to act as negative regulator of the melanization response, since silencing Rel2 triggered the melanization of a significant number of *Plasmodium* ookinetes (Frolet et al., 2006; Meister et al., 2005). How this negative regulation is established is still not clear, but it was recently shown that Rel2 is required for the infection-induced upregulation of *CLIPA2*, a cSPH that negatively regulates TEP1 activation (Kamareddine et al., 2016). Hence, CLIPA2 might fulfil the functional link between Imd/Rel2 and melanization.

5.2 Melanization and Complement Pathway

Evidence for the existence of a functional link between melanization and complement came from studies in the mosquito *A. gambiae*. Initially, it was shown that TEP1 is required for the melanization of *P. berghei* ookinetes as they egress into the basal labyrinth of the midgut epithelium in the refractory *A. gambiae* strain L3–5 (Blandin et al., 2004). Later, TEP1 was reported to be equally required for the melanization of *P. berghei* ookinetes in *CTL4* kd susceptible mosquitoes (Povelones et al., 2011) which are known to exhibit strong melanotic refractoriness to *Plasmodium* ookinetes (Osta et al., 2004); Cosilencing *CTL4* and *TEP1* in the susceptible N'gousso strain abolished melanization leading to the normal development of oocysts. While these studies indicate that TEP1 is required for parasite melanization irrespective of the mosquito genetic background, they do not necessarily link complement to melanization as the latter may also be perceived as a clearance mechanism triggered indirectly in response to the dead ookinetes killed by TEP1, rather than directly by TEP1 itself. In this context, encasing the dead ookinetes in a melanotic capsule would isolate them from the tissues and possibly avoid continuous stimulus to the mosquito immune system by parasite debris. Subsequent studies using bacterial infections of mosquitoes provided irrefutable evidence on the key role of TEP1 in controlling PPO activation. In fact, *TEP1* kd abolished PO activity in the hemolymph in response to *E. coli* systemic infections (Povelones et al., 2013). CLIPA8 is an essential positive regulator of PPO activation in response to systemic infections, which is cleaved soon after the injection of bacteria into the mosquito hemolymph (Schnitger et al., 2007). *TEP1* kd abolished CLIPA8 cleavage suggesting that TEP1 acts upstream of the PPO-activating cascade. CLIPA2 is another cSPH that negatively regulates TEP1 consumption during systemic infections. *CLIPA2* kd enhanced TEP1 activity leading to an exaggerated PO activity in the hemolymph following *E. coli* infections (Yassine et al., 2012). Furthermore, *TEP1* kd abolished the recruitment of PPO to *B. bassiana* hyphae invading the mosquito hemocoel, hence inhibiting melanization (Yassine et al., 2012). Altogether, these studies indicate that mosquito complement imposes a tight control upon the infection-induced melanization response. As a matter of fact, our current perception is that melanization in the mosquito is one of the effector branches initiated by TEP1 in parallel to those involved in lysis and opsonization. However, these results raise an important question as to the role of mosquito PRRs, if any, in activating melanization. This response, whose activation in insects has been classically attributed to PRRs belonging to different gene families including βGRP (Jiang et al., 2004; Ma and Kanost, 2000) and PGRPs (Park et al., 2006), seems to obey different rules in *A. gambiae*, at least so far. Could it be that TEP1 is the point of convergence of several PRRs? Several mosquito PRRs exhibit a similar role to TEP1 with respect to providing resistance to *Plasmodium* ookinete stages. These include LRRD7, a member of the leucine-rich immune protein family (Dong et al., 2006a); GNBPA2, GNBPB3 and GNBPB4 (Warr et al., 2008); the fibrinogen-related proteins FBN8, FBN9 and FBN39 (Dong et al., 2006a) and AgDscam, a hypervariable receptor of the immunoglobulin-domains superfamily (Dong et al., 2006b). However, it is not known whether these PRRs are required for PPO activation nor whether they act upstream or downstream of TEP1.

6. CONCLUDING REMARKS

The melanization reaction has been observed long time ago during the 19th century in diseased silk worms; however, it was only during the last two decades that significant progress has been made in characterizing PPO activation pathways in insects. There is no doubt that the main complexity in these pathways resides in understanding the fine regulation of the serine protease cascades that control PPO activation. This complexity is indirectly reflected in the large number of genes encoding clip-domain serine proteases in several insects, especially in mosquitoes. A rigorous biochemical approach in amenable insects such as *Manduca* and *Tenebrio* has succeeded in defining functional modules of these proteases as described in this review. Though convincing, it is worth noting that these studies have been conducted largely in vitro using recombinant and hemolymph purified proteins and may not reflect accurately the complex regulation of these proteases in vivo. For instance, if two serine proteinases are able to cleave the same substrate in vitro does it mean that they do so with a similar efficiency in vivo? What kind of interactions fine tune the specificity of a proteinase to its substrate to dilute the effect of functional redundancy expected from the significant expansion of this gene family in several insects? Furthermore, are these cascades truly linear in vivo as we usually like to envision them or they form branched networks whereby more than one protease may act on the same substrate but with different efficiencies? Does the latter scenario provide more sensitivity in tuning PPO activation in response to the class of the microbe and infectious dose by creating several inputs on a same node in the pathway that can be regulated independently? Finally, what components of these cascades are localized to microbial surfaces in response to recognition by PRRs and how are they recruited, especially PPO? What are the roles of clip domains and their spectrum of interactions during that process? These provocative questions are intended to challenge the way PPO activation cascades are perceived and to stimulate thinking of original non-reductionist approaches to decipher the molecular pathways controlling PPO activation.

REFERENCES

Altincicek, B., et al., 2013. Next generation sequencing based transcriptome analysis of septic-injury responsive genes in the beetle Tribolium castaneum. PLoS One 8. e52004.

An, C., Kanost, M.R., 2010. Manduca sexta serpin-5 regulates prophenoloxidase activation and the Toll signaling pathway by inhibiting hemolymph proteinase HP6. Insect Biochem. Mol. Biol. 40, 683–689.

An, C., Ishibashi, J., Ragan, E.J., Jiang, H., Kanost, M.R., 2009. Functions of Manduca sexta hemolymph proteinases HP6 and HP8 in two innate immune pathways. J. Biol. Chem. 284, 19716–19726.

An, C., Budd, A., Kanost, M.R., Michel, K., 2011. Characterization of a regulatory unit that controls melanization and affects longevity of mosquitoes. Cell. Mol. Life Sci. 68, 1929–1939.

An, C., Zhang, M., Chu, Y., Zhao, Z., 2013. Serine protease MP2 activates prophenoloxidase in the melanization immune response of Drosophila melanogaster. PLoS One 8. e79533.

Andersen, S.O., 2010. Insect cuticular sclerotization: a review. Insect Biochem. Mol. Biol. 40, 166–178.

Ayres, J.S., Schneider, D.S., 2008. A signaling protease required for melanization in Drosophila affects resistance and tolerance of infections. PLoS Biol. 6, 2764–2773.

Beck, M.H., Strand, M.R., 2007. A novel polydnavirus protein inhibits the insect prophenoloxidase activation pathway. Proc. Natl. Acad. Sci. U.S.A. 104, 19267–19272.

Binggeli, O., Neyen, C., Poidevin, M., Lemaitre, B., 2014. Prophenoloxidase activation is required for survival to microbial infections in Drosophila. PLoS Pathog. 10. e1004067.

Blandin, S., et al., 2004. Complement-like protein TEP1 is a determinant of vectorial capacity in the malaria vector *Anopheles gambiae*. Cell 116, 661–670.

Boissiere, A., et al., 2012. Midgut microbiota of the malaria mosquito vector Anopheles gambiae and interactions with Plasmodium falciparum infection. PLoS Pathog. 8. e1002742.

Buchon, N., et al., 2009. A single modular serine protease integrates signals from pattern-recognition receptors upstream of the Drosophila Toll pathway. Proc. Natl. Acad. Sci. U.S.A. 106, 12442–12447.

Cao, X., et al., 2015. Sequence conservation, phylogenetic relationships, and expression profiles of nondigestive serine proteases and serine protease homologs in Manduca sexta. Insect Biochem. Mol. Biol. 62, 51–63.

Christen, J.M., Hiromasa, Y., An, C., Kanost, M.R., 2012. Identification of plasma proteinase complexes with serpin-3 in Manduca sexta. Insect Biochem. Mol. Biol. 42, 946–955.

Davis, M.M., Primrose, D.A., Hodgetts, R.B., 2008. A member of the p38 mitogen-activated protein kinase family is responsible for transcriptional induction of Dopa decarboxylase in the epidermis of Drosophila melanogaster during the innate immune response. Mol. Cell. Biol. 28, 4883–4895.

De Gregorio, E., et al., 2002. An immune-responsive Serpin regulates the melanization cascade in *Drosophila*. Dev. Cell 3, 581–592.

Dong, Y., et al., 2006a. *Anopheles gambiae* immune responses to human and rodent *Plasmodium* parasite species. PLoS Pathog. 2. e52.

Dong, Y., Taylor, H.E., Dimopoulos, G., 2006b. AgDscam, a hypervariable immunoglobulin domain-containing receptor of the *Anopheles gambiae* innate immune system. PLoS Biol. 4. e229.

Dong, Y., Manfredini, F., Dimopoulos, G., 2009. Implication of the mosquito midgut microbiota in the defense against malaria parasites. PLoS Pathog. 5. e1000423.

Dunstone, M.A., Whisstock, J.C., 2011. Crystallography of serpins and serpin complexes. Methods Enzymol. 501, 63–87.

Eleftherianos, I., Revenis, C., 2011. Role and importance of phenoloxidase in insect hemostasis. J. Innate Immun. 3, 28–33.

Eleftherianos, I., et al., 2007. An antibiotic produced by an insect-pathogenic bacterium suppresses host defenses through phenoloxidase inhibition. Proc. Natl. Acad. Sci. U.S.A. 104, 2419–2424.

Eleftherianos, I., et al., 2009. A single locus from the entomopathogenic bacterium Photorhabdus luminescens inhibits activated Manduca sexta phenoloxidase. FEMS Microbiol. Lett. 293, 170–176.

Forst, S., Dowds, B., Boemare, N., Stackebrandt, E., 1997. Xenorhabdus and Photorhabdus spp.: bugs that kill bugs. Annu. Rev. Microbiol. 51, 47–72.

Fraiture, M., et al., 2009. Two mosquito LRR proteins function as complement control factors in the TEP1-mediated killing of *Plasmodium*. Cell Host Microbe 5, 273–284.

Frolet, C., Thoma, M., Blandin, S., Hoffmann, J.A., Levashina, E.A., 2006. Boosting NF-kappaB-dependent basal immunity of *Anopheles gambiae* aborts development of *Plasmodium berghei*. Immunity 25, 677–685.

Fuchs, S., Behrends, V., Bundy, J.G., Crisanti, A., Nolan, T., 2014. Phenylalanine metabolism regulates reproduction and parasite melanization in the malaria mosquito. PLoS One 9. e84865.

Futahashi, R., Fujiwara, H., 2005. Melanin-synthesis enzymes coregulate stage-specific larval cuticular markings in the swallowtail butterfly, Papilio xuthus. Dev. Genes Evol. 215, 519–529.

Futahashi, R., Fujiwara, H., 2007. Regulation of 20-hydroxyecdysone on the larval pigmentation and the expression of melanin synthesis enzymes and yellow gene of the swallowtail butterfly, Papilio xuthus. Insect Biochem. Mol. Biol. 37, 855–864.

Garver, L.S., Dong, Y., Dimopoulos, G., 2009. Caspar controls resistance to *Plasmodium falciparum* in diverse anopheline species. PLoS Pathog. 5. e1000335.

Gerttula, S., Jin, Y.S., Anderson, K.V., 1988. Zygotic expression and activity of the Drosophila Toll gene, a gene required maternally for embryonic dorsal-ventral pattern formation. Genetics 119, 123–133.

Gimonneau, G., et al., 2014. Composition of Anopheles coluzzii and Anopheles gambiae microbiota from larval to adult stages. Infect. Genet. Evol. 28, 715–724.

Gorman, M.J., An, C., Kanost, M.R., 2007a. Characterization of tyrosine hydroxylase from Manduca sexta. Insect Biochem. Mol. Biol. 37, 1327–1337.

Gorman, M.J., Wang, Y., Jiang, H., Kanost, M.R., 2007b. Manduca sexta hemolymph proteinase 21 activates prophenoloxidase-activating proteinase 3 in an insect innate immune response proteinase cascade. J. Biol. Chem. 282, 11742–11749.

Gulley, M.M., Zhang, X., Michel, K., 2013. The roles of serpins in mosquito immunology and physiology. J. Insect Physiol. 59, 138–147.

Gupta, S., Wang, Y., Jiang, H., 2005. Manduca sexta prophenoloxidase (proPO) activation requires proPO-activating proteinase (PAP) and serine proteinase homologs (SPHs) simultaneously. Insect Biochem. Mol. Biol. 35, 241–248.

Hiruma, K., Riddiford, L.M., 1985. Hormonal regulation of dopa decarboxylase during a larval molt. Dev. Biol. 110, 509–513.

Hiruma, K., Riddiford, L.M., Hopkins, T.L., Morgan, T.D., 1985. Roles of dopa decarboxylase and phenoloxidase in the melanization of the tobacco hornworm and their control by 20-hydroxyecdysone. J. Comp. Physiol. B 155, 659–669.

Huntington, J.A., Read, R.J., Carrell, R.W., 2000. Structure of a serpin-protease complex shows inhibition by deformation. Nature 407, 923–926.

Infanger, L.C., et al., 2004. The role of phenylalanine hydroxylase in melanotic encapsulation of filarial worms in two species of mosquitoes. Insect Biochem. Mol. Biol. 34, 1329–1338.

Ji, C., Wang, Y., Guo, X., Hartson, S., Jiang, H., 2004. A pattern recognition serine proteinase triggers the prophenoloxidase activation cascade in the tobacco hornworm, Manduca sexta. J. Biol. Chem. 279, 34101–34106.

Jiang, H., Kanost, M.R., 2000. The clip-domain family of serine proteinases in arthropods. Insect Biochem. Mol. Biol. 30, 95–105.

Jiang, H., Wang, Y., Kanost, M.R., 1998. Pro-phenol oxidase activating proteinase from an insect, *Manduca sexta*: a bacteria-inducible protein similar to *Drosophila* easter. Proc. Natl. Acad. Sci. U.S.A. 95, 12220–12225.

Jiang, H., Wang, Y., Yu, X.Q., Kanost, M.R., 2003a. Prophenoloxidase-activating proteinase-2 from hemolymph of Manduca sexta. A bacteria-inducible serine proteinase containing two clip domains. J. Biol. Chem. 278, 3552–3561.

Jiang, H., Wang, Y., Yu, X.Q., Zhu, Y., Kanost, M., 2003b. Prophenoloxidase-activating proteinase-3 (PAP-3) from *Manduca sexta* hemolymph: a clip-domain serine proteinase

regulated by serpin-1J and serine proteinase homologs. Insect Biochem. Mol. Biol. 33, 1049–1060.
Jiang, H., Ma, C., Lu, Z.Q., Kanost, M.R., 2004. Beta-1,3-glucan recognition protein-2 (betaGRP-2)from Manduca sexta; an acute-phase protein that binds beta-1,3-glucan and lipoteichoic acid to aggregate fungi and bacteria and stimulate prophenoloxidase activation. Insect Biochem. Mol. Biol. 34, 89–100.
Jiang, R., et al., 2009. Three pairs of protease-serpin complexes cooperatively regulate the insect innate immune responses. J. Biol. Chem. 284, 35652–35658.
Johnson, J.K., et al., 2003. A potential role for phenylalanine hydroxylase in mosquito immune responses. Insect Biochem. Mol. Biol. 33, 345–354.
Kamareddine, L., Nakhleh, J., Osta, M.A., 2016. Functional interaction between apolipophorins and complement regulate the mosquito immune response to systemic infections. J. Innate Immun. 8, 314–326.
Kan, H., et al., 2008. Molecular control of phenoloxidase-induced melanin synthesis in an insect. J. Biol. Chem. 283, 25316–25323.
Kanost, M.R., Jiang, H., 2015. Clip-domain serine proteases as immune factors in insect hemolymph. Curr. Opin. Insect Sci. 11, 47–55.
Kellenberger, C., et al., 2011. Structure-function analysis of grass clip serine protease involved in Drosophila Toll pathway activation. J. Biol. Chem. 286, 12300–12307.
Kim, M.S., et al., 2002. A new easter-type serine protease cleaves a masquerade-like protein during prophenoloxidase activation in *Holotrichia diomphalia* larvae. J. Biol. Chem. 277, 39999–40004.
Kim, C.H., et al., 2008. A three-step proteolytic cascade mediates the activation of the peptidoglycan-induced toll pathway in an insect. J. Biol. Chem. 283, 7599–7607.
Koch, P.B., Behnecke, B., Weigmann-Lenz, M., Ffrench-Constant, R.H., 2000. Insect pigmentation: activities of beta-alanyldopamine synthase in wing color patterns of wild-type and melanic mutant swallowtail butterfly Papilio glaucus. Pigment Cell Res. 13 (Suppl. 8), 54–58.
Leclerc, V., et al., 2006. Prophenoloxidase activation is not required for survival to microbial infections in *Drosophila*. EMBO Rep. 7, 231–235.
Lee, S.Y., Soderhall, K., 2001. Characterization of a pattern recognition protein, a masquerade-like protein, in the freshwater crayfish Pacifastacus leniusculus. J. Immunol. 166, 7319–7326.
Lee, K.Y., et al., 2002. A zymogen form of masquerade-like serine proteinase homologue is cleaved during pro-phenoloxidase activation by Ca2+ in coleopteran and *Tenebrio molitor* larvae. Eur. J. Biochem. 269, 4375–4383.
Lemaitre, B., et al., 1995. Functional analysis and regulation of nuclear import of dorsal during the immune response in Drosophila. EMBO J. 14, 536–545.
Li, J., Ma, L., Lin, Z., Zou, Z., Lu, Z., 2016. Serpin-5 regulates prophenoloxidase activation and antimicrobial peptide pathways in the silkworm, Bombyx mori. Insect Biochem. Mol. Biol. 73, 27–37.
Ligoxygakis, P., et al., 2002. A serpin mutant links Toll activation to melanization in the host defence of *Drosophila*. EMBO J. 21, 6330–6337.
Lu, Z., Jiang, H., 2008. Expression of Manduca sexta serine proteinase homolog precursors in insect cells and their proteolytic activation. Insect Biochem. Mol. Biol. 38, 89–98.
Lu, Z., Beck, M.H., Wang, Y., Jiang, H., Strand, M.R., 2008. The viral protein Egf1.0 is a dual activity inhibitor of prophenoloxidase-activating proteinases 1 and 3 from *Manduca sexta*. J. Biol. Chem. 283, 21325–21333.
Lu, Z., Beck, M.H., Strand, M.R., 2010. Egf1.5 is a second phenoloxidase cascade inhibitor encoded by Microplitis demolitor bracovirus. Insect Biochem. Mol. Biol. 40, 497–505.

Ma, C., Kanost, M.R., 2000. A beta1,3-glucan recognition protein from an insect, Manduca sexta, agglutinates microorganisms and activates the phenoloxidase cascade. J. Biol. Chem. 275, 7505–7514.

Matskevich, A.A., Quintin, J., Ferrandon, D., 2010. The Drosophila PRR GNBP3 assembles effector complexes involved in antifungal defenses independently of its Toll-pathway activation function. Eur. J. Immunol. 40, 1244–1254.

Meister, S., et al., 2005. Immune signaling pathways regulating bacterial and malaria parasite infection of the mosquito *Anopheles gambiae*. Proc. Natl. Acad. Sci. U.S.A. 102, 11420–11425.

Michel, K., Budd, A., Pinto, S., Gibson, T.J., Kafatos, F.C., 2005. *Anopheles gambiae* SRPN2 facilitates midgut invasion by the malaria parasite *Plasmodium berghei*. EMBO Rep. 6, 891–897.

Nam, H.J., Jang, I.H., You, H., Lee, K.A., Lee, W.J., 2012. Genetic evidence of a redox-dependent systemic wound response via Hayan protease-phenoloxidase system in Drosophila. EMBO J. 31, 1253–1265.

Nappi, A.J., Christensen, B.M., 2005. Melanogenesis and associated cytotoxic reactions: applications to insect innate immunity. Insect Biochem. Mol. Biol. 35, 443–459.

Niare, O., et al., 2002. Genetic loci affecting resistance to human malaria parasites in a West African mosquito vector population. Science 298, 213–216.

Oduol, F., Xu, J., Niare, O., Natarajan, R., Vernick, K.D., 2000. Genes identified by an expression screen of the vector mosquito Anopheles gambiae display differential molecular immune response to malaria parasites and bacteria. Proc. Natl. Acad. Sci. U.S.A. 97, 11397–11402.

Olson, S.T., Gettins, P.G., 2011. Regulation of proteases by protein inhibitors of the serpin superfamily. Prog. Mol. Biol. Transl. Sci. 99, 185–240.

Osta, M.A., Christophides, G.K., Kafatos, F.C., 2004. Effects of mosquito genes on *Plasmodium* development. Science 303, 2030–2032.

Park, J.W., et al., 2006. A synthetic peptidoglycan fragment as a competitive inhibitor of the melanization cascade. J. Biol. Chem. 281, 7747–7755.

Park, J.W., et al., 2007. Clustering of peptidoglycan recognition protein-SA is required for sensing lysine-type peptidoglycan in insects. Proc. Natl. Acad. Sci. U.S.A. 104, 6602–6607.

Piao, S., et al., 2005. Crystal structure of a clip-domain serine protease and functional roles of the clip domains. EMBO J. 24, 4404–4414.

Povelones, M., Waterhouse, R.M., Kafatos, F.C., Christophides, G.K., 2009. Leucine-rich repeat protein complex activates mosquito complement in defense against *Plasmodium* parasites. Science 324, 258–261.

Povelones, M., Upton, L.M., Sala, K.A., Christophides, G.K., 2011. Structure-function analysis of the Anopheles gambiae LRIM1/APL1C complex and its interaction with complement C3-like protein TEP1. PLoS Pathog. 7. e1002023.

Povelones, M., et al., 2013. The CLIP-domain serine protease homolog SPCLIP1 regulates complement recruitment to microbial surfaces in the malaria mosquito Anopheles gambiae. PLoS Pathog. 9. e1003623.

Qiu, P., Pan, P.C., Govind, S., 1998. A role for the Drosophila Toll/Cactus pathway in larval hematopoiesis. Development 125, 1909–1920.

Riehle, M.M., et al., 2006. Natural malaria infection in *Anopheles gambiae* is regulated by a single genomic control region. Science 312, 577–579.

Riehle, M.M., et al., 2008. Anopheles gambiae APL1 is a family of variable LRR proteins required for Rel1-mediated protection from the malaria parasite, *Plasmodium berghei*. PLoS One 3. e3672.

Rodriguez-Andres, J., et al., 2012. Phenoloxidase activity acts as a mosquito innate immune response against infection with Semliki Forest virus. PLoS Pathog. 8. e1002977.

Roh, K.B., et al., 2009. Proteolytic cascade for the activation of the insect toll pathway induced by the fungal cell wall component. J. Biol. Chem. 284, 19474–19481.
Ross, J., Jiang, H., Kanost, M.R., Wang, Y., 2003. Serine proteases and their homologs in the Drosophila melanogaster genome: an initial analysis of sequence conservation and phylogenetic relationships. Gene 304, 117–131.
Scherfer, C., et al., 2008. Drosophila Serpin-28D regulates hemolymph phenoloxidase activity and adult pigmentation. Dev. Biol. 323, 189–196.
Schnitger, A.K., Kafatos, F.C., Osta, M.A., 2007. The melanization reaction is not required for survival of *Anopheles gambiae* mosquitoes after bacterial infections. J. Biol. Chem. 282, 21884–21888.
Sideri, M., Tsakas, S., Markoutsa, E., Lampropoulou, M., Marmaras, V.J., 2008. Innate immunity in insects: surface-associated dopa decarboxylase-dependent pathways regulate phagocytosis, nodulation and melanization in medfly haemocytes. Immunology 123, 528–537.
Silverman, G.A., et al., 2010. Serpins flex their muscle: I. Putting the clamps on proteolysis in diverse biological systems. J. Biol. Chem. 285, 24299–24305.
Strand, M.R., Burke, G.R., 2013. Polydnavirus-wasp associations: evolution, genome organization, and function. Curr. Opin. Virol. 3, 587–594.
Sutthangkul, J., et al., 2015. Suppression of shrimp melanization during white spot syndrome virus infection. J. Biol. Chem. 290, 6470–6481.
Suwanchaichinda, C., Ochieng, R., Zhuang, S., Kanost, M.R., 2013. Manduca sexta serpin-7, a putative regulator of hemolymph prophenoloxidase activation. Insect Biochem. Mol. Biol. 43, 555–561.
Takahashi, D., Dai, H., Hiromasa, Y., Krishnamoorthi, R., Kanost, M.R., 2014. Self-association of an insect beta-1,3-glucan recognition protein upon binding laminarin stimulates prophenoloxidase activation as an innate immune response. J. Biol. Chem. 289, 28399–28410.
Takahashi, D., Garcia, B.L., Kanost, M.R., 2015. Initiating protease with modular domains interacts with beta-glucan recognition protein to trigger innate immune response in insects. Proc. Natl. Acad. Sci. U.S.A. 112, 13856–13861.
Takehana, A., et al., 2002. Overexpression of a pattern-recognition receptor, peptidoglycan-recognition protein-LE, activates imd/relish-mediated antibacterial defense and the prophenoloxidase cascade in *Drosophila* larvae. Proc. Natl. Acad. Sci. U.S.A. 99, 13705–13710.
Takehana, A., et al., 2004. Peptidoglycan recognition protein (PGRP)-LE and PGRP-LC act synergistically in Drosophila immunity. EMBO J. 23, 4690–4700.
Tang, H., Kambris, Z., Lemaitre, B., Hashimoto, C., 2006. Two proteases defining a melanization cascade in the immune system of *Drosophila*. J. Biol. Chem. 281, 28097–28104.
Tang, H., Kambris, Z., Lemaitre, B., Hashimoto, C., 2008. A serpin that regulates immune melanization in the respiratory system of Drosophila. Dev. Cell 15, 617–626.
Tong, Y., Kanost, M.R., 2005. *Manduca sexta* serpin-4 and serpin-5 inhibit the prophenol oxidase activation pathway: cDNA cloning, protein expression, and characterization. J. Biol. Chem. 280, 14923–14931.
Tong, Y., Jiang, H., Kanost, M.R., 2005. Identification of plasma proteases inhibited by *Manduca sexta* serpin-4 and serpin-5 and their association with components of the prophenol oxidase activation pathway. J. Biol. Chem. 280, 14932–14942.
Vavricka, C.J., Christensen, B.M., Li, J., 2010. Melanization in living organisms: a perspective of species evolution. Protein Cell 1, 830–841.
Veillard, F., Troxler, L., Reichhart, J.M., 2016. Drosophila melanogaster clip-domain serine proteases: structure, function and regulation. Biochimie 122, 255–269.

Volz, J., Muller, H.M., Zdanowicz, A., Kafatos, F.C., Osta, M.A., 2006. A genetic module regulates the melanization response of *Anopheles* to *Plasmodium*. Cell. Microbiol. 8, 1392–1405.

Wang, Y., Jiang, H., 2004a. Prophenoloxidase (proPO) activation in Manduca sexta: an analysis of molecular interactions among proPO, proPO-activating proteinase-3, and a cofactor. Insect Biochem. Mol. Biol. 34, 731–742.

Wang, Y., Jiang, H., 2004b. Purification and characterization of Manduca sexta serpin-6: a serine proteinase inhibitor that selectively inhibits prophenoloxidase-activating proteinase-3. Insect Biochem. Mol. Biol. 34, 387–395.

Wang, Y., Jiang, H., 2006. Interaction of beta-1,3-glucan with its recognition protein activates hemolymph proteinase 14, an initiation enzyme of the prophenoloxidase activation system in Manduca sexta. J. Biol. Chem. 281, 9271–9278.

Wang, Y., Jiang, H., 2007. Reconstitution of a branch of the Manduca sexta prophenoloxidase activation cascade in vitro: snake-like hemolymph proteinase 21 (HP21) cleaved by HP14 activates prophenoloxidase-activating proteinase-2 precursor. Insect Biochem. Mol. Biol. 37, 1015–1025.

Wang, Y., Jiang, H., 2008. A positive feedback mechanism in the Manduca sexta prophenoloxidase activation system. Insect Biochem. Mol. Biol. 38, 763–769.

Wang, Y., Lu, Z., Jiang, H., 2014. Manduca sexta proprophenoloxidase activating proteinase-3 (PAP3) stimulates melanization by activating proPAP3, proSPHs, and proPOs. Insect Biochem. Mol. Biol. 50, 82–91.

Wang, Y.H., et al., 2015. A critical role for CLSP2 in the modulation of antifungal immune response in mosquitoes. PLoS Pathog. 11. e1004931.

Warr, E., Das, S., Dong, Y., Dimopoulos, G., 2008. The *Gram-negative bacteria-binding protein* gene family: its role in the innate immune system of *Anopheles gambiae* and in anti-*Plasmodium* defence. Insect Mol. Biol. 17, 39–51.

Waterhouse, R.M., et al., 2007. Evolutionary dynamics of immune-related genes and pathways in disease-vector mosquitoes. Science 316, 1738–1743.

Yang, F., Wang, Y., He, Y., Jiang, H., 2016. In search of a function of Manduca sexta hemolymph protease-1 in the innate immune system. Insect Biochem. Mol. Biol. 76, 1–10.

Yassine, H., Kamareddine, L., Osta, M.A., 2012. The mosquito melanization response is implicated in defense against the entomopathogenic fungus Beauveria bassiana. PLoS Pathog. 8. e1003029.

Yassine, H., Kamareddine, L., Chamat, S., Christophides, G.K., Osta, M.A., 2014. A serine protease homolog negatively regulates TEP1 consumption in systemic infections of the malaria vector Anopheles gambiae. J. Innate Immun. 6, 806–818.

Yu, X.Q., Jiang, H., Wang, Y., Kanost, M.R., 2003. Nonproteolytic serine proteinase homologs are involved in prophenoloxidase activation in the tobacco hornworm, Manduca sexta. Insect Biochem. Mol. Biol. 33, 197–208.

Zhang, R., et al., 2003. Characterization and properties of a 1,3-beta-D-glucan pattern recognition protein of Tenebrio molitor larvae that is specifically degraded by serine protease during prophenoloxidase activation. J. Biol. Chem. 278, 42072–42079.

Zhang, X., An, C., Sprigg, K., Michel, K., 2016. CLIPB8 is part of the prophenoloxidase activation system in Anopheles gambiae mosquitoes. Insect Biochem. Mol. Biol. 71, 106–115.

Zhao, M., et al., 2005. A novel 43-kDa protein as a negative regulatory component of phenoloxidase-induced melanin synthesis. J. Biol. Chem. 280, 24744–24751.

Zhao, P., Li, J., Wang, Y., Jiang, H., 2007. Broad-spectrum antimicrobial activity of the reactive compounds generated in vitro by Manduca sexta phenoloxidase. Insect Biochem. Mol. Biol. 37, 952–959.

Zou, Z., et al., 2008. Mosquito RUNX4 in the immune regulation of PPO gene expression and its effect on avian malaria parasite infection. Proc. Natl. Acad. Sci. U.S.A. 105, 18454–18459.
Zou, Z., Shin, S.W., Alvarez, K.S., Kokoza, V., Raikhel, A.S., 2010. Distinct melanization pathways in the mosquito Aedes aegypti. Immunity 32, 41–53.

CHAPTER FOUR

Microbiota, Gut Physiology, and Insect Immunity

Ji-Hoon Lee, Kyung-Ah Lee, Won-Jae Lee
School of Biological Science, Seoul National University and National Creative Research Initiative Center for hologenomics, Seoul, South Korea

Contents

1. Introduction 112
2. Gut Structure and Gut Microbes 113
   2.1 Gut Structure and Function 113
   2.2 The Microbial Environment of the Gut Epithelium 115
3. Gut Immunity 119
   3.1 The IMD Pathway 120
   3.2 The DUOX Pathway 124
4. Gut Renewal 129
   4.1 ISC Self-Renewal, Differentiation, and Proliferation 129
   4.2 Gut Renewal and Innate Immune Systems 131
5. Concluding Remarks 132
Acknowledgements 133
References 133

Abstract

This chapter aims to visualize insect gut as a life-sustaining organ that is resilient yet interactive with the changing environment to maintain its immunological and physiological homeostasis. As in all metazoans, insect gut is where the organism interacts most actively with the external ecosystem. A healthy gut epithelium properly controls incoming foodborne microbes as well as microbiota while maintaining its structural and functional integrity. Novel insights into gut immunity and physiology have been made using the fruit fly *Drosophila melanogaster* as a model system. Here, we begin our discussion with the gut architecture and the microbial environment the gut faces. Then, we review the current understanding of the immune responses of the gut epithelium involving the immune deficiency and dual oxidase pathways to restrict unwanted microbial colonization. We also discuss how the gut epithelium maintains its functionality by utilizing controlled proliferation and differentiation of intestinal stem cells despite damage-causing gut environment.

Advances in Insect Physiology, Volume 52
ISSN 0065-2806
http://dx.doi.org/10.1016/bs.aiip.2016.11.001

1. INTRODUCTION

The metazoan gut is one of the innermost organs based on its physical location, but it is where the most active interaction with environment takes place in the organism. For a healthy life, the gut processes ingested substances with efficiency to supply enough nutrients for the organism while preventing invasion of foodborne germs to keep internal organs aseptic by acting as a physicochemical barrier (Buchon et al., 2013a). Furthermore, the gut epithelium interacts with various species of microorganisms constituting the gut microbiota (Lee and Brey, 2013). Therefore, the gut epithelium must have a delicate strategy to prevent pathogens, control microbiota, and maintain its structural and functional integrity.

Insects like many other metazoans share common aspects of gut–microbe interactions and gut physiology, including host-gut microbe symbiosis, controlling pathogens using the mucosal immune systems, and maintenance of gut homeostasis (You et al., 2014). The diversity of insects and their niche ecology makes it difficult to generalize gut microbiota and gut physiology. In this review, we will focus on an insect, the fruit fly *Drosophila melanogaster*. With powerful genetic tools and rich resources, *Drosophila* has been a leading in vivo model system and has made seminal contributions in this field (Imler, 2014). Specifically, pathogens and commensal microbes are well defined in *Drosophila*, and mechanistic studies are possible with the genetics of both the host and the microorganisms (Ryu et al., 2008; Shin et al., 2011).

In this review, we discuss the functionality of the gut epithelium to control microbes and to maintain its integrity. First, we describe the microbial environment that the gut epithelium faces. In the gut lumen, there is continuous flow of incoming microorganisms with ingested food, including pathogens and opportunistic pathogens. There is also the gut microbiota colonized in the gut. Therefore, it is evident that there are intensive interactions between the gut and microbes. We will also present recently discovered interesting aspects of mutualism such as microbiota-modulated animal development and behaviour. Second, we discuss the strategies of the gut epithelium to sweep unwanted microorganisms while preserving commensal microbes. Unlike systemic innate immunity where Toll and immune deficiency (IMD) pathways are essential, the immune response in mucosal epithelium depends on the IMD pathway and dual oxidase (DUOX) pathway. We will present the basic regulatory mechanisms of these defence systems and their physiological roles. Finally, we will discuss the gut renewal programme, a fundamental aspect of gut physiology for gut cell homeostasis.

2. GUT STRUCTURE AND GUT MICROBES

2.1 Gut Structure and Function

Although, at first sight, it may look like a long flexible tube with little structural features, insect gut is a regionalized organ both structurally and functionally (Buchon and Osman, 2015). Adult *Drosophila* gut is composed of structurally, functionally, and developmentally distinct three primary domains: the foregut, the midgut, and the hindgut (Lawrence and Johnston, 1986; Lemaitre and Miguel-Aliaga, 2013; Murakami et al., 1999) (Fig. 1). The foregut is at the anteriormost region and is originated from the ectoderm. The foregut includes the pharynx and the oesophagus for the passage of ingested food and crop for food storage (Buchon et al., 2013a). The midgut is the region from the cardia (proventriculus) to the junction of the midgut and hindgut and is originated from the endoderm. The cardia serves as a valve for food passage regulation, and the midgut–hindgut junction is where the Malpighian tubules, the functional analog of the mammalian kidney, are attached (Cagan, 2003; King, 1988). The primary function of the midgut is food digestion and nutrient absorption. Following the midgut is the ectoderm-derived hindgut (Murakami et al., 1999), which is responsible for the absorption of water and ions and extends to the rectum (Lengyel and Iwaki, 2002; Murakami and Shiotsuki, 2001). The midgut is one of the largest organs in insects. We will focus on the pivotal role of the midgut in host physiology, such as metabolism and immunity, in our discussion later in this chapter.

The midgut is a single layer of epithelium and a visceral muscle layer with circular and longitudinal muscles wraps around the epithelium with the basal membrane in between (Micchelli and Perrimon, 2006; Ohlstein and

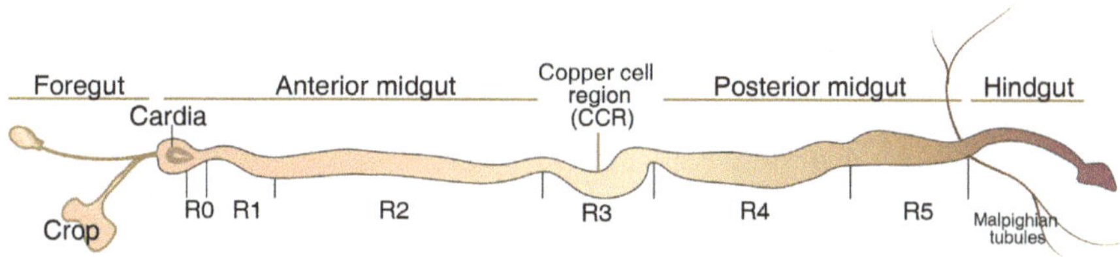

Fig. 1 Gut anatomy of adult *Drosophila*. See text for details.

Spradling, 2006; Sandborn et al., 1967). The midgut epithelium contains four different types of cells: enterocytes (ECs), enteroendocrine cells (EECs), ISCs, and enteroblasts (EBs) (Micchelli and Perrimon, 2006; Ohlstein and Spradling, 2006). The ECs are fully differentiated cells and most abundant in the epithelium. ECs are large polyploid cells with the primary function of secreting digestive enzymes and absorbing nutrients. The EECs are the other differentiated cells, and their role is to secrete hormones. The ISCs are the dividing progenitor cells. Finally, the EBs are more restricted progenitor cells produced by ISC divisions. EBs further differentiate to generate either ECs or EECs. The luminal side of the midgut is covered with the peritrophic matrix, a noncellular chitin polymer layer with chitin-binding proteins, to protect the epithelial cells from direct contact with ingested food and gut microbes. A mucus layer fills the area between the peritrophic matrix and the epithelium. The peritrophic matrix and mucus layer as well as the tight cell-to-cell junctions in the epithelium constitute the physical barrier to protect the organism from pathogen invasion and toxic molecules (Buchon et al., 2013a). The proliferation and differentiation programme in the gut epithelium is the central player of gut homeostasis regulation. We will further discuss these issues in the following sections.

The midgut is further highly regionalized along the anterior–posterior axis based on its structure and function. Roughly, the midgut can be divided into three regions, anterior and posterior regions relative to the middle-located copper cell region (CCR) with an acidic pH due to the activity of vacuolar H+ ATPase pump proteins on the apical surface of differentiated copper cells (Shanbhag and Tripathi, 2009). In this regard, the CCR is often compared with the mammalian stomach (Dubreuil, 2004). Functionally, the anterior midgut is responsible for the breakdown of food by secreted digestive enzymes, the CCR is for further digestion with its low pH, and the posterior midgut is where absorption of nutrients begins (Marianes and Spradling, 2013). The three regions of the midgut can be further divided based on the anatomy. When the radius was measured along the midgut, six major constrictions forming boundaries of six subregions, R0–R5, were found (Buchon et al., 2013b) (Fig. 1). In this manner, R0 corresponds to the cardia, R1–2 to the anterior midgut, R3 to the CCR, and R4–5 to the posterior midgut. Moreover, based on the morphology, physiology, and genetic properties of the cells, the midgut is further compartmentalized into 10 or 14 subregions (Buchon et al., 2013b; Marianes and Spradling, 2013). Since epithelial cells in each subregion seem to exhibit distinct functions and genes with essential roles in midgut physiology are differentially expressed in the subregions, understanding the distinct functions of each subregion might significantly expand our knowledge of midgut function.

2.2 The Microbial Environment of the Gut Epithelium

The *Drosophila* gut epithelium begins to interact with microorganisms as soon as the insect begins its first oral feeding after embryogenesis. Since then, the gut epithelium is exposed to constant flow of microbial environment throughout the life of the insect. Among the various microbes in the insect gut, some have stronger impacts on host physiology. For instance, foodborne pathogens are harmful microbes causing host pathology or even death, while some microbes in the microbiota make mutualistic relationships and provide benefits to host physiology. In this section, we will discuss the pathogenic and beneficial interactions of *Drosophila* with gut pathogens and gut microbiota, respectively.

2.2.1 Gut Pathogens

Pathogens are infectious microorganisms and often lead to host pathology or even death. A small number of enteric pathogens in *Drosophila* has been characterized, although pathogenic host–microbe interactions in *Drosophila* have been widely studied as a model system to understand the innate immune system after septic injury (Mistry et al., 2016). One of the natural *Drosophila* gut pathogens is *Erwinia carotovora carotovora 15* (*ECC15*). *ECC15* is a Gram-negative phytopathogen causing plant pathology and uses insects as vectors (Basset et al., 2000). *ECC15* induces strong immune activation in the *Drosophila* gut epithelium but is not a lethal pathogen in normal wild-type adult flies as *ECC15* is efficiently cleared by the host immune system. However, *ECC15* stays in the gut longer and causes pathology or lethality when the host immune system is compromised. Thus, *ECC15* is a representative opportunistic pathogen in adult *Drosophila* and a useful resource to study gut immunity and to test immune gene functions (Buchon et al., 2009b; Ha et al., 2005a). *Pseudomonas entomophila* is another important natural *Drosophila* gut pathogen (Vodovar et al., 2005). It is an entomopathogen causing pathology in insects including *Drosophila* and silkworm. *P. entomophila* is known to produce virulent factors such as insecticidal pore-forming toxins, proteases, and hemolysin (Vodovar et al., 2006). As such, high-dose oral infection causes lethality in adult *Drosophila*. The interaction of *ECC15* and *P. entomophila* with the local immune system in the adult *Drosophila* gut epithelium will be further discussed in the following sections.

2.2.2 Gut Microbiota

Most of the microorganisms that have entered into the gut lumen transiently pass through or are eliminated by the host immune system, but some of them are allowed to stay longer and proliferate in the healthy host gut lumen. These microbes closely associated with the gut ecosystem constitute the gut microbiota. Since each species of the gut microbiota has a distinctive ability to metabolize nutrients, produce second metabolites, and interact with the host immune system, prolonged interaction with a specific member of the gut microbiota may have an impact on a diverse ranges of host physiology (Fig. 2).

Obligate mutualism between lower termites and their gut microbiota has been discovered back in the 19th century (Brune and Dietrich, 2015; Leidy, 1881). These termites are absolutely dependent on the microbiota dwelling in their hindgut to convert lignocellulose in ingested wood particles to short-chain fatty acids, which are absorbed by hindgut cells as energy sources. Without gut microbiota, termites still ingest wood but die of starvation eventually. Since then, more examples of insect–microbiota mutualism have been found (Wang et al., 2013). Nevertheless, these obligate mutualisms are just a small portion of the widespread host-gut microbe symbiosis, and our knowledge of the role of microbiota in host physiology is yet very limited.

A vast majority of microbes constituting the gut microbiota are referred to as commensal microbes. By definition, commensals gain benefits from the host but are neither beneficial nor harmful to the host. However, commensals are renamed as mutualists if a hidden benefit of the commensal bacteria is revealed. Therefore, in a sense, commensals can be viewed as a pool of gut-interacting microbes with potential regulatory functions in host physiology. For example, as gut-associated microorganisms, the microbiota often

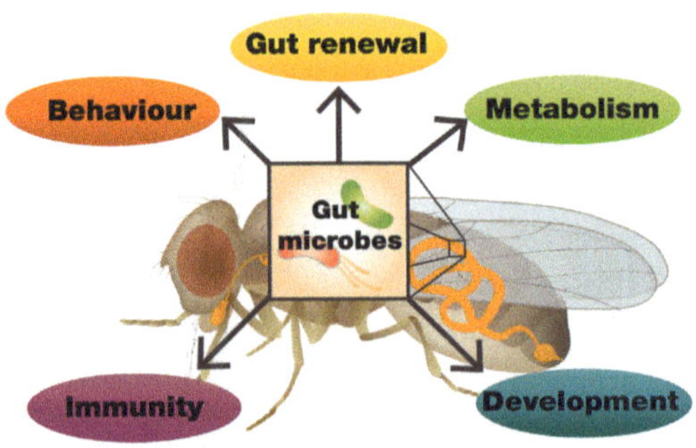

Fig. 2 The role of gut microbiota on *Drosophila* physiology. See text for details.

influences the physiology of the host gut. Compared with germ-free *Drosophila*, conventionally reared animals with microbiota have increased rate of ISC proliferation and gut epithelial turnover (Buchon et al., 2009a). Furthermore, recent studies done by different research groups have revealed that the effect of gut microbiota is not only restricted to gut physiology but is also extended to other remote organs as well as the systemic physiology of the host animal (Hsiao et al., 2013; Sharon et al., 2010; Shin et al., 2011; Storelli et al., 2011).

In *Drosophila*, the gut microbiota is dominated by approximately 5–20 species of commensal bacteria of the Acetobacteraceae and Enterobacteriaceae families and the Lactobacillale order, with variation between laboratories (Chandler et al., 2011; Cox and Gilmore, 2007; Ren et al., 2007; Ryu et al., 2008; Storelli et al., 2011). For instance, *Lactobacillus plantarum*, *L. brevis*, *Acetobacter pomorum*, *Gluconobacter morbifer*, and *Commensalibacter intestini* were the five dominant species in the gut microbiota of flies grown in our laboratory (Ryu et al., 2008). These indigenous gut bacteria can be grown in vitro. The relatively simple composition of the gut microbiota, the accessibility of these gut microbes for in vitro culture and manipulation, and the ease of gnotobiotic research are some of the good reasons to use *Drosophila* model to study the host-gut microbe interaction.

Using a gnotobiotic *Drosophila* model, it has been shown that commensal gut microbes play a pivotal role in host development by regulating host metabolism (Shin et al., 2011; Storelli et al., 2011). While conventionally reared *Drosophila* took less than 7 days to develop from eggs to pupae, germ-free animals took about 9 days under a standard laboratory diet condition (Shin et al., 2011). The role of microbiota became more obvious when the flies were grown in medium with reduced amount of yeast from 2% to 0.1%. Whereas it took about 9 days for egg-to-pupa development for conventionally reared animals with 0.1% yeast, all axenic ones died as first instar larvae (Shin et al., 2011). The lethality of axenic larvae under malnutrition condition was rescued by mono-association with any of *L. plantarum*, *L. brevis*, *G. morbifer*, *C. intestini*, and *A. pomorum*. Strikingly, *A. pomorum* alone could fully recover the delayed development as well (Shin et al., 2011). Further genetic studies revealed that the metabolic pathway of *Acetobacter* to produce acetic acid from ethanol (involving pyrroloquinoline quinone-dependent alcohol dehydrogenase (PQQ-ADH)-dependent oxidative respiratory chain) is in part responsible for the growth-promoting activity of *A. pomorum*. A PQQ-ADH mutant of *A. pomorum*, which could not fully rescue the developmental defect on its own, rescued the defect when acetic acid was supplemented, although acetic acid alone was unable to rescue the defect (Shin et al., 2011). The regulatory role of *A. pomorum* in host development is in part through inducing systemic activation of the insulin signalling pathway (Shin et al., 2011). Similarly, Storelli et al. (2011) also uncovered the growth-promoting role of a species of commensal bacteria, *L. plantarum*, in part via systemic activation of the insulin signalling pathway. Although the precise mechanism of the growth-promoting function of these commensal bacteria is yet to be determined, these reports support that commensal microbes can serve as a regulator of host growth and development through their specific metabolic functions.

It is interesting to note that commensal bacteria are found to regulate host behaviour such as mating preference (Sharon et al., 2010). *Drosophila* grown on a molasses medium preferred to mate with flies from the same molasses medium and flies grown on a starch medium preferred those from the same starch medium, although both groups of flies were originated from the same stock. The microbiota may have caused this mating preference as the preference disappeared when they were treated with antibiotics. Consistently, the two groups had different microbiota compositions, i.e., molasses flies had mixed population of microbes, whereas starch flies had the dominant *L. plantarum*. Furthermore, this homogamic mating preference was recapitulated by the mono-association of *L. plantarum*. Sharon et al. (2010) found that there are differences in the levels of cuticular sex pheromones between the two groups and the differences decreased with antibiotic treatment, suggestive of a regulatory role of microbiota in the production of the sex pheromone. A regulatory function of microbiota for host behaviour has also been identified in the mouse model of autism spectrum disorder (Hsiao et al., 2013). Controlling host behaviour was a previously unexpected function of microbiota, leading to the speculation that there might be a microbiota–gut–brain axis to control animal physiology.

Furthermore, the immunoprotective role of gut microbiota during enteric infection was also illustrated. Compared with normal *Drosophila* with microbiota, germ-free animals were more susceptible to oral infection with *Candida albicans*, *Serratia marcescens*, and *P. aeruginosa* (Blum et al., 2013; Glittenberg et al., 2011). Furthermore, higher population of *L. plantarum* provided an additional protection (Blum et al., 2013). Although the mechanism of the protective function of microbiota against pathogens is not fully understood, one possibility is through the ability of gut microbes to compete directly with pathogens as presented in other insect models (Scott et al., 2008). Another possibility is that gut microbes interact with the host immune system and indirectly control infectious microbes. Recently, it has been shown that *A. pomorum* is responsible for antiviral activity in the fly gut by stimulating the expression of Pvf2, an ERK pathway ligand, via NF-κB-dependent gut immune system (Sansone et al., 2015).

As discussed earlier, gut microbiota is an active regulator of several aspects of host physiology, including intestinal turnover, metabolism, development, behaviour, and protection against pathogens. Therefore, maintaining a healthy microbiota is an important task for hosts. Then, how are the composition and proportion of gut microbiota determined? Since microbiota forms from what the insect ingested, the composition and proportion of gut microbiota often represent the insect's living environment and feeding behaviour. For example, the diversity of the gut microbiota of wild-caught *Drosophila* (different *Drosophila* species having distinct natural habitats) is higher than that of laboratory *Drosophila* (Chandler et al., 2011). Therefore, food-associated microbes are the initial source for microbiota composition. The feeding behaviour including the oral–faecal cycle might be a player in fortifying the composition and proportion of gut microbiota. However, obviously, not all environmental microbes mixed in food act as gut microbiota. Furthermore, the amount of gut microbiota is also regulated, and uncontrolled increase of microbiota causes host pathology (Guo et al., 2014). In this regard, the host immune system is an important factor in shaping gut microbiota community (Ryu et al., 2008). However, the mechanism by which host immunity controls gut microbiota consortium is not fully understood, which will be a fascinating field of research.

3. GUT IMMUNITY

Like all living organisms, discrimination of self and nonself is an essential ability in the life of insects and is made possible through the immune system. Insects are devoid of adaptive immunity and solely depend on the innate immune system. *Drosophila* is one of the best-studied model organisms in the research on innate immunity. Genetic analysis of *Drosophila* has revealed the existence of two conserved innate immune systems: the Toll and IMD pathways (Lemaitre et al., 1995, 1996). These innate immune systems guard the whole organism from systemic infections by directing the synthesis of various effector molecules known as antimicrobial peptides primarily in the fat body, a functional equivalent of the mammalian liver and adipose tissue. However, the local immune response in the gut epithelium is quite different from systemic immunity because it is in constant contact with microbiota. The primary immune system in the midgut epithelium is the DUOX pathway (Ha et al., 2005a, 2009a,b). The IMD pathway is also responsible for antimicrobial peptide expression in the midgut (Tzou et al., 2000). However, it has been shown that the Toll pathway is not operating and is dispensable for the host in the gut epithelium. An important feature of the DUOX and IMD pathways is that these local immune pathways have mechanisms to defend against pathogens while preserving commensal microbes. Improper control of these immune pathways contributes to host pathogenesis. We will discuss the regulatory mechanism and the physiological role of the DUOX and IMD pathways in the *Drosophila* gut epithelium.

3.1 The IMD Pathway

The IMD pathway was first discovered with the identification of the essential role of the *IMD* gene in systemic immunity (Lemaitre et al., 1995). It was found that *imd* mutant animals exhibited the phenotype of increased susceptibility to septic injury, which was due to impaired expression of antimicrobial peptide genes, the effectors of immune response (Boman et al., 1972). The IMD pathway is involved in the following (1) the recognition of bacterial-derived peptidoglycans, (2) the intracellular signalling cascade leading to the activation of transcription factor Relish, a nuclear factor κB (NF-κB) homolog, (3) the expression of antimicrobial peptides, and (4) negative regulatory mechanisms. We will cover this signalling pathway briefly in this section with an emphasis on the role of IMD in gut physiology. More in-depth reviews on the signalling components of the IMD pathway are available elsewhere (Imler, 2014; Myllymaki et al., 2014).

3.1.1 Positive and Negative Regulation of Intestinal IMD Pathway

Drosophila gut cells recognize gut-associated microbiota by sensing peptidoglycans. Peptidoglycan is a mesh-like polymer forming the bacterial cell wall. It is made up of sugar chains of β-1,4-linked *N*-acetylglucosamine and *N*-acetylmuramic acid and peptide chains cross-linking the sugar chains. The composition of peptides in the peptide chain is a source of modification in the peptidoglycan. The peptidoglycan of *Drosophila* gut-associated microbiota (mostly Gram-negative bacteria and Gram-positive bacilli) is characterized by the peptide chain consisting of four amino acids, including meso-diaminopimelic acid and thus known as DAP-type peptidoglycan. The host cell recognizes peptidoglycans with a family of peptidoglycan-recognition proteins (PGRPs) characterized by the existence of a PGRP domain. Genes encoding PGRP proteins can be classified based on their length of transcripts, long (L) and short (S). Among the long PGRPs, PGRP-LC is a transmembrane receptor that binds to DAP-type peptidoglycans (Choe et al., 2002; Gottar et al., 2002; Ramet et al., 2002). Peptidoglycans can also be recognized with PGRP-LE. Full-length PGRP-LE resides in the cytoplasm, recognizes internalized DAP-type peptidoglycans, and activates the IMD pathway (Bosco-Drayon et al., 2012).

The proximal event at the downstream of PGRP-LC activation is not fully understood, but right after peptidoglycan binding, a signalling complex composed of IMD, Dredd, and FADD is recruited to the plasma membrane and activated (Georgel et al., 2001; Naitza et al., 2002) (Fig. 3). Dredd is a caspase-8 homolog, and activated Dredd cleaves IMD and Relish for their

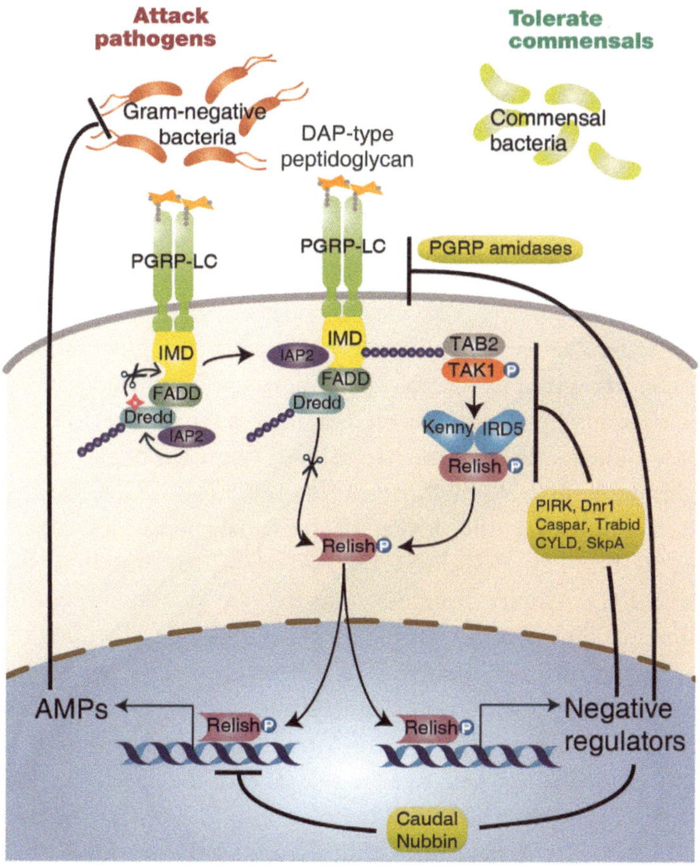

Fig. 3 The regulation of the IMD pathway. See text for details.

activation. Ultimately, the N-terminal part of cleaved Relish, after further modifications, translocates into the nucleus for transcription of target genes (Khush et al., 2001).

Activation of the IMD pathway requires ubiquitinations and phosphorylations of the signal components (Fig. 3). First, Dredd activation requires its K63-ubiquitination by the E3-ligase inhibitor of apoptosis 2 (IAP2) (Meinander et al., 2012). IAP2 also mediates ubiquitination of IMD. Cleavage of IMD at its N-terminus by Dredd exposes an IAP2-binding motif, which interacts with BIR domain in IAP2. Thus, IAP2 is recruited to IMD and generates K63-polyubiquitination, which is required for the recruitment of the TAK1/TAB2 complex (Vidal et al., 2001). TAB2 binds on K63-conjugated ubiquitin, and TAK1 is a MAPKKK kinase for the activation of the IKK complex. The IKK complex consists of two subunits, immune response deficient 5 (IRD5), which has catalytic activity, and Kenny, the regulatory subunit. The activated IKK complex, then, phosphorylates Relish on multiple sites, which is required for Relish activity as a transcription factor, such as recruitment of RNA polymerase II (Erturk-Hasdemir et al., 2009; Silverman et al., 2000). It is known that Relish induces a diverse range of immunity genes involved in nonself recognition, signalling pathways, proteolysis, and antibacterial activity. Especially, rapid induction of antimicrobial peptides capable of killing bacteria and/or fungi is a representative readout of IMD pathway activation following bacterial challenges.

In contrast to several aseptic internal organs, the gut epithelium is in constant contact with gut microbiota. Therefore, it should have a strategy to avoid chronic IMD activation that may have deleterious effects on health. It was shown that the activity of IMD signal is negatively regulated at multiple steps (Fig. 3). First, negative regulation of IMD signal can be achieved by reducing the levels of bacterial peptidoglycans. PGRPs including PGRP-LB, PGRP-SC1a, PGRP-SC1b, and PGRP-SC2 are characterized by their amidase activity to degrade peptidoglycans (Bischoff et al., 2006; Guo et al., 2014; Paredes et al., 2011). Second, negative regulation of the pathway is also achieved by the expression of negative regulators, which downregulate the activity of the IMD pathway components. For instance, PIRK is a transcriptional target of the IMD pathway and a strong negative regulator of the pathway (Aggarwal et al., 2008; Kleino et al., 2008; Lhocine et al., 2008). As PIRK interacts with PGRP-LC, PGRP-LE, and IMD, PIRK may disrupt the formation and activation of the IMD-containing signalling complex

(Aggarwal et al., 2008). There are more inhibitory proteins that negatively regulate the pathway, including Defense repressor 1 (Dnr1) for Dredd inhibition (Foley and O'Farrell, 2004; Guntermann et al., 2009); Caspar for Dredd-dependent Relish cleavage inhibition (Kim et al., 2006); Trabid at the level of TAK1 (Fernando et al., 2014); the deubiquitinating enzyme cylindromatosis (CYLD) (Tsichritzis et al., 2007); SkpA, a subunit of SCF-E3 ubiquitin ligase targeting Relish (Khush et al., 2002); and transcriptional repressors including caudal (Ryu et al., 2008) and the Oct1 homolog Nubbin (Dantoft et al., 2013). In addition to evading chronic activation of the pathway, the essential role of these negative regulatory modules is to preserve the healthy community of the gut microbiota (see later).

3.1.2 Physiological Roles of the Intestinal IMD Pathway

The essential role of the IMD pathway has been demonstrated mostly in the context of systemic infections. However, the role of IMD pathway in host survival during enteric infection is quite modest. The essential role of IMD pathway in host survival is visible only with enteric infection with certain types of pathogens. For example, in wild-type *Drosophila*, oral infection of *P. entomophila* with a mutation in *AprA*, which encodes a virulent protease, activates the local IMD pathway and the pathogen is efficiently removed from the gut. However, in Relish-mutant flies, the mutant *P. entomophila* persists (Liehl et al., 2006). It has also been shown that *Drosophila* with a mutation in the IMD pathway genes is more susceptible to oral infection with *S. marcescens* (Nehme et al., 2007). However, in most enteric infections including *ECC15* infection, the IMD pathway-mutant animals are almost equally viable when compared with control animals (Ha et al., 2005b). Interestingly, it has been shown that the number of gut microbiota increases in the IMD pathway-mutant animals, indicating that the IMD pathway is required to antagonize the excess proliferation of gut microbiota (Guo et al., 2014). This suggests a differential role of the IMD signal activity in gut immunity, although the precise role of the IMD pathway in controlling gut infections remains to be determined.

Recent studies indicate the significance of negative regulation of the IMD pathway in gut immune homeostasis. The insect gut is already occupied with microbiota, which are sources of peptidoglycans triggering the IMD pathway. Therefore, to preserve its microbiota, the gut epithelium suppresses the IMD pathway utilizing its negative regulators. However, if the gut epithelium fails to repress the IMD pathway, hyperactivation of the pathway causes pathologic symptoms including commensal dysbiosis and dysplasia (Bosco-Drayon et al., 2012; Guo et al., 2014; Lhocine et al., 2008; Ryu et al., 2008). For instance, caudal is a gut-specific transcriptional repressor of antimicrobial peptide genes. Knockdown of caudal or over-expression of antimicrobial peptide genes in the gut causes gut cell apoptosis, decreased survival rate, and most interestingly, change in gut commensal community leading to the dominance of *G. morbifer*, a pathobiont (Ryu et al., 2008). Oral infection with *G. morbifer* in axenic flies causes apoptosis in the gut and lethality, suggesting that the shifting in gut commensal community to the *G. morbifer* dominance (or dysbiosis) is responsible for the gut pathology (Ryu et al., 2008). A genetic study of PGRP-SC2 also demonstrates the significance of negative regulation of the IMD pathway for gut homeostasis (Guo et al., 2014). The authors showed that knockdown of PGRP-SC2, which negatively regulates the IMD pathway through its amidase activity, increased dysbiosis and dysplasia (Guo et al., 2014). Taken together, it is likely that an important function of the IMD pathway in the gut is to control the quality and quantity of gut microbiota.

3.2 The DUOX Pathway

Compared with systemic immunity where the IMD pathway is absolutely required, the function of the IMD pathway in the gut epithelial immunity (e.g. clearing invading pathogens) is less significant (Ha et al., 2005a). As the gut epithelium can efficiently eliminate microorganisms from microbe-contaminated food, there might be another defence mechanism in the gut epithelium (Ha et al., 2005a). Infection-induced reactive oxygen species (ROS) molecule was suggested as a candidate of bactericidal effector in the gut epithelium. Flies with reduced activity of secretory immune-regulated catalase (IRC) exhibited lethality when they were subjected to enteric infection, while flies lacking AMP expression (e.g. IMD pathway mutant flies) were unaffected (Ha et al., 2005b). These observations support an interesting scenario that when the fly gut was infected with bacteria, gut cells generate the ROS molecule as a bactericidal effector and that ROS should be quickly removed by IRC, as ROS is also harmful to gut cells. Lethality in IRC-knockdown flies, thus, is due to excess bacterial-induced ROS levels in the gut lumen, which should have been controlled by IRC (Ha et al., 2005a). How does the gut epithelium produce bactericidal ROS? It was found that the DUOX, a member of the nicotinamide adenine dinucleotide phosphate oxidase (NOX) family is responsible for microbicidal ROS generation in the gut. In this section, we will review the current understanding of the DUOX pathway, which acts as a first line of host defence in the gut epithelium.

3.2.1 DUOX in Gut Immunity

NOX/DUOX family proteins share a catalytic gp91phox domain, while DUOX contains an additional peroxidase homology domain (PHD). The general function of NOX enzyme is to generate superoxide anion in the extracellular space by electron transfer from NADPH in the cytosol to oxygen across the membrane. The superoxide anion is then converted to H_2O_2 spontaneously or enzymatically. In mammalian phagocytes, H_2O_2 generated from NOX2 is further converted to the bactericidal ROS effector HOCl by secreted myeloperoxidase (i.e. NOX2-myeloperoxidase system for oxidative burst in the phagosome).

In *Drosophila*, there are one NOX homolog and one DUOX homolog (Ha et al., 2005a). Interestingly, like a mammalian NOX2-myeloperoxidase system, the *Drosophila* gut epithelium is able to produce HOCl in a DUOX-dependent manner (Ha et al., 2005a). However, it requires further investigation whether the PHD of DUOX is directly responsible for the in vivo HOCl production or an additional protein with a myeloperoxidase-like activity is involved in HOCl production (Fig. 4).

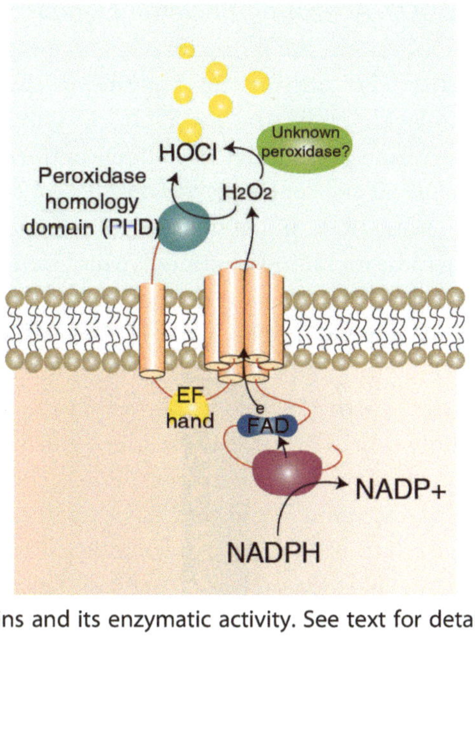

Fig. 4 DUOX domains and its enzymatic activity. See text for details.

Genetic studies using *Drosophila* have demonstrated that DUOX is required for host survival during enteric infection. Flies with knockdown of DUOX gene either ubiquitously or specifically in the gut epithelium showed increased lethality after feeding with bacteria- or yeast-contaminated foods (Ha et al., 2005a, 2009a). Furthermore, DUOX-knockdown flies were unable to clear gut-infected GFP-tagged pathogens or yeast, which persisted and proliferated in DUOX-knockdown fly guts (Ha et al., 2005a, 2009a).

If DUOX constitutes an essential gut immune system to kill pathogens and yeasts, how does microbiota colonize in the gut where DUOX exists? A striking feature of the DUOX signalling system is that DUOX is activated only with transient microorganisms but not with indigenous gut commensal bacteria. ROS, specifically HOCl, the bactericidal effector generated by DUOX, can be monitored in vivo utilizing the rhodamine-based R19S sensor (Chen et al., 2016). Oral infection with *ECC15* induced ROS generation in the gut, whereas oral feeding with indigenous bacteria such as *C. intestini*, *A. pomorum*, or *L. plantarum* maintained the ROS production at basal levels (Lee et al., 2013). This observation suggests the existence of a bacterial-derived substance that the DUOX signalling system utilizes to differentiate between indigenous and transient microorganisms.

3.2.2 Uracil as a DUOX-Activating Bacterial Ligand

How does the DUOX system differentially recognize indigenous and transient microorganisms? To answer this question, the identity of the bacterial-derived molecule activating the DUOX system has to be determined, and a series of research studies have been performed in this aspect. First, using a *Drosophila* S2 cell reporter system to detect DUOX-dependent intracellular ROS generation, it has been determined that well-known bacterial-derived ligands, including peptidoglycans, are unable to stimulate ROS generation, whereas whole bacteria efficiently induce ROS (Ha et al., 2009a). This suggests that a novel bacterial-derived molecule induces DUOX activity. To identify the bacterial-derived DUOX-activating molecule, bacterial culture supernatant was fractionated using high-performance liquid chromatography (HPLC), and each fraction with a peak was tested if it induces ROS (Lee et al., 2013). In this manner, uracil, the nucleobase in uridine, was identified as a specific ligand for DUOX activation. It was shown that uracil is secreted from noncommensal bacteria such as *ECC15* but not from the gut commensal bacteria such as *C. intestini* (Lee et al., 2013). Uracil was also secreted from several other pathogens, including *Vibrio*, *Klebsiella*,

Shigella, *Pseudomonas*, and *Serratia*, which suggests that secretion of high amount of uracil is rather a general feature of pathogens (Lee et al., 2013). Furthermore, uracil treatment alone is sufficient to induce ROS generation in the gut, indicating that uracil is a specific agonist for DUOX activation (Lee et al., 2013). Several important questions remain to be answered. For example, it is unclear why and how noncommensal pathogens secrete uracil while commensal bacteria do not. What is the molecular mechanism of uracil secretion? Do pathogens produce molecules other than uracil capable of inducing DUOX activity? All the future studies in this direction will provide a clear picture of how the gut epithelium distinguishes and discriminates unwanted bacteria while retaining the beneficial bacteria.

3.2.3 Complex Regulation of Signal Pathways Leading to DUOX Activation

How is the signal triggered by uracil transduced in the cell to activate DUOX? The first clue came from the observation that superoxide anion production from the membrane fraction of a dissected fly gut requires calcium ion, suggestive of calcium ion as a key regulator of DUOX activity (Ha et al., 2005a). Indeed, microorganisms are able to mobilize intracellular calcium levels from the ER in *Drosophila* S2 cells (Ha et al., 2009a). This observation led to test if DUOX activity is regulated by phospholipase C-β (PLCβ) and G protein α q (Gαq) because calcium release from the ER is under the control of inositol 1,4,5-triphosphate (IP3) receptor and IP3 generation requires PLCβ activity at the downstream of Gαq. Observations have been made in support of this model. First, IP3 generation was induced in S2 cells treated with microorganisms, for which Gαq and PLCβ were required. Second, increase of intracellular calcium in vivo was detected in microbe-fed gut epithelium using a fluorescence resonance energy transfer (FRET) assay. Third, in the absence of Gαq or PLCβ activity, the gut ROS level was decreased and calcium mobilization was blocked, which were rescued by transgenes expressing these genes. In addition, microbe-fed flies with impaired function of Gαq or PLCβ displayed increased host mortality (Ha et al., 2009a). All these observations indicate that Gαq- and PLCβ-dependent calcium mobilization from the ER is required for DUOX enzymatic activity (Fig. 5).

As gut infection induces DUOX expression along with DUOX enzymatic activity, DUOX activity is also regulated at its transcription level (Ha et al., 2005a). There is a putative binding site for activating transcription factor 2 (ATF2) at the near upstream of DUOX translation start site, raising a

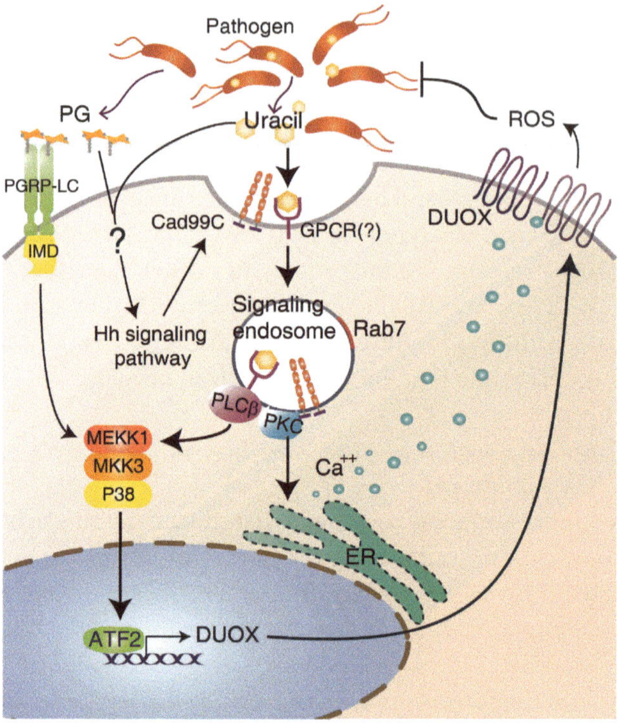

Fig. 5 The regulation of the DUOX pathway. See text for details.

possibility that DUOX transcription is regulated by ATF2 and its upstream p38 pathway. A series of experiments tested this idea and revealed that the signalling cascade involving MEKK1, MKK3, p38, and ATF2 indeed regulates DUOX transcription (Ha et al., 2009b) (Fig. 5).

Interestingly, DUOX expression was induced by either peptidoglycan or bacterial ligand other than peptidoglycan (possibly uracil) (Ha et al., 2009b). This observation led to the identification that transcriptional activation of DUOX is achieved through two independent pathways (Fig. 5): (1) peptidoglycan-dependent pathway with the cascade of PGRP-LC/IMD/MEKK1/MKK3/p38/ATF2 and (2) uracil-dependent PLCβ pathway. Notably, this PLCβ pathway does not involve IMD activation but merges with the peptidoglycan-dependent pathway at the downstream of IMD to induce MEKK1/MKK3/p38 activation (Ha et al., 2009b).

In conventionally reared fly guts, p38 stays largely inactive despite the presence of microbiota, suggesting that there is a negative regulation mechanism for the peptidoglycan-dependent p38 pathway. Indeed, DUOX transcription is further controlled by negative regulation of peptidoglycan-dependent p38 activation, for which PLCβ, calcineurin B (a calcium- and calmodulin-dependent serine/threonine protein phosphatase), and MAP kinase phosphatase 3 (MKP3) are required (Ha et al., 2009b). The important implication of this negative regulation is that massive DUOX activation via induction of DUOX transcription occurs only in infection conditions with a surge of peptidoglycans, and the gut epithelium tolerates certain amount of peptidoglycans from commensal bacteria maintaining basal levels of DUOX expression. The beneficial effect of this negative control of DUOX expression is demonstrated by the increased gut pathology in flies with MKP3 knockdown (Ha et al., 2009b).

Further investigations for the regulatory mechanism of DUOX activation revealed a novel function of the Hedgehog signalling pathway in DUOX-mediated gut immunity (Lee et al., 2015). An RNA-Seq analysis identified induction of genes involved in the Hedgehog pathway, including Hedgehog ligand and Cad99C, a cadherin-like transmembrane protein. Functional analyses uncovered their essential roles in the DUOX pathway. First, uracil activates Hedgehog expression and subsequent activation of the Hedgehog pathway, which is required for DUOX activation (Lee et al., 2015). Next, uracil-dependent Hedgehog signal activation induces the expression of Cad99C in the apical membrane of ECs. Upon enteric infection, ECs generate multiple $Cad99^{+}/Rab7^{+}$ endosomes around the brush border membrane of ECs. Formation of the Cad99C + endosome is essential for DUOX-dependent ROS production as well (Lee et al., 2015). Colocalization of Cad99C and PLCβ suggests that the $Cad99C^{+}$ foci are the signalling endosome to recruit these proteins and serve as a platform for PLCβ-dependent calcium mobilization for DUOX pathway activation (Lee et al., 2015) (Fig. 5). Flies with reduced Hedgehog signalling activity are highly susceptible to enteric infection, which can be rescued by ectopic expression of Cad99C in the ECs, demonstrating the essential role of Hedgehog signalling and Cad99C expression for DUOX activity.

4. GUT RENEWAL

4.1 ISC Self-Renewal, Differentiation, and Proliferation

The midgut is a dynamic organ. In adult *D. melanogaster*, the intestinal epithelial cells are renewed every 1 week (Micchelli and Perrimon, 2006). This dynamic gut renewal process is dependent on the continuous supply of new gut cells by asymmetric division of ISC, which gives rise to two daughter cells, one to be self-renewed ISC and the other to be EB for further differentiation into EC or EEC. The fate decision of ISC daughters requires the Delta-Notch signalling pathway (Micchelli and Perrimon, 2006; Ohlstein and Spradling, 2006, 2007) and the BMP signalling pathway that antagonizes Delta-Notch signalling (Tian and Jiang, 2014). Delta-Notch signalling is also important for the terminal differentiation into ECs and EECs (Ohlstein and Spradling, 2007; Perdigoto et al., 2011).

In addition to the cell fate decision process after ISC division, the regulation of ISC proliferation rate is crucial for the maintenance of the structural and functional integrity of the gut epithelium. If the rate of ISC proliferation is too fast, unwanted cells accumulate. In ageing guts, accumulation of overproduced abnormal cells causes pathology (Biteau et al., 2008). On the other hand, if the rate of ISC proliferation is too slow, damaged cells cannot be replaced resulting in the loss of gut integrity and organismal death. Controlled ISC proliferation is especially important when the epithelium is under the condition of acute damage due to enteric pathogens or toxic materials (Buchon et al., 2009a). Therefore, ISC should be equipped with signalling systems to read epithelial damages to adjust the rate of its division. Currently, multiple signalling pathways involved in the regulation of ISC proliferation have been identified. Jak–Stat, EGFR, Hippo, JNK, or Wingless pathway activation in ISCs is sufficient to induce ISC proliferation (Biteau et al., 2008; Cordero et al., 2012; Jiang et al., 2011; Karpowicz et al., 2010; Lee et al., 2009). Myc may serve as a common downstream effector of Jak–Stat, EGFR, Hippo, and Wingless pathways for ISC proliferation (Ren et al., 2013). Insulin Receptor signalling in ISC is also required for ISC proliferation (Amcheslavsky et al., 2009). Nearby injured cells produce signalling ligands and, therefore, are responsible for the activation of these pathways in ISC. In stressed ECs, Jak–Stat pathway ligands, especially Upd3, and EGFR pathway ligands, especially Keren, are expressed under the control of JNK and Hippo pathways (Jiang et al., 2009, 2011; Ren et al., 2010) (Fig. 6). EGFR pathway ligands, vein and spitz, are expressed in visceral muscles and progenitors, respectively (Jiang et al., 2011). In stressed conditions, EBs produce Upd2 through the activation of Hedgehog pathway (Tian et al., 2015). EBs also express Wingless ligand under the control of JNK pathway (Cordero et al., 2012). Activity of Hippo pathway in ISC might be under the control of intercellular interaction between two atypical cadherins, Fat in ISC and Dachsous (DS) in EC (Karpowicz et al., 2010) (Fig. 6). Therefore, all cellular components in the gut epithelium are capable to control ISC

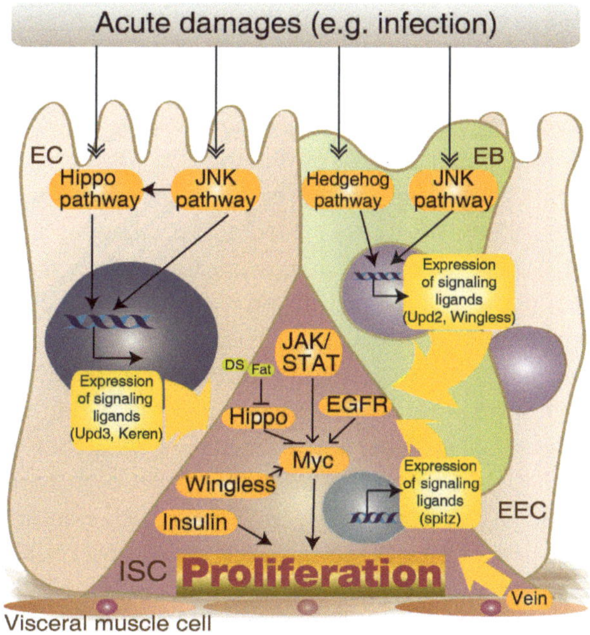

Fig. 6 Environmental signal-dependent regulation of gut cell renewal. See text for details.

proliferation and thereby serve as delicate sensors for environmental damage-induced gut renewal programme.

4.2 Gut Renewal and Innate Immune Systems

Accelerated ISC division rate was observed in the gut epithelium lacking IMD pathway activity, suggesting a possible relationship between gut cell renewal and IMD pathway (Buchon et al., 2009a). However, it was found that IMD pathway does not control the ISC division rate directly, but indirectly by modulating the number of gut bacteria. For example, the number of gut bacteria in the IMD pathway mutant flies was 10-fold higher than that of wild type (Buchon et al., 2009a), and this increased bacterial load in IMD mutant flies is the direct cause of accelerated ISC division rate.

A possible relationship between gut cell renewal and DUOX system was also proposed. It was found that bacterial-induced ISC division was impaired in the absence of DUOX activity (Buchon et al., 2009a), suggesting that ROS molecule generated by DUOX is the key factor for the induction of ISC division. In support of this notion, oral administration of H_2O_2 or paraquat, which produces ROS, also induced ISC mitosis (Buchon et al.,

2009a). How are ROS able to activate the signalling pathways for ISC proliferation introduced in the previous section? One possibility is that ROS indirectly activate signalling pathways controlling ISC proliferation by acting as tissue-damaging agents. This argument is supported by the observation that tissue-damaging agents unrelated to ROS production such as SDS, DSS, and bleomycin also induce ISC proliferation (Buchon et al., 2009a; Karpowicz et al., 2010; Ren et al., 2010, 2013; Shaw et al., 2010; Staley and Irvine, 2010). Another possibility is that ROS directly activate these signalling pathways by targeting redox-sensitive components of these signalling cascades (Lee, 2009). For instance, ROS are known to activate Jak–Stat pathway by targeting redox-sensitive protein tyrosine phosphatases (Liu et al., 2004), JNK pathway by targeting thioredoxin (Junn et al., 2000), and Wnt pathway by targeting nucleoredoxin (Funato et al., 2006). As ROS production and tissue damage are reciprocal events where ROS cause tissue damage and tissue damage induces ROS production, it is difficult to dissect these two possibilities. Nevertheless, it is clear that DUOX-generated ROS molecule mediates infection-induced gut cell homeostasis.

5. CONCLUDING REMARKS

In this chapter, we discussed structural features of insect gut, gut pathogens and microbiota, gut immunity, and gut cell homeostasis, based on recent landmark discoveries made with *D. melanogaster* model. Interconnections among these various aspects of gut physiology have become more evident with recent findings introduced here. For example, fine regulation of gut immune systems, used for defence against exogenous pathogens, turned out to have an important role in preserving healthy gut microbiota. Also, gut microbiota has various roles in host physiology including host immune regulation and gut stem cell homeostasis. Furthermore, the structural and functional integrity of gut under various environmental stresses is dependent on the dynamicity of gut renewal process controlled by complex signalling pathways. Although these interconnected topics collectively provide insights leading to a more integrated view of gut physiology, many important questions remain unanswered. How does gut immune system control the composition and ratio of gut microbial species? What are the exact mechanisms by which gut microbiota dysbiosis occurs? What are the relationships between gut microbiota and animal-level of physiology (e.g. animal longevity or behaviour)? Current understanding in *Drosophila* gut–microbiota interactions will serve as the framework to foster these investigations toward the core knowledge of 'microbiota–gut immunity–animal physiology'.

ACKNOWLEDGEMENTS

This work was supported by the National Creative Research Initiative Programme (Grant No. 2015R1A3A2033475 to W.-J.L.). K.-A.L. and J.-H.L. are supported by Basic Science Research Programme (NRF-2016R1C1B2016287 and NRF-2015R1D1A1A01058837, respectively).

REFERENCES

Aggarwal, K., Rus, F., Vriesema-Magnuson, C., Erturk-Hasdemir, D., Paquette, N., Silverman, N., 2008. Rudra interrupts receptor signaling complexes to negatively regulate the IMD pathway. PLoS Pathog. 4, e1000120.

Amcheslavsky, A., Jiang, J., Ip, Y.T., 2009. Tissue damage-induced intestinal stem cell division in Drosophila. Cell Stem Cell 4, 49–61.

Basset, A., Khush, R.S., Braun, A., Gardan, L., Boccard, F., Hoffmann, J.A., Lemaitre, B., 2000. The phytopathogenic bacteria Erwinia carotovora infects Drosophila and activates an immune response. Proc. Natl. Acad. Sci. U.S.A. 97, 3376–3381.

Bischoff, V., Vignal, C., Duvic, B., Boneca, I.G., Hoffmann, J.A., Royet, J., 2006. Down-regulation of the Drosophila immune response by peptidoglycan-recognition proteins SC1 and SC2. PLoS Pathog. 2, e14.

Biteau, B., Hochmuth, C.E., Jasper, H., 2008. JNK activity in somatic stem cells causes loss of tissue homeostasis in the aging Drosophila gut. Cell Stem Cell 3, 442–455.

Blum, J.E., Fischer, C.N., Miles, J., Handelsman, J., 2013. Frequent replenishment sustains the beneficial microbiome of Drosophila melanogaster. MBio 4, e00860-13.

Boman, H.G., Nilsson, I., Rasmuson, B., 1972. Inducible antibacterial defence system in Drosophila. Nature 237, 232–235.

Bosco-Drayon, V., Poidevin, M., Boneca, I.G., Narbonne-Reveau, K., Royet, J., Charroux, B., 2012. Peptidoglycan sensing by the receptor PGRP-LE in the Drosophila gut induces immune responses to infectious bacteria and tolerance to microbiota. Cell Host Microbe 12, 153–165.

Brune, A., Dietrich, C., 2015. The gut microbiota of termites: digesting the diversity in the light of ecology and evolution. Annu. Rev. Microbiol. 69, 145–166.

Buchon, N., Osman, D., 2015. All for one and one for all: regionalization of the Drosophila intestine. Insect Biochem. Mol. Biol. 67, 2–8.

Buchon, N., Broderick, N.A., Chakrabarti, S., Lemaitre, B., 2009a. Invasive and indigenous microbiota impact intestinal stem cell activity through multiple pathways in Drosophila. Genes Dev. 23, 2333–2344.

Buchon, N., Broderick, N.A., Poidevin, M., Pradervand, S., Lemaitre, B., 2009b. Drosophila intestinal response to bacterial infection: activation of host defense and stem cell proliferation. Cell Host Microbe 5, 200–211.

Buchon, N., Broderick, N.A., Lemaitre, B., 2013a. Gut homeostasis in a microbial world: insights from Drosophila melanogaster. Nat. Rev. Microbiol. 11, 615–626.

Buchon, N., Osman, D., David, F.P., Fang, H.Y., Boquete, J.P., Deplancke, B., Lemaitre, B., 2013b. Morphological and molecular characterization of adult midgut compartmentalization in Drosophila. Cell Rep. 3, 1725–1738.

Cagan, R., 2003. The signals that drive kidney development: a view from the fly eye. Curr. Opin. Nephrol. Hypertens. 12, 11–17.

Chandler, J.A., Lang, J.M., Bhatnagar, S., Eisen, J.A., Kopp, A., 2011. Bacterial communities of diverse Drosophila species: ecological context of a host-microbe model system. PLoS Genet. 7, e1002272.
Chen, X., Lee, K.A., Ren, X., Ryu, J.C., Kim, G., Ryu, J.H., Lee, W.J., Yoon, J., 2016. Synthesis of a highly HOCl-selective fluorescent probe and its use for imaging HOCl in cells and organisms. Nat. Protoc. 11, 1219–1228.
Choe, K.M., Werner, T., Stoven, S., Hultmark, D., Anderson, K.V., 2002. Requirement for a peptidoglycan recognition protein (PGRP) in Relish activation and antibacterial immune responses in Drosophila. Science 296, 359–362.
Cordero, J.B., Stefanatos, R.K., Scopelliti, A., Vidal, M., Sansom, O.J., 2012. Inducible progenitor-derived Wingless regulates adult midgut regeneration in Drosophila. EMBO J. 31, 3901–3917.
Cox, C.R., Gilmore, M.S., 2007. Native microbial colonization of Drosophila melanogaster and its use as a model of Enterococcus faecalis pathogenesis. Infect. Immun. 75, 1565–1576.
Dantoft, W., Davis, M.M., Lindvall, J.M., Tang, X., Uvell, H., Junell, A., Beskow, A., Engstrom, Y., 2013. The Oct1 homolog Nubbin is a repressor of NF-kappaB-dependent immune gene expression that increases the tolerance to gut microbiota. BMC Biol. 11, 99.
Dubreuil, R.R., 2004. Copper cells and stomach acid secretion in the Drosophila midgut. Int. J. Biochem. Cell Biol. 36, 745–752.
Erturk-Hasdemir, D., Broemer, M., Leulier, F., Lane, W.S., Paquette, N., Hwang, D., Kim, C.H., Stoven, S., Meier, P., Silverman, N., 2009. Two roles for the Drosophila IKK complex in the activation of Relish and the induction of antimicrobial peptide genes. Proc. Natl. Acad. Sci. U.S.A. 106, 9779–9784.
Fernando, M.D., Kounatidis, I., Ligoxygakis, P., 2014. Loss of Trabid, a new negative regulator of the drosophila immune-deficiency pathway at the level of TAK1, reduces life span. PLoS Genet. 10, e1004117.
Foley, E., O'Farrell, P.H., 2004. Functional dissection of an innate immune response by a genome-wide RNAi screen. PLoS Biol. 2, E203.
Funato, Y., Michiue, T., Asashima, M., Miki, H., 2006. The thioredoxin-related redox-regulating protein nucleoredoxin inhibits Wnt-beta-catenin signalling through dishevelled. Nat. Cell Biol. 8, 501–508.
Georgel, P., Naitza, S., Kappler, C., Ferrandon, D., Zachary, D., Swimmer, C., Kopczynski, C., Duyk, G., Reichhart, J.M., Hoffmann, J.A., 2001. Drosophila immune deficiency (IMD) is a death domain protein that activates antibacterial defense and can promote apoptosis. Dev. Cell 1, 503–514.
Glittenberg, M.T., Kounatidis, I., Christensen, D., Kostov, M., Kimber, S., Roberts, I., Ligoxygakis, P., 2011. Pathogen and host factors are needed to provoke a systemic host response to gastrointestinal infection of Drosophila larvae by Candida albicans. Dis. Model. Mech. 4, 515–525.
Gottar, M., Gobert, V., Michel, T., Belvin, M., Duyk, G., Hoffmann, J.A., Ferrandon, D., Royet, J., 2002. The Drosophila immune response against Gram-negative bacteria is mediated by a peptidoglycan recognition protein. Nature 416, 640–644.
Guntermann, S., Primrose, D.A., Foley, E., 2009. Dnr1-dependent regulation of the Drosophila immune deficiency signaling pathway. Dev. Comp. Immunol. 33, 127–134.
Guo, L., Karpac, J., Tran, S.L., Jasper, H., 2014. PGRP-SC2 promotes gut immune homeostasis to limit commensal dysbiosis and extend lifespan. Cell 156, 109–122.
Ha, E.M., Oh, C.T., Bae, Y.S., Lee, W.J., 2005a. A direct role for dual oxidase in Drosophila gut immunity. Science 310, 847–850.
Ha, E.M., Oh, C.T., Ryu, J.H., Bae, Y.S., Kang, S.W., Jang, I.H., Brey, P.T., Lee, W.J., 2005b. An antioxidant system required for host protection against gut infection in Drosophila. Dev. Cell 8, 125–132.

Ha, E.M., Lee, K.A., Park, S.H., Kim, S.H., Nam, H.J., Lee, H.Y., Kang, D., Lee, W.J., 2009a. Regulation of DUOX by the Galphaq-phospholipase Cbeta-Ca2+ pathway in Drosophila gut immunity. Dev. Cell 16, 386–397.
Ha, E.M., Lee, K.A., Seo, Y.Y., Kim, S.H., Lim, J.H., Oh, B.H., Kim, J., Lee, W.J., 2009b. Coordination of multiple dual oxidase-regulatory pathways in responses to commensal and infectious microbes in drosophila gut. Nat. Immunol. 10, 949–957.
Hsiao, E.Y., McBride, S.W., Hsien, S., Sharon, G., Hyde, E.R., McCue, T., Codelli, J.A., Chow, J., Reisman, S.E., Petrosino, J.F., Patterson, P.H., Mazmanian, S.K., 2013. Microbiota modulate behavioral and physiological abnormalities associated with neurodevelopmental disorders. Cell 155, 1451–1463.
Imler, J.L., 2014. Overview of Drosophila immunity: a historical perspective. Dev. Comp. Immunol. 42, 3–15.
Jiang, H., Patel, P.H., Kohlmaier, A., Grenley, M.O., McEwen, D.G., Edgar, B.A., 2009. Cytokine/Jak/Stat signaling mediates regeneration and homeostasis in the Drosophila midgut. Cell 137, 1343–1355.
Jiang, H., Grenley, M.O., Bravo, M.J., Blumhagen, R.Z., Edgar, B.A., 2011. EGFR/Ras/MAPK signaling mediates adult midgut epithelial homeostasis and regeneration in Drosophila. Cell Stem Cell 8, 84–95.
Junn, E., Han, S.H., Im, J.Y., Yang, Y., Cho, E.W., Um, H.D., Kim, D.K., Lee, K.W., Han, P.L., Rhee, S.G., Choi, I., 2000. Vitamin D3 up-regulated protein 1 mediates oxidative stress via suppressing the thioredoxin function. J. Immunol. 164, 6287–6295.
Karpowicz, P., Perez, J., Perrimon, N., 2010. The Hippo tumor suppressor pathway regulates intestinal stem cell regeneration. Development 137, 4135–4145.
Khush, R.S., Leulier, F., Lemaitre, B., 2001. Drosophila immunity: two paths to NF-kappaB. Trends Immunol. 22, 260–264.
Khush, R.S., Cornwell, W.D., Uram, J.N., Lemaitre, B., 2002. A ubiquitin-proteasome pathway represses the Drosophila immune deficiency signaling cascade. Curr. Biol. 12, 1728–1737.
Kim, M., Lee, J.H., Lee, S.Y., Kim, E., Chung, J., 2006. Caspar, a suppressor of antibacterial immunity in Drosophila. Proc. Natl. Acad. Sci. U.S.A. 103, 16358–16363.
King, D.G., 1988. Cellular organization and peritrophic membrane formation in the cardia (proventriculus) of Drosophila melanogaster. J. Morphol. 196, 253–282.
Kleino, A., Myllymaki, H., Kallio, J., Vanha-Aho, L.M., Oksanen, K., Ulvila, J., Hultmark, D., Valanne, S., Ramet, M., 2008. Pirk is a negative regulator of the Drosophila Imd pathway. J. Immunol. 180, 5413–5422.
Lawrence, P.A., Johnston, P., 1986. Observations on cell lineage of internal organs of Drosophila. J. Embryol. Exp. Morphol. 91, 251–266.
Lee, W.J., 2009. Bacterial-modulated host immunity and stem cell activation for gut homeostasis. Genes Dev. 23, 2260–2265.
Lee, W.J., Brey, P.T., 2013. How microbiomes influence metazoan development: insights from history and Drosophila modeling of gut-microbe interactions. Annu. Rev. Cell Dev. Biol. 29, 571–592.
Lee, W.C., Beebe, K., Sudmeier, L., Micchelli, C.A., 2009. Adenomatous polyposis coli regulates Drosophila intestinal stem cell proliferation. Development 136, 2255–2264.
Lee, K.A., Kim, S.H., Kim, E.K., Ha, E.M., You, H., Kim, B., Kim, M.J., Kwon, Y., Ryu, J.H., Lee, W.J., 2013. Bacterial-derived uracil as a modulator of mucosal immunity and gut-microbe homeostasis in Drosophila. Cell 153, 797–811.
Lee, K.A., Kim, B., Bhin, J., Kim, D.H., You, H., Kim, E.K., Kim, S.H., Ryu, J.H., Hwang, D., Lee, W.J., 2015. Bacterial uracil modulates Drosophila DUOX-dependent gut immunity via Hedgehog-induced signaling endosomes. Cell Host Microbe 17, 191–204.
Leidy, J., 1881. The parasites of the termites. J. Acad. Nat. Sci. (Phila.) 8, 425–447.

Lemaitre, B., Miguel-Aliaga, I., 2013. The digestive tract of Drosophila melanogaster. Annu. Rev. Genet. 47, 377–404.

Lemaitre, B., Kromer-Metzger, E., Michaut, L., Nicolas, E., Meister, M., Georgel, P., Reichhart, J.M., Hoffmann, J.A., 1995. A recessive mutation, immune deficiency (imd), defines two distinct control pathways in the Drosophila host defense. Proc. Natl. Acad. Sci. U.S.A. 92, 9465–9469.

Lemaitre, B., Nicolas, E., Michaut, L., Reichhart, J.M., Hoffmann, J.A., 1996. The dorsoventral regulatory gene cassette spatzle/Toll/cactus controls the potent antifungal response in Drosophila adults. Cell 86, 973–983.

Lengyel, J.A., Iwaki, D.D., 2002. It takes guts: the Drosophila hindgut as a model system for organogenesis. Dev. Biol. 243, 1–19.

Lhocine, N., Ribeiro, P.S., Buchon, N., Wepf, A., Wilson, R., Tenev, T., Lemaitre, B., Gstaiger, M., Meier, P., Leulier, F., 2008. PIMS modulates immune tolerance by negatively regulating Drosophila innate immune signaling. Cell Host Microbe 4, 147–158.

Liehl, P., Blight, M., Vodovar, N., Boccard, F., Lemaitre, B., 2006. Prevalence of local immune response against oral infection in a Drosophila/Pseudomonas infection model. PLoS Pathog. 2, e56.

Liu, T., Castro, S., Brasier, A.R., Jamaluddin, M., Garofalo, R.P., Casola, A., 2004. Reactive oxygen species mediate virus-induced STAT activation: role of tyrosine phosphatases. J. Biol. Chem. 279, 2461–2469.

Marianes, A., Spradling, A.C., 2013. Physiological and stem cell compartmentalization within the Drosophila midgut. Elife 2, e00886.

Meinander, A., Runchel, C., Tenev, T., Chen, L., Kim, C.H., Ribeiro, P.S., Broemer, M., Leulier, F., Zvelebil, M., Silverman, N., Meier, P., 2012. Ubiquitylation of the initiator caspase DREDD is required for innate immune signalling. EMBO J. 31, 2770–2783.

Micchelli, C.A., Perrimon, N., 2006. Evidence that stem cells reside in the adult Drosophila midgut epithelium. Nature 439, 475–479.

Mistry, R., Kounatidis, I., Ligoxygakis, P., 2016. Exploring interactions between pathogens and the Drosophila gut. Dev. Comp. Immunol. 64, 3–10.

Murakami, R., Shiotsuki, Y., 2001. Ultrastructure of the hindgut of Drosophila larvae, with special reference to the domains identified by specific gene expression patterns. J. Morphol. 248, 144–150.

Murakami, R., Takashima, S., Hamaguchi, T., 1999. Developmental genetics of the Drosophila gut: specification of primordia, subdivision and overt-differentiation. Cell. Mol. Biol. (Noisy-le-Grand) 45, 661–676.

Myllymaki, H., Valanne, S., Ramet, M., 2014. The Drosophila imd signaling pathway. J. Immunol. 192, 3455–3462.

Naitza, S., Rosse, C., Kappler, C., Georgel, P., Belvin, M., Gubb, D., Camonis, J., Hoffmann, J.A., Reichhart, J.M., 2002. The Drosophila immune defense against gram-negative infection requires the death protein dFADD. Immunity 17, 575–581.

Nehme, N.T., Liegeois, S., Kele, B., Giammarinaro, P., Pradel, E., Hoffmann, J.A., Ewbank, J.J., Ferrandon, D., 2007. A model of bacterial intestinal infections in Drosophila melanogaster. PLoS Pathog. 3, e173.

Ohlstein, B., Spradling, A., 2006. The adult Drosophila posterior midgut is maintained by pluripotent stem cells. Nature 439, 470–474.

Ohlstein, B., Spradling, A., 2007. Multipotent Drosophila intestinal stem cells specify daughter cell fates by differential notch signaling. Science 315, 988–992.

Paredes, J.C., Welchman, D.P., Poidevin, M., Lemaitre, B., 2011. Negative regulation by amidase PGRPs shapes the Drosophila antibacterial response and protects the fly from innocuous infection. Immunity 35, 770–779.

Perdigoto, C.N., Schweisguth, F., Bardin, A.J., 2011. Distinct levels of Notch activity for commitment and terminal differentiation of stem cells in the adult fly intestine. Development 138, 4585–4595.
Ramet, M., Manfruelli, P., Pearson, A., Mathey-Prevot, B., Ezekowitz, R.A., 2002. Functional genomic analysis of phagocytosis and identification of a Drosophila receptor for E. coli. Nature 416, 644–648.
Ren, C., Webster, P., Finkel, S.E., Tower, J., 2007. Increased internal and external bacterial load during Drosophila aging without life-span trade-off. Cell Metab. 6, 144–152.
Ren, F., Wang, B., Yue, T., Yun, E.Y., Ip, Y.T., Jiang, J., 2010. Hippo signaling regulates Drosophila intestine stem cell proliferation through multiple pathways. Proc. Natl. Acad. Sci. U.S.A. 107, 21064–21069.
Ren, F., Shi, Q., Chen, Y., Jiang, A., Ip, Y.T., Jiang, H., Jiang, J., 2013. Drosophila Myc integrates multiple signaling pathways to regulate intestinal stem cell proliferation during midgut regeneration. Cell Res. 23, 1133–1146.
Ryu, J.H., Kim, S.H., Lee, H.Y., Bai, J.Y., Nam, Y.D., Bae, J.W., Lee, D.G., Shin, S.C., Ha, E.M., Lee, W.J., 2008. Innate immune homeostasis by the homeobox gene caudal and commensal-gut mutualism in Drosophila. Science 319, 777–782.
Sandborn, E.B., Duclos, S., Messier, P.E., Roberge, J.J., 1967. Atypical intestinal striated muscle in Drosophila melanogaster. J. Ultrastruct. Res. 18, 695–702.
Sansone, C.L., Cohen, J., Yasunaga, A., Xu, J., Osborn, G., Subramanian, H., Gold, B., Buchon, N., Cherry, S., 2015. Microbiota-dependent priming of antiviral intestinal immunity in drosophila. Cell Host Microbe 18, 571–581.
Scott, J.J., Oh, D.C., Yuceer, M.C., Klepzig, K.D., Clardy, J., Currie, C.R., 2008. Bacterial protection of beetle-fungus mutualism. Science 322, 63.
Shanbhag, S., Tripathi, S., 2009. Epithelial ultrastructure and cellular mechanisms of acid and base transport in the Drosophila midgut. J. Exp. Biol. 212, 1731–1744.
Sharon, G., Segal, D., Ringo, J.M., Hefetz, A., Zilber-Rosenberg, I., Rosenberg, E., 2010. Commensal bacteria play a role in mating preference of Drosophila melanogaster. Proc. Natl. Acad. Sci. U.S.A. 107, 20051–20056.
Shaw, R.L., Kohlmaier, A., Polesello, C., Veelken, C., Edgar, B.A., Tapon, N., 2010. The Hippo pathway regulates intestinal stem cell proliferation during Drosophila adult midgut regeneration. Development 137, 4147–4158.
Shin, S.C., Kim, S.H., You, H., Kim, B., Kim, A.C., Lee, K.A., Yoon, J.H., Ryu, J.H., Lee, W.J., 2011. Drosophila microbiome modulates host developmental and metabolic homeostasis via insulin signaling. Science 334, 670–674.
Silverman, N., Zhou, R., Stoven, S., Pandey, N., Hultmark, D., Maniatis, T., 2000. A Drosophila IkappaB kinase complex required for Relish cleavage and antibacterial immunity. Genes Dev. 14, 2461–2471.
Staley, B.K., Irvine, K.D., 2010. Warts and Yorkie mediate intestinal regeneration by influencing stem cell proliferation. Curr. Biol. 20, 1580–1587.
Storelli, G., Defaye, A., Erkosar, B., Hols, P., Royet, J., Leulier, F., 2011. Lactobacillus plantarum promotes Drosophila systemic growth by modulating hormonal signals through TOR-dependent nutrient sensing. Cell Metab. 14, 403–414.
Tian, A., Jiang, J., 2014. Intestinal epithelium-derived BMP controls stem cell self-renewal in Drosophila adult midgut. Elife 3, e01857.
Tian, A., Shi, Q., Jiang, A., Li, S., Wang, B., Jiang, J., 2015. Injury-stimulated Hedgehog signaling promotes regenerative proliferation of Drosophila intestinal stem cells. J. Cell Biol. 208, 807–819.
Tsichritzis, T., Gaentzsch, P.C., Kosmidis, S., Brown, A.E., Skoulakis, E.M., Ligoxygakis, P., Mosialos, G., 2007. A Drosophila ortholog of the human cylindromatosis tumor suppressor gene regulates triglyceride content and antibacterial defense. Development 134, 2605–2614.

Tzou, P., Ohresser, S., Ferrandon, D., Capovilla, M., Reichhart, J.M., Lemaitre, B., Hoffmann, J.A., Imler, J.L., 2000. Tissue-specific inducible expression of antimicrobial peptide genes in Drosophila surface epithelia. Immunity 13, 737–748.

Vidal, S., Khush, R.S., Leulier, F., Tzou, P., Nakamura, M., Lemaitre, B., 2001. Mutations in the Drosophila dTAK1 gene reveal a conserved function for MAPKKKs in the control of rel/NF-kappaB-dependent innate immune responses. Genes Dev. 15, 1900–1912.

Vodovar, N., Vinals, M., Liehl, P., Basset, A., Degrouard, J., Spellman, P., Boccard, F., Lemaitre, B., 2005. Drosophila host defense after oral infection by an entomopathogenic Pseudomonas species. Proc. Natl. Acad. Sci. U.S.A. 102, 11414–11419.

Vodovar, N., Vallenet, D., Cruveiller, S., Rouy, Z., Barbe, V., Acosta, C., Cattolico, L., Jubin, C., Lajus, A., Segurens, B., Vacherie, B., Wincker, P., Weissenbach, J., Lemaitre, B., Medigue, C., Boccard, F., 2006. Complete genome sequence of the entomopathogenic and metabolically versatile soil bacterium Pseudomonas entomophila. Nat. Biotechnol. 24, 673–679.

Wang, J., Weiss, B.L., Aksoy, S., 2013. Tsetse fly microbiota: form and function. Front. Cell. Infect. Microbiol. 3, 69.

You, H., Lee, W.J., Lee, W.J., 2014. Homeostasis between gut-associated microorganisms and the immune system in Drosophila. Curr. Opin. Immunol. 30, 48–53.

CHAPTER FIVE

Intestinal Stem Cells: A Decade of Intensive Research in *Drosophila* and the Road Ahead

Yiorgos Apidianakis, Vasilia Tamamouna, Savvas Teloni, Chrysoula Pitsouli
University of Cyprus, Nicosia, Cyprus

Contents

1. Introduction 140
2. The Evolutionary Origin of ISCs 141
3. Primary Reports on *Drosophila* ISCs 143
4. The Molecular Characterization of *Drosophila* ISCs 144
5. Other SC Lineages in the *Drosophila* Gastrointestinal Tract: SCs of the Malpighian Tubules, the Hindgut, the Gastric Region and Stomach, and the Midgut Copper Cell Region 147
6. The Developmental Origin of *Drosophila* ISCs 149
7. The ISC Niche (Autocrine, Paracrine) and Hormonal (Endocrine) Signals Regulate ISC Activity 152
8. Ageing and Oxidative Stress Modulate ISC Activity and Vice Versa 153
9. Epithelial Regeneration (Regenerative Inflammation) as a Mechanism of Host Defence 155
10. Conserved Signalling Pathways Controlling Midgut ISC Proliferation (Notch, Wg/Wnt, PVR/PDGFR-VEGFR, JAK/STAT, JNK, p38, Hpo, EGFR, TOR, BMP, Slit-Robo, Bursicon-dLGR2, Hh, Ca^{2+}) 157
11. ISCs and Intestinal Tumours 160
12. Dietary Control of ISCs 162
13. Molecularly Defined Midgut Compartmentalization and ISCs 164
14. Drug Testing and *Drosophila* ISCs 165
15. How to Kill ISCs 167
16. Sexual Identity of ISCs 168
17. Conclusions 169

References 169

Advances in Insect Physiology, Volume 52
ISSN 0065-2806
http://dx.doi.org/10.1016/bs.aiip.2017.03.002

Abstract

Since the molecular characterization of *Drosophila* midgut progenitors in 2006, a few hundred articles studying fly intestinal stem cells have already been published. There was a relative lag phase in creating new knowledge until 2009, when at least 20 papers per year started being published on the subject and at least 40 per year since 2013. Here, we ponder on the substantial literature prior to intestinal stem cell molecular identification, including intestinal stem cell development and evolutionary origin, and describe the milestones achieved since then with an emphasis on their impact on biomedical research. The existing literature illuminates aspects of intestinal stem cell function in terms of homeostasis and disease. We discuss key findings on (a) the genetic markers of stem cells, their asymmetric or symmetric divisions and their progeny, (b) signalling pathways or networks and organ communication, (c) bacterial infections and microbiota, (d) dietary factors and drugs and (e) ageing. The accumulated knowledge provides lessons relevant to intestinal hyperplasia, dysplasia and intestinal cell metastasis, signalling pathway integration, the role of regenerative inflammation in host defence and tumorigenesis, the role of diet and the potential for translational therapeutics through drugs against intestinal inflammation, tumours and ageing.

1. INTRODUCTION

Somatic stem cells (SCs) of adult organisms are characterized by their self-renewal capacity and their ability to generate differentiated progeny. The first report about somatic SCs abiding to this rule is found in the literature of 1896 describing the precursor cells of the haematopoietic system, which are able to generate both red and white blood cells (Pappenheim, 1896; Ramalho-Santos and Willenbring, 2007). There is an earlier report about self-renewal of the mammalian intestinal epithelium in 1893, noting mitoses in the crypts of Lieberkuhn that fuel the intestinal surface epithelium of the villi with differentiated daughters (Bizzozero, 1893). In 1947 the rate and mechanism of intestinal renewal was more clearly described (Stevens and Leblond, 1947). Nevertheless, the precise locations of intestinal stem cells (ISCs) in the crypt bottoms (the Crypt Base Columnar Stem Cell, CBC) and directly above the Paneth cells (the Position +4 Stem Cell) were found in 1974 and 1978, respectively (Cheng and Leblond, 1974; Potten et al., 1978). Since then, modern lineage tracing, and other genetic and genomic methodologies have greatly contributed to the refinement of the molecular mechanism of

ISC renewal and differentiation. Moreover, the recent development of "miniguts" from CBCs in organoid cultures and the recellularization of a human colon scaffold with epithelial and other cell types promise to answer questions relevant to ISC behaviour and therapeutics (Chen et al., 2016; Clevers, 2013). Herein, we take a fresh look from the evolutionary, the developmental and the homeostasis vs disease perspective into the seminal work done in *Drosophila* and its contribution to our knowledge on ISCs.

2. THE EVOLUTIONARY ORIGIN OF ISCs

Processing of food and absorption of nutrients are keys to organismal growth and maintenance. In animals, ranging from flatworms to humans both functions are performed by the digestive system, which exhibits anterior to posterior functional regionalization. Even animals with simple body organization, such as the cnidarians (jelly fish, corals) and the ctenophores (comb jellies) that lack mesodermal tissues, have an endodermally derived tubular gut. Furthermore, sponges that lack internal organs altogether use specialized cells with motile microvilli resembling enterocytes (ECs), the choanocytes, to phagocytose food particles and absorb the nutrients (Takashima et al., 2013a). Obviously, intestinal cell functions are necessary and, thus, present in all animalia.

The intestine is constantly damaged due to the passage of food causing normal wear and tear or the ingestion of pathogens. To maintain its long-term fitness and functionality, the intestine of most animals undergoes constant cell renewal. Intestinal regeneration mediated by resident ISCs is an effective strategy to maintain a healthy organ and it is well characterized in mammals and *Drosophila*, but other long-lived organisms use alternative, yet powerful, strategies to achieve intestinal maintenance, i.e. dedifferentiation of intestinal cells or mobilization and recruitment of mitosis-capable neoblasts to the damaged gut. Below is a summary of what we know about regenerative cells contributing to intestinal maintenance in different animal groups.

In cnidarians, and specifically *Hydra*, the ECs of the gastrodermis continuously divide to keep their numbers constant, whereas gland cells divide in response to a signal from the I-cells, which are migratory ectodermal cells attracted to the gut. In ctenophores gut regeneration utilizes cells of the mesoglea located between the gastrodermis and the ectoderm (Takashima et al., 2013a). In protostomes, there is a great variety in the mechanisms mediating intestinal regeneration: bryozoa (lophophorates) and bivalves (moluscs) use mitosis of differentiated digestive tract cells, flatworms (planaria, *Schmitea mediterranea*) use migrating multipotent SCs, the neoblasts, to replace lost gut cells and annelids (segmented worms), which can regenerate every part of their body, use either neoblasts or mitotic endoderm cells for gut regeneration, depending on the species (Bely, 2014; Vogt, 2012). Interestingly, nematodes, like *Caenorhabditis elegans*, do not contain ISCs nor do they mobilize other mitotic cells to regenerate their intestine. However, the other clade of ecdysozoa, the arthropods, use either ISCs (in the insects, like *Drosophila*) or other resident SCs of the intestinal tract, such as the E cells of the hepatopancreas and mitotic cells of the midgut cecae (in decapods, like the lobster). Thus, it seems that intestinal regeneration via SC proliferation is lost in nematodes (Takashima et al., 2013a; Vogt, 2012). Echinoderms and chordates (constituting the deuterostome group) have impressive intestinal regenerative abilities. For example, studies in the echinoderm sea cucumber *Holothuria glaberrina* have shown that during autotomy or surgical removal of the gut, specialized cells closely associated with the mature gut, the mesothelial cells, dedifferentiate, activate mitosis and redifferentiate to achieve tissue regeneration (Mashanov et al., 2005, 2014b). Interestingly, these cells are positive for mammalian stemness factors (Mashanov et al., 2015). The simplest chordates, the tunicates or sea squirts (e.g. *Ciona intestinalis*) can regenerate their surgically removed oral siphon and gut by dedifferentiation and activation of mitosis of differentiated cells (Mashanov et al., 2014a; Tiozzo et al., 2008). Lastly, amphibians (*Xenopus laevis*), fish (Medaka, Zebrafish), birds (chick) and mammals (mouse, rat, human) have well-characterized resident ISCs, which are responsible for intestinal maintenance and regeneration (Mashanov et al., 2014a; Tiozzo et al., 2008).

Impact: To achieve intestinal homeostasis only arthropods (e.g. *Drosophila*) among protostomes and deuterostomes (e.g. human) use resident ISCs. All other protostomes and radiata use either cell dedifferentiation or multipotent neoblast-like cells (Fig. 1). Given the evolutionary distance between arthropods and deuterostomes, it is likely that the ISCs appeared independently in flies and mammals as the result of convergent evolution. If this is the case, important differences should exist at the molecular and cellular level, but also similarities in the effect of environmental factors (e.g. microbes and diet) and ageing between the insect and the mammalian intestine.

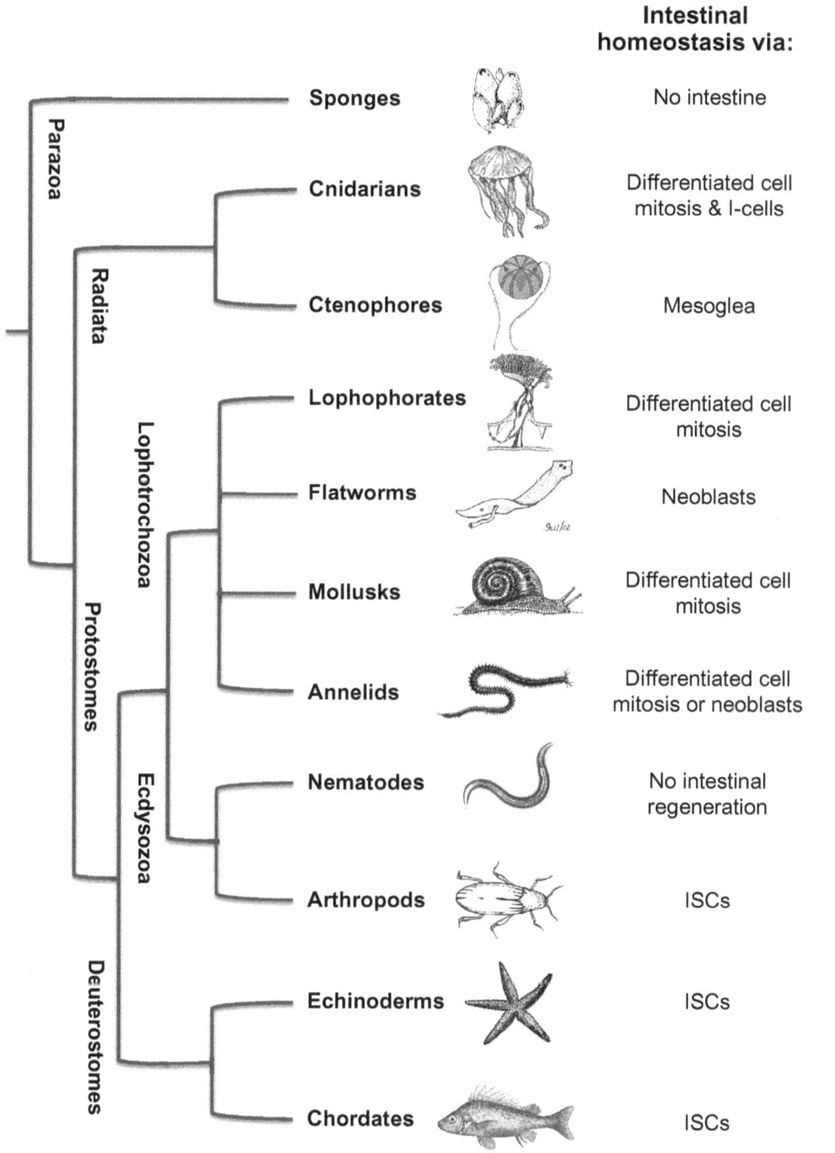

Fig. 1 Phylogenetic tree of the mechanisms of intestinal regeneration in animals.

3. PRIMARY REPORTS ON *DROSOPHILA* ISCs

The first reference about the existence of ISCs in *Drosophila melanogaster* is found in the Biology of *Drosophila* monograph edited by Demerec (1950). In the relevant chapter, the now known ISCs of the middle intestine (the midgut) are described as the "regenerative cells" that exhibit small size and are interdispersed in the adult intestinal epithelium. Nevertheless, no molecular or functional characterization was noted (Miller, 1950). This observation was corroborated in Baumann (2001) who noticed small potentially regenerative cells with distinct subcellular characteristics and staining markers in the adult posterior midgut (Baumann, 2001). The speculation about the identity of these cells was based on the regenerative cells with similar morphology in other insects (Baldwin and Hakim, 1991; Day and Powning, 1949; Snodgrass, 1935; Strasburger, 1932; Wigglesworth, 1965). Nevertheless, previous studies are conflicting over the existence of cell proliferation in the adult *Drosophila* gut (Bozcuk, 1972). One such study indicates that DNA synthesis and endoreplication, but not mitosis, are taking place upon ageing (Bozcuk, 1972). Thus, despite the intensive use of *Drosophila* as a powerful model system to understand human biology, it took more than half a century for scientists to establish the existence and molecularly characterize ISCs in the adult *Drosophila* midgut (Micchelli and Perrimon, 2006; Ohlstein and Spradling, 2006). Since then, an expanding body of literature has placed *Drosophila* in the forefront of ISC research. The chronological advances stemming from studies of *Drosophila* ISCs and their impact on our understanding of ISC biology will be the focus of this chapter (Fig. 2).

4. THE MOLECULAR CHARACTERIZATION OF *DROSOPHILA* ISCs

The existence of *Drosophila* ISCs was established in 2006, when two studies published in Nature characterized the adult *Drosophila* midgut ISCs and their differentiated progeny molecularly (Micchelli and Perrimon, 2006; Ohlstein and Spradling, 2006). Using mosaic analyses and molecular markers to label the intestinal cells, the authors collectively showed that the naturally lost mature cells of the adult *Drosophila* midgut are continuously replaced by pluripotent ISCs that have a simple lineage: each ISC divides asymmetrically to produce itself and a transient enteroblast (EB), which will eventually differentiate as an absorptive EC or as a secretory enteroendocrine cell (EE). Both ISCs and EBs are diploid cells that reside basally in the epithelium and express the transcription factor Escargot (Esg); the EBs

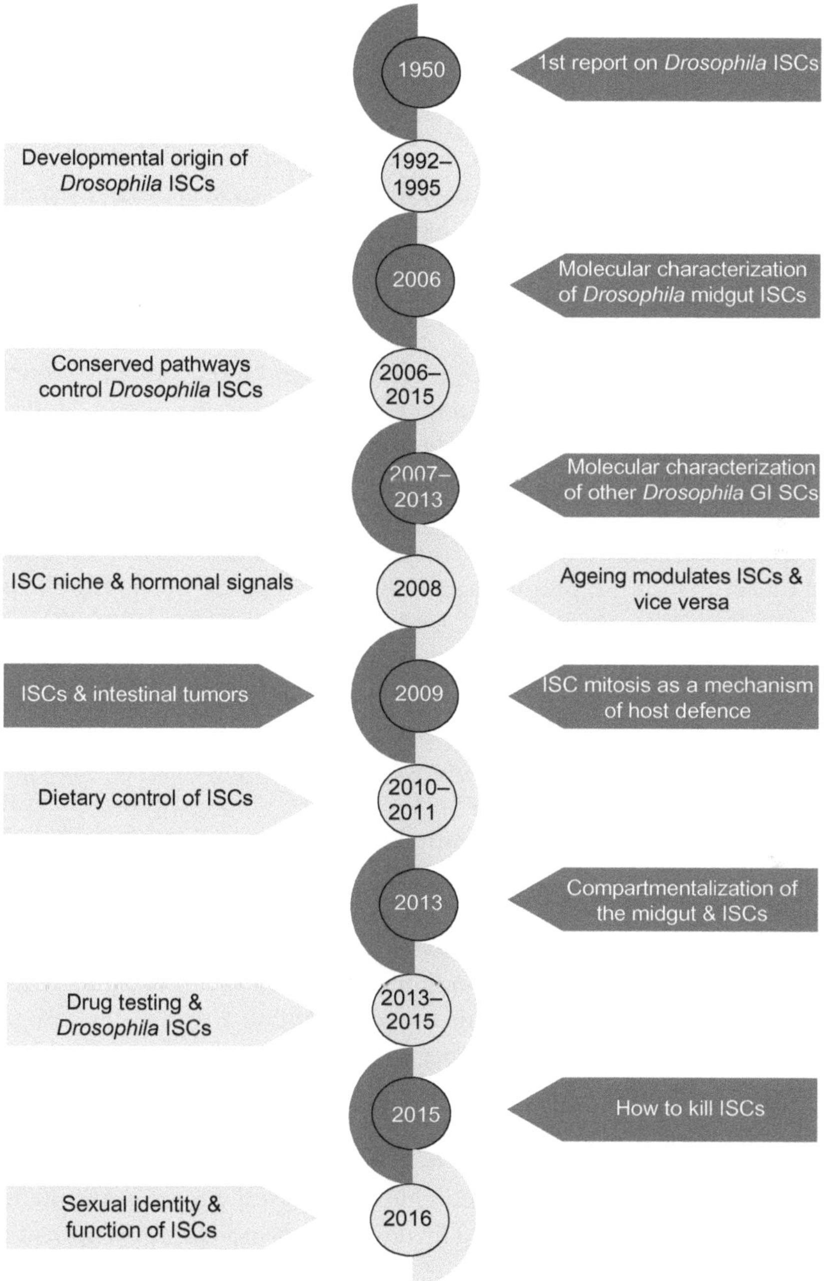

Fig. 2 The chronology of research milestones on *Drosophila* intestinal stem cells (ISCs).

are also positive for Suppressor of Hairless [Su(H)], a reporter of Notch pathway activation. The ECs are epithelial polyploid cells with cellular junctions, exhibit apicobasal polarity, have an apical actin-rich brush border that faces the gut lumen and express *Myo1A*, a gene encoding the nonclassical myosin Myo1A (Jiang and Edgar, 2009). EEs are small diploid cells labelled by the transcription factor Prospero (Pros) and express different endocrine peptides, such as tachykinin or allatostatin (Micchelli and Perrimon, 2006; Ohlstein and Spradling, 2006). The list of hormones produced locally by EEs in the *Drosophila* midgut has since been extensively expanded (Veenstra and Ida, 2014; Veenstra et al., 2008).

Interestingly, both studies discovered that the Notch pathway is critical for differentiation of the adult midgut ISCs. Specifically, loss of Notch in mosaic clones or downregulation of Notch via tissue-specific RNAi leads to increased numbers of ISCs. A closer look of the mosaics shows that mutant cells may develop as a group of ISCs or as a group of EEs, but not ECs, suggesting that Notch is required to balance ISC proliferation and differentiation by promoting the EC fate. Follow-up studies showed that the Delta (Dl) protein, the ligand for Notch is highly endocytosed in ISCs and activates the Notch receptor in the EB. Depending on the strength of the signal, the EB will become an EC (high Notch, 90% of the time) or an EE (low Notch, 10% of the time) (Biteau and Jasper, 2014; Ohlstein and Spradling, 2007). Further experiments have implicated various genes involved in the Notch pathway in ISC biology and showed that distinct levels of Notch signalling are required for commitment and differentiation of ISCs (Bardin et al., 2010; Perdigoto et al., 2011). In addition, a closer look at lineage dynamics of ISCs during homeostasis refined the original model of asymmetric ISC divisions and underscored a role of Notch in the decision of an ISC to divide not only asymmetrically but also symmetrically, which, similarly to mammals, involves neutral competition between ISC progeny (De Navascues et al., 2012). Further evidence for symmetric ISC divisions in the midgut came from the study of ISC behaviour in emerging female adults: when adult flies eclose, their midgut is not yet mature, but grows adaptively to feeding to achieve its final size. Specifically, upon feeding the ISCs of newly emerged flies divide symmetrically to generate the final number of adult midgut ISCs and finally bring the midgut to its normal size (O'Brien et al., 2011).

A closer look at EE differentiation with the use of a novel lineage system in the *Drosophila* midgut showed that the Su(H)-positive EB cells almost never produced EEs, indicating that EEs are generated by a distinct pre-EE progenitor (Beehler-Evans and Micchelli, 2015; Biteau and Jasper, 2014; Zeng and Hou, 2015). There are three ways that an ISC can produce an EE: an ISC can differentiate to EE, it can asymmetrically divide to self-renew and produce an EE or it can divide symmetrically to generate two EEs. EE differentiation requires the transcription factor Prospero, which acts downstream of the *achaete-scute* complex (AS-C) to determine ISC commitment to the EE fate (Zeng and Hou, 2015). Interestingly, EEs secrete Slit, which activates its receptor Robo2 in the ISCs, which in turn downregulates Prospero in the ISCs establishing a negative-feedback loop for ISC differentiation towards the EE fate (Biteau and Jasper, 2014; Zeng and Hou, 2015).

Impact: The molecular characterization of *Drosophila* ISCs in 2006 marked the beginning of a whole new era in fruit fly research, allowing studies focusing on basic biology questions of somatic SCs in a model organism with great genetics. The conservation of intestinal maintenance via ISCs between flies and mammals offers exciting possibilities for studies of conserved genes and pathways.

5. OTHER SC LINEAGES IN THE *DROSOPHILA* GASTROINTESTINAL TRACT: SCs OF THE MALPIGHIAN TUBULES, THE HINDGUT, THE GASTRIC REGION AND STOMACH, AND THE MIDGUT COPPER CELL REGION

The adult *Drosophila* intestine presents anterior–posterior differences in structure and function and is broadly separated in the anterior intestine or foregut, the middle intestine or midgut and the posterior intestine or hindgut. The foregut and hindgut have an ectodermal origin, whereas the midgut is endodermal. Food intake and initial processing takes place in the foregut (mouth, oesophagus and crop), further processing and nutrient absorption take place in the midgut and final absorption and waste removal take place in the hindgut (pylorus, ileum, rectum) (Pitsouli and Perrimon, 2008; Takashima and Hartenstein, 2012).

Following the molecular characterization of the midgut ISCs in *Drosophila*, a series of studies have identified SCs in other organs associated with the gastrointestinal (GI) tract: the malpighian tubules that function in excretion and water-ion balance, the hindgut where final nutrient absorption and waste disposal takes place and the foregut, the site of initial food processing (Fox and Spradling, 2009; Singh et al., 2007, 2011; Takashima et al., 2008).

In addition, specialized ISCs were characterized in the midgut copper cell region (CCR), which functionally resembles the mammalian stomach due to its acidity (Strand and Micchelli, 2013).

The malpighian tubules are connected to the *Drosophila* intestinal tube at the junction between the midgut and hindgut, they are dedicated to excretion and they functionally resemble the mammalian kidney. Before the malpighian tubule SC identification in 2007, the insect adult malpighian tubules were thought to be very stable organs that did not require renewal (Singh et al., 2007). Using lineage tracing and molecular marker labelling, the authors showed that the adult malpighian tubule differentiated cells arise from multipotent SCs located in the ureters and the lower tubules, the renal/nephric SCs (RNSCs). Adult RNSCs express the *esg* gene, similar to midgut ISCs, they divide symmetrically to self-renew and generate a renalblast, which will eventually differentiate to the Type 1 Cut-positive renal cell or the Type 2 Tsh-positive renal cell. RNSC self-renewal is controlled by autocrine Jak/Stat signalling (Singh et al., 2007). The Notch pathway acts with Jak/Stat to regulate RNSC self-renewal (Li et al., 2014b), and the EGFR/MAPK pathway controls RNSC proliferation (Li et al., 2015).

In 2008 a study published in Nature identified the SCs of the developing and regenerating hindgut epithelium in the pylorus, the anterior part of the hindgut, the domain adjacent to the midgut (Takashima et al., 2008). The authors showed that the developing and adult hindgut ECs are generated by progenitor cells residing in a ring of slowly proliferating cells, the "hindgut proliferation zone" (HPZ). Within this region, the self-renewal and proliferation of ISCs is controlled by the antagonistic actions of the conserved Wingless (Wg)/Wnt and Hedgehog (Hh) signalling pathways. Specifically, Wg/Wnt promotes self-renewal and proliferation, whereas Hh inhibits proliferation to promote differentiation (Takashima et al., 2008). Interestingly, this is reminiscent of the function of Wg/Wnt and Hh during mammalian ISC homeostasis in the crypts and villi of the intestine (Pitsouli and Perrimon, 2008). Nevertheless, a study published in 2009 (Fox and Spradling, 2009) showed that the hindgut is not constitutively renewed, as the midgut is, because the mature hindgut ECs are not continuously lost due to apoptosis. Instead, ISCs of the HPZ function as facultative SCs and they are activated upon acute damage to repair the hindgut (Fox and Spradling, 2009).

In 2011 another multipotent SC population was identified in the anterior-most part of the adult *Drosophila* midgut, the cardia (Singh et al., 2011). The cardia is a structure that functions as the gastric valve and is located at the foregut–midgut boundary. Using lineage analysis and marker expression, SCs were found in the cardia to contribute daughters to two directions: the anterior midgut and the oesophagus and crop. These cells were named gastric stem cells (GaSCs) since the cardia with the crop together function as the mammalian stomach. In addition, it was shown that the Wg/Wnt, Jak/Stat and Hh pathways regulate proliferation, self-renewal and differentiation of GaSCs (Singh et al., 2011).

Finally, a paper published in 2013 studied a subset of adult midgut ISCs, those renewing the acidic CCR of the intestine (Strand and Micchelli, 2013). The CCR corresponds to a structurally and functionally characteristic part of the midgut that encompasses *esg*-expressing SCs, EEs expressing allatostatin or neuropeptide F, small interstitial cells and specialized gland cells that secrete acid and store dietary copper, thus called copper cells. Given its lower pH, the CCR functionally mimics the mammalian stomach. Interestingly, the ISCs of the CCR, which the authors named gastric stem cells (GSSCs), are not dividing as fast as ISCs in other parts of the midgut, but they are more quiescent. Interestingly, the EGFR pathway is differentially regulated in the CCR and plays a key role in activation of GSSCs from quiescence when needed (Strand and Micchelli, 2013).

Impact: The presence of SCs in different parts of the GI tract in flies underscores the compartmentalization of ISC function in the *Drosophila* intestine and is reminiscent of the presence of resident ISCs in various parts of the mammalian GI tract. Similarities and differences in the biology of these ISCs will be relevant to the physiology of the different intestinal parts.

6. THE DEVELOPMENTAL ORIGIN OF *DROSOPHILA* ISCs

The adult *Drosophila* intestine derives during metamorphosis from the larval intestine after dramatic remodelling that ensures that larval cells are removed and all adult cell types are present (Takashima and Hartenstein, 2012).

The *Drosophila* larval GI tract derives during embryogenesis from an anterior invagination (the stomodeal invagination) that gives rise to the foregut and the anterior-most part of the midgut, the anterior tip of the ventral furrow that gives rise to most of the anterior midgut and a posterior invagination (the proctodeal invagination) that gives rise to the posterior midgut and hindgut. The larval midgut is composed of the intestinal epithelium inner layer, which develops from the endoderm, and the visceral muscle outer layer of mesodermal origin that surrounds it. The midgut intestinal epithelium develops from mesenchymal cells that undergo mesenchymal–epithelial transition and is thus considered a secondary epithelium. Specifically, at embryonic stage 10, the endoderm forms two mesenchymal cell masses, the anterior and the posterior midgut rudiments, which lie adjacent to the ectodermal foregut and hindgut primordia, respectively. The anterior rudiment encompasses principle midgut epithelial cells (PMECs) and adult midgut precursors (AMPs), whereas the posterior rudiment, in addition to PMECs and AMPs, also encompasses interstitial cell precursors (ICPs). At embryonic stage 11, the midgut rudiments establish contact with the mesodermal visceral muscle layer. Interactions of PMECs with the visceral muscle induce migration of the midgut rudiments, mesenchymal–epithelial transition of PMECs and their differentiation that by embryonic stage 13 will allow formation of the larval midgut monolayer epithelium. At this stage, the AMPs and ICPs retain their mesenchymal character and are associated with the apical surface of the monolayer, with AMPs distributed along the length of the epithelium and ICPs clustered in the region where the anterior and posterior midgut rudiments fuse. Later in development, ICPs integrate into the epithelium, whereas AMPs move to the basal surface of the epithelium (Simon and Gordon, 1995; Tepass and Hartenstein, 1994).

The progenitors of all adult midgut cells are the AMPs, which are specified in the endoderm during embryogenesis and increase in numbers during larval stages (Hartenstein and Jan, 1992; Tepass and Hartenstein, 1994, 1995). The AMPs form nests of proliferating cells dispersed in the differentiated larval midgut and divide seven to eight times to generate the adult midgut (Hartenstein, 1993; Micchelli, 2012; Strand and Micchelli, 2011; Tepass and Hartenstein, 1994). Symmetric divisions of AMPs ensure increase of their numbers and asymmetric divisions produce lineage-restricted progeny during larval and pupal stages (Mathur et al., 2010; Strand and Micchelli, 2011; Takashima et al., 2016).

During metamorphosis, dramatic tissue remodelling takes place and the differentiated larval midgut cells are shed concomitantly with the expansion and fusion of the AMP nests and their progeny that give rise to the adult midgut. The future adult midgut ISCs arise from precursors of ISCs (presumptive ISCs, pISCs), which are first detected as motile mesenchymal cells in early pupae at the onset of metamorphosis, divide two to three times to increase their numbers and finally also generate differentiated EEs; the EC progeny of pISCs arise only at the time of eclosion of the adult (Micchelli, 2012; Takashima et al., 2011, 2016). Interestingly, it was recently shown that the intestinal visceral muscle is maintained during metamorphosis, acts as a scaffold for intestinal elongation and is necessary for the normal development of the ISCs (Aghajanian et al., 2016; Klapper, 2000). It has been shown that during embryonic development the proneural genes of the AS-C give competence to the AMPs and removal of the genes leads to loss of AMPs without other disruptions of the midgut epithelium, whereas Notch signalling controls AMP number and loss of Notch leads to increased numbers of AMPs at the expense of ECs (Micchelli et al., 2011). In addition, the EGFR/Ras/MAPK pathway is crucial for AMP proliferation and acts as a potent mitogen upon activation through its ligand Vein, which is expressed in the visceral muscle (Jiang and Edgar, 2009; Micchelli, 2012).

In addition to the midgut ISCs, SCs that regenerate the hindgut, the foregut and the malpighian tubules have been described (Fox and Spradling, 2009; Singh et al., 2007, 2011; Takashima et al., 2008). During development, the foregut and the hindgut arise from two ring-shaped domains of dividing cells, appearing at the anterior and posterior end of the larval midgut, respectively. The posterior ring corresponds to the HPZ and encompasses the SCs of the hindgut. Cells of the HPZ move posteriorly to generate the mature hindgut ECs and different subsets of these progenitors generate distinct populations of ECs in the different parts of the hindgut, the pylorus, the ileum and the rectum (Fox and Spradling, 2009; Takashima et al., 2008). Surprisingly, a subset of HPZ progenitors moves towards the anterior and invade the adjacent midgut epithelium to form the ECs of the posterior end of the midgut. The Wg/Wnt pathway balances the proportion of HPZ progenitors that contribute to the hindgut vs the midgut and the GATAe transcription factor instructs the cells that migrate anteriorly to form midgut ECs (Takashima et al., 2013b). Furthermore, during early stages of metamorphosis, a subset of midgut AMPs migrate posteriorly to form the ureters (at the midgut/hindgut junction) and later on subsets of AMPs migrate from the ureters to the malpighian tubules to establish the adult renal stem cell population (Takashima et al., 2013b). Thus, although the boundaries of the ectoderm and endoderm are thought to be clearly defined in the embryo, it seems that during development and metamorphosis there is significant migration of progenitors from the endodermal midgut to the ectodermal renal tubules and from the ectodermal hindgut to the endodermal midgut.

Impact: Similarly to the mouse *Achaete scute-like 2* (*Ascl2*) gene that drives ISC multiplication in the small intestine (van der Flier et al., 2009), the *Drosophila* AS-C genes drive AMP expansion via symmetric divisions during embryogenesis. In the adult, the *Drosophila* AS-C genes appear necessary only for the EE cell fate specification (Amcheslavsky et al., 2014; Bardin et al., 2010). Thus, there are more commonalities in terms of symmetric divisions and the role of AS-C genes between the adult mouse ISCs during homeostasis and *Drosophila* midgut AMPs during development.

7. THE ISC NICHE (AUTOCRINE, PARACRINE) AND HORMONAL (ENDOCRINE) SIGNALS REGULATE ISC ACTIVITY

To maintain their numbers during homeostasis and achieve activation during regeneration, the ISCs need to receive and respond to autocrine, paracrine and hormonal signals arising from themselves, neighbouring cells or tissues and distant organs, respectively. The local microenvironment that produces autocrine and paracrine signals to ensure SC self-renewal and promote their differentiation during physiological homeostasis constitutes the "SC niche". The prototypical SC niche (i.e. in the female or male *Drosophila* germline) corresponds to an anatomical location in the tissue in contact with the SCs that produces self-renewal signals. When a SC divides, its daughter located closer to the niche retains its self-renewal property and is maintained as a SC, whereas the one located further from the niche cannot receive the niche signals and differentiates (Spradling et al., 2001). In the *Drosophila* midgut where ISCs are interdispersed throughout the epithelium, many paracrine signal-producing cells may control ISC activity along the length of the gut, that is, being part of the ISC niche.

The intestinal epithelial cells of the *Drosophila* midgut are ensheathed by a layer of visceral muscle, which is connected to the tracheal system to get oxygen, and in some parts of the intestine, it is contacted by neurons. Between the basal side of the epithelium and the visceral muscle, there is a basement membrane layer through which molecules can travel to and from the epithelium (Kux and Pitsouli, 2014). The first report of a functional ISC niche in the midgut came in 2008, when it was shown that Wg/Wnt produced by the visceral muscle reaches the ISCs through the basement membrane to control their self-renewal during homeostasis (Lin et al., 2008). In addition to Wg/Wnt, several studies have shown that during physiological homeostasis the visceral muscle is a source of EGFs, Dpp/BMP and insulin-like peptides (Biteau and Jasper, 2011; Buchon et al., 2010; Guo et al., 2013; Jiang et al., 2011; O'Brien et al., 2011; Veenstra et al., 2008; Xu et al., 2011); the trachea is a source of Dpp/BMP (Li et al., 2013a), the neurons produce insulin-like peptides (Cognigni et al., 2011; Linneweber et al., 2014), the ECs produce EGFs, BMPs and Upds (Buchon et al., 2010; Jiang et al., 2011; Li et al., 2013b; Osman et al., 2012; Xu et al., 2011), the EBs are a source of Wg/Wnt and EGFs (Cordero et al., 2012a; Xu et al., 2011) and the EEs produce hormones like Tachykinin and Bursicon (Amcheslavsky et al., 2014; Scopelliti et al., 2014). All these signals are integrated to control ISC maintenance and differentiation. Further description of the pathways activated in response to these signals and their integration during maintenance and regeneration is described in a separate section later. Importantly, differentiation of ISC progeny is not a result of departure from the niche, but due to asymmetric ISC divisions.

8. AGEING AND OXIDATIVE STRESS MODULATE ISC ACTIVITY AND VICE VERSA

In 2008 two independent studies found that aged flies exhibit an increase in ISCs and EBs due to PDGF/VEGF receptor (PVR) pathway activation (Biteau et al., 2008; Choi et al., 2008). According to these studies, the PVR ligand, Pvf2, is induced in the midgut progenitors during ageing and is necessary for the overaccumulation of ISCs. Similarly, oxidative stress induced by paraquat feeding or by the loss of catalase function mimics the ageing phenotype (Choi et al., 2008). Moreover, MAPK/p38b acts downstream of PVR/Pvf2 in stressed and aged intestines to induce misdifferentiation of ECs (Park et al., 2009). Biteau et al. (2008) showed that the cytoprotective Jun N-terminal kinase (JNK) pathway is activated in the old and stressed intestine causing proliferation of ISCs and accumulation of misdifferentiated progeny. Interestingly, the old age phenotype can be reversed by ectopic expression of the Notch/Delta differentiation pathway underscoring the need for balanced JNK-Notch activity for intestinal homeostasis during ageing (Biteau et al., 2008).

A multitude of subsequent studies uncovered additional aspects of the biology of the aged and stressed intestine. For example, the aged *Drosophila* midgut presents various pathological phenotypes, such as leakiness (Rera et al., 2012), increased DNA damage and centrosome amplification in the ISCs (Park et al., 2014), loss of compartmentalization (Li et al., 2016) and increased Redox signalling in ECs (Albrecht et al., 2011). The ATR DNA damage response pathway is necessary for ISC maintenance during ageing (Park et al., 2015) and metformin can reverse the DNA damage and the ISC centrosome amplification in aged midguts (Na et al., 2013, 2015).

Interestingly, the ISC proliferation phenotype of the ageing intestine can be rescued by overexpression of the *dPGC-1/spargel*, a regulator of energy metabolism, mitochondrial biogenesis and respiration, in the ISCs/EBs of the *Drosophila* midgut. Reducing the ageing phenotypes in the intestine extends fly lifespan indicating that ISC health has a direct effect on animal health (Rera et al., 2011). A similar phenotype has been observed when the yeast NDI1 mitochondrial gene encoding a NADPH dehydrogenase is overexpressed in ISCs/EBs indicating that ISC mitochondrial health is critical for lifespan regulation (Hur et al., 2013). Strikingly, perturbations of the ISC dysplasia phenotype of the aged *Drosophila* intestine underscored the protective role of ISC proliferation in healthy ageing and lifespan. Moderate reduction of the ISC proliferation phenotype in the aged intestine leads to increased lifespan, whereas excessive ISC proliferation restricts lifespan, highlighting the need for moderate ISC proliferation for healthy ageing (Biteau et al., 2010).

Furthermore, the *Drosophila* midgut is populated by resident microbes, which live harmonically with their host. During ageing, the balance between the populations of commensal microbes is perturbed leading to dysbiosis and increased ISC proliferation. Reducing dysbiosis lengthens lifespan, for example, by improving gut immune homeostasis via PGRP-SC2 expression (Guo et al., 2014) or gut compartmentalization (Li et al., 2016). In addition, Rapamycin, which inhibits the target-of-Rapamycin (TOR) pathway, has been recently shown to extend healthspan and lifespan by maintaining gut stability during ageing, inhibiting age-induced ISC proliferation and reducing intestinal ROS produced by the Duox pathway (Fan et al., 2015).

Interestingly, Lamin B is progressively lost during ageing in fat body cells leading to systemic inflammation and secretion of peptidoglycan recognition proteins (Chen et al., 2014). These secreted proteins reach the midgut where they suppress the immune deficiency (Imd) pathway leading to midgut hyperplasia due to dysbiosis. Mechanistically, loss of Lamin B in aged fat bodies contributes to heterochromatin loss and derepression of genes involved in immune response. Similar functions are mediated by mammalian lamins highlighting parallels with the fly system and the potential to understand immunosenescence in a powerful model organism (Chen et al., 2014).

The aged *Drosophila* intestine was also found to be prone to tumorigenesis. In 2008 a histopathological study reported tumour accumulation upon ageing in the male fly testis and midgut (Salomon and Jackson, 2008). Using the Oregon-R fly strain and screening ~150 males per time point the authors found that tumours accumulate in the testis and midgut starting the 3rd and 4th week of age, respectively. On the 5th week 18% and 1.3% of flies contained tumours in the testis and midgut (Salomon and Jackson, 2008). Despite the low tumour frequency in the Oregon-R midgut, dysplastic epithelium appears as early as 2 weeks of adult life. While most adult fly tissues lack mitotic activity, the testis and midgut are constantly renewed in adult males. The authors speculated that adult SCs might be susceptible to neoplastic transformation, because the effectiveness of DNA repair declines with age (Salomon and Jackson, 2008).

Tumour accumulation in the adult midgut was further studied and molecularly explained in a 2015 study showing frequent somatic mutations in ISCs and tumour formation during ageing (Siudeja et al., 2015). Mechanisms driving mutagenesis include somatic deletions and large chromosomal rearrangements in hemizygous males and the loss of heterozygosity by mitotic homologous recombination in heterozygous midgut SCs. Both mechanisms may result in the inactivation of important genes, such as the tumour suppressor *Notch*. *Notch* is located on the X chromosome, thus, in male flies is only present in one copy and its deletion leads to spontaneous neoplasias in approximately 10% of 6-week-old males (Siudeja et al., 2015).

Impact: The *Drosophila* midgut has been successfully used to address the mechanisms of ageing. Interestingly, increased ROS and commensal microbe dysbiosis combined with DNA instability lead to the aged gut phenotype. Rescuing these phenotypes genetically or chemically leads to extended healthspan and lifespan. Furthermore, it is very likely that tumorigenesis due to spontaneous mutations occurring during ageing is not limited to Notch. Therefore, the aforementioned studies pave the way for modelling and studying age-related spontaneous neoplasia in flies and the microbial, environmental, genetic and epigenetic factors that may affect it.

9. EPITHELIAL REGENERATION (REGENERATIVE INFLAMMATION) AS A MECHANISM OF HOST DEFENCE

The contribution of *Drosophila* in the field of innate immunity was established initially with the work of Bruno Lemaitre, Jules Hoffman, and colleagues in the mid-1990 and subsequently with the pertinent Nobel prize in "Physiology and Medicine" in 2011 (Lemaitre and Hoffmann, 2007). Through the years, it became apparent that *Drosophila*, similar to mammals, can defend itself from microbes entering the host circulation via the Toll and

Imd NF-κB pathways, whereas in barrier epithelia, such as the intestine, the Imd/NF-κB, the JAK/STAT and the Duox pathways drive the production of antimicrobial peptides and reactive oxygen species (Buchon et al., 2013a). Beyond innate immunity, another part of *Drosophila* host defence was elucidated in 2009, that is, epithelial regeneration. Following infection with *Pseudomonas entomophila, Pseudomonas aeruginosa, Erwinia carotovora* and *Serratia marcescens* the JNK and JAK–STAT pathways induce regeneration in the *Drosophila* midgut, which in turn promotes fly survival to infection (Apidianakis et al., 2009; Buchon et al., 2009a,b; Cronin et al., 2009). To the contrary, *Vibrio cholerae* infection and cholera toxin decrease ISC division to reduce host defence, and genetic manipulations that increase ISC mitosis promote host survival to infection (Wang et al., 2013). Thus, *V. cholerae* may actively decrease ISC division to circumvent this defence response of *Drosophila*. Suboptimal ISC response has also been noted, when flies are fed with very high doses of *P. entomophila*, because regeneration is blocked (Buchon et al., 2009b; Chakrabarti et al., 2012). Lack of regeneration via ISCs has also been noted upon *P. aeruginosa* oral infection (Limmer et al., 2011), apparently, when infection is less virulent to the flies. In addition, feeding on *S. marcescens* appears more lethal when regeneration is activated (Cronin et al., 2009). On the other hand, not only bacterial pathogens, but also microbiota can induce regeneration in the midgut (Buchon et al., 2009b). For example, symbiotic *Lactobacilli* induce regeneration as part of the host defence response (Jones et al., 2013). Finally, ageing alters intestinal microbiota in *Drosophila* resulting in intestinal dysplasia (Biteau et al., 2008; Guo et al., 2014). Theoretically, intestinal deterioration over time may involve the induction of the EGFR pathway, which has been shown to suppress the Imd/NF-κB pathway in various cells including ISCs (Moreira et al., 2011). Alternatively, chronic activation of the transcription factor Foxo upon ageing reduces expression of PGRP-SC2, a negative regulator of Imd/NF-κB signalling, resulting in commensal dysbiosis, ISC hyperproliferation and epithelial dysplasia (Guo et al., 2014).

In mammals, mucosal infection may accelerate the rate of regeneration and induce inflammatory cytokines and concomitant infiltration of immune cells, such as neutrophils, macrophages and T cells in the mucosa that may cause further damage and induce regeneration of the epithelium (Christofi and Apidianakis, 2015). Similarly, phagocytosis by *Drosophila* haemocytes is critical for the control of bacteria escaping from the intestine and into the haemolymph (Kocks et al., 2005; Nehme et al., 2007), and haemocytes accumulate in the infected gut or act at a distance to assist regeneration

(Ayyaz et al., 2015; Chakrabarti et al., 2016). Strikingly, in both *Drosophila* and mammals, local and remote signals drive regenerative inflammation, a process that encompasses not only infiltrating immune cells, but also ISC microenvironment signals able to regenerate the epithelium (Karin and Clevers, 2016; Panayidou and Apidianakis, 2013).

Impact: Taking into account the various types of regeneration in the *Drosophila* intestine (Guo et al., 2016), with or without transit-amplifying cells or via endoreplication, it will be interesting to characterize the genetic, environmental and microbial factors that lead to optimal/beneficial as opposed to suboptimal/detrimental regeneration. This is also important in terms of translational research, because many human pathogens can be effectively modelled in flies (Panayidou et al., 2014). Future research may also explore how regeneration pertains to other aspects of host defence, namely, permeability, mucus secretion, peristalsis, intestinal and faecal pH, microbiota and production of antimicrobials, such as antimicrobial peptides, lysozymes and reactive oxygen species.

10. CONSERVED SIGNALLING PATHWAYS CONTROLLING MIDGUT ISC PROLIFERATION (NOTCH, WG/WNT, PVR/PDGFR-VEGFR, JAK/STAT, JNK, P38, HPO, EGFR, TOR, BMP, SLIT-ROBO, BURSICON-dLGR2, HH, CA^{2+})

Intestinal homeostasis is coordinated by the action of highly conserved signalling pathways. Most of them have been extensively described and reviewed. Here, we refer to the pace of their discoveries, while commenting on prominent similarities and discrepancies among studies. In 2006 the articles that first characterized the midgut ISCs molecularly, also showed that the Notch signalling pathway controls ISC proliferation and differentiation (Micchelli and Perrimon, 2006; Ohlstein and Spradling, 2006). One year later, the ligand of Notch, Delta was established as a marker of ISCs (Ohlstein and Spradling, 2007). In 2008 the growth factor Pvf2 (PVR/PDGFR-VEGFR pathway) was found necessary for ISC proliferation upon ageing or oxidative stress (Choi et al., 2008). The same year Wg/Wnt was demonstrated as necessary for ISC maintenance (Belenkaya et al., 2008; Lin et al., 2008) and later on as necessary for ISC proliferation during homeostasis and upon tissue injury (Cordero et al., 2012b).

2009 was a year of major discoveries in the field, such as the elucidation of the role of the insulin-like receptor (InR), JAK/STAT, JNK and p38 signalling pathways in midgut ISC regulation. The earliest report that year demonstrated the role of systemic insulin in ISC proliferation (Amcheslavsky et al., 2009). Two years later, the same authors showed that the TOR pathway, which can be controlled by the insulin pathway, is implicated in ISC proliferation (Amcheslavsky et al., 2011). In 2009 the JAK/STAT pathway was found induced in the ISCs upon Upd cytokine secretion from stressed/damaged ECs (Buchon et al., 2009a; Cronin et al., 2009; Jiang and Edgar, 2009; Jiang et al., 2009). A few months later the JNK pathway was shown to be necessary for ISC proliferation, when EC damage is induced by stress, infection or microbiota (Apidianakis et al., 2009; Buchon et al., 2009b). Finally, p38b was found necessary upon ageing or oxidative stress to induce aberrant proliferation of ISCs (Park et al., 2009).

In 2010 four articles described the role of Hippo (Hpo) pathway as an EC stress sensor in the intestine and responsive to changes that disrupt the integrity of the epithelium (Karpowicz et al., 2010; Ren et al., 2010; Shaw et al., 2010; Staley and Irvine, 2012). In addition to revealing a unique noncell autonomous role of Hpo signalling in stressed ECs to block ISC proliferation, Ren et al. (2010) showed that the Hpo pathway regulates the transcription factor Yorkie (Yki) in EBs, which in turn promotes ISC proliferation in response to the tissue-damaging reagent dextran sulphate sodium (DSS) (Ren et al., 2010).

In 2010 and 2011 four papers described the pivotal role of the EGFR pathway in ISC mitosis. The work collectively shows that EC damage or stress induces multiple EGFR ligands that activate EGFR in ISCs and concomitant proliferation (Biteau and Jasper, 2011; Buchon et al., 2010; Jiang and Edgar, 2011; Xu et al., 2011). The EGFR ligand Vein is specifically expressed in visceral muscle cells and is important for ISC maintenance and proliferation. Two additional EGFR ligands, Spitz and Keren, act redundantly to promote ISC maintenance and proliferation. Interestingly, the EGFR pathway contributes to gut morphogenesis cell autonomously in ECs to properly coordinate the delamination and anoikis of damaged cells (Buchon et al., 2010). To restore midgut homeostasis upon infection, ISC proliferation, differentiation, incorporation and morphogenesis of new ECs and expulsion of damaged ones is required. It appears that the EGFR pathway is involved in all these stages and the synchronization of the cellular events leading to midgut repair.

In 2013 and the following years the BMP (Dpp/Gbb) pathway was found to both positively and negatively regulate midgut ISC proliferation. Three different studies claimed that BMP ligands emanate from the trachea or muscle to inhibit ISC mitosis or from ECs to promote CCR GSSC proliferation and differentiation (Guo et al., 2013; Li et al., 2013a; Tian and Jiang, 2014). A forth study showed that during oral infection Dpp is induced in the visceral muscle and subsequently Gbb is induced in ECs to signal progenitor cells towards EB differentiation or EC maturation (Zhou et al., 2015). Interestingly, the Iroquois/IRX family protein Mirror downregulates the Dpp pathway, which acts as a tumour suppressor against Ras oncogenic midgut tumours of flies and human colorectal cancer cells (Martorell et al., 2014a). Finally, haemocytes were found to accumulate in the infected gut secreting Dpp to induce regeneration as early as 4 h upon intestinal infection with *E. carotovora* by activating its receptor Saxophone and its downstream target Smox (Ayyaz et al., 2015). A few hours later, activated Dpp is received by ISCs through the receptors Saxophone/Thickveins to reestablish ISC quiescence by activating its downstream target Mad.

In 2014 Biteau and Jasper found that the secreted ligand Slit is secreted by the EEs cells to signal via its receptor Robo2 in ISCs to limit their commitment to differentiate into EE cells (Biteau and Jasper, 2014). Bursicon is an additional signal secreted by midgut EE cells, which acts in a paracrine manner on the visceral muscle. Bursicon binds to its receptor, DLGR2, in the visceral muscle to repress the EGFR pathway ligand Vein and concomitant ISC proliferation (Scopelliti et al., 2014).

In 2014 the growth factor Hh was found expressed in multiple midgut cell types to induce ISC proliferation during tissue homeostasis (Li et al., 2014a). A 2015 study extended these results showing that the tissue-damaging agent DSS induces Hh in ISCs, EBs and ECs to activate Hh signalling and Upd2 expression in EBs (Tian et al., 2015). Upd2 activates the JAK/STAT pathway in ISCs for their proliferation (Tian et al., 2015). Without Hh signalling in progenitor cells, ISCs do not respond to DSS exposure, indicating the necessity of one more pathway in ISC damage-induced proliferation.

Given that multiple pathways may control ISC proliferation and differentiation, an important issue is signalling pathway integration. Towards this, Ren and colleagues showed that midgut damage caused by DSS feeding stimulates dMyc expression via the Hpo pathway, whereas bleomycin feeding activates dMyc through the JAK/STAT and EGFR pathways. dMyc integrates signals from multiple signalling pathways in the midgut ISCs and is required for optimal proliferation upon tissue damage (Ren et al., 2013). Another level of signal integration appears to be the Ca^{2+} oscillation modes within ISCs. In response to a wide range of dietary and stress stimuli,

ISCs reversibly transit between Ca^{2+}oscillation states that represent poised or activated modes of proliferation. The authors propose that the dynamic regulation of intracellular Ca^{2+} levels in ISCs allows effective integration of diverse mitogenic signals in ISCs to adjust their proliferative activity according to the needs of the tissue (Deng et al., 2015).

Impact: All the aforementioned pathways have homologues in mammals, suggesting analogous ways of mammalian ISC regulation. It is important though to also pinpoint the differences in these pathways between flies and mammals. For example, the Notch pathway acts as a tumour suppressor in the *Drosophila* midgut, as opposed to tumour inducer in mammals (Fre et al., 2005). Similarly, the Wg/Wnt signalling controls fly ISC proliferation and maintenance, although in mammals it also controls ISC differentiation (Crosnier et al., 2006). Further studies are likely to refine the mode of action of the various pathways, but also assess the role of additional pathways in regeneration, as well as the contribution of genetic, environmental and microbial stimuli that act selectively through some of them.

Notwithstanding differences with mammals, a strong message comes from *Drosophila* studies: all midgut cell types studied so far, namely, the EBs, ECs, muscle cells and EEs may serve as delicate sensors for the homeostatic and the damage-induced gut renewal program. Despite intensive efforts to integrate signalling pathways in a meaningful way, for example, hypothesizing that most of them converge on dMyc or the Ca^{2+} oscillation modes within ISCs, the type of damaging agent (e.g. DSS vs bleomycin or *E. carotovora* vs *P. entomophila*) and the protocol of its administration may reveal different aspects of the puzzle.

11. ISCs AND INTESTINAL TUMOURS

Most studies on *Drosophila* carcinogenesis use oncogene expression or loss of function mutations of tumour suppressors to assess tumour development during the larval stage or in adult flies (Gonzalez, 2013). To mimic molecular events similar to those of sporadic tumorigenesis in mammals, clones of *Apc* mutant cells that overexpress the *Ras* oncogene were created (Martorell et al., 2014b). Nevertheless, loss of *Apc* (Wg/Wnt pathway) or *Ras* oncogene (EGFR pathway) expression in the *Drosophila* midgut progenitors suffices to induce tumours (Apidianakis et al., 2009; Lee et al., 2009).

The Wg/Wnt pathway is required for intestinal homeostasis and tumorigenesis in flies, mice and humans (Cordero and Sansom, 2012; Cordero et al., 2012b; Lee et al., 2009; Lin et al., 2008; Pinto et al., 2003). In a 2009 seminal study it was shown that mutations in the homologues of the *APC* gene, *Apc1* and *Apc2*, lead to hyperproliferation of ISCs that induce hyperplasia characterized by multilayering of the epithelium (Lee et al., 2009). Importantly, loss of *Apc1* from ISCs leads to induction of the *Myc* oncogene (Cordero et al., 2012a). *Myc* induction via Wg/Wnt signalling in turn induces the Spitz/EGF ligand cell autonomously in progenitor cells, but also Upd2/3/IL6 nonautonomously in ECs. Both events lead to ISC-mediated hyperplasia (Cordero et al., 2012b). The mechanism of tumour-autonomous Spitz/EGF and tumour niche nonautonomous Upd2/3/IL6 secretion has been corroborated by an independent study on *Notch* loss of function-derived tumours (Patel et al., 2015).

In 2009 another seminal study modelled *Ras* oncogene and the tumour suppressors *Notch* and *dlg* in ISC-mediated intestinal dysplasia (Apidianakis et al., 2009). Concurrently, it was shown that JNK and JAK/STAT pathways are induced in the midgut of *Drosophila* upon intestinal infection (Apidianakis et al., 2009; Buchon et al., 2009a,b; Cronin et al., 2009; Jiang et al., 2009). Strikingly, when infection is combined with a latent *Ras1* oncogenic background, normal regeneration is diverted towards intestinal dysplasia characterized by multilayering of the epithelium, loss of apicobasal polarity, reduction of lumen diameter, enhancement of faecal output and higher mortality (Apidianakis et al., 2009; Christofi and Apidianakis, 2013). Not only *Ras* oncogene expression, but also loss of tumour suppressors exemplified by *Notch* and *dlg*, synergize with infection to induce ISC-emanating tumours (Apidianakis et al., 2009). Nevertheless, midgut *Ras* tumours never become invasive. This was corroborated by an independent assessment of *Ras* oncogene, *Apc* loss and *Ras-Apc* midgut clones, which never become invasive despite the prominent dysplasia caused by the combination of *Ras* and *Apc* mutations (Martorell et al., 2014b). Contrary to midgut, sustained oral infection synergizes with the *Ras1* oncogene to induce basal invasion and dissemination of hindgut cells to distant sites (Bangi et al., 2012). In the hindgut, the *Ras* oncogene suffices to induce EC dissemination, and this effect is boosted by an innate immune response (Bangi et al., 2012).

More recently Perrimon and colleagues developed a model for organ wasting of adult *Drosophila* ovary, fat body and muscle by activating the Hpo pathway target, Yki, in midgut progenitors (Kwon et al., 2015). Wasting was attributed to the secretion of insulin/IGF antagonist ImpL2 from the expanded midgut progenitors and its action on remote organs

(Kwon et al., 2015). A synchronous publication corroborated these results showing that malignant, but not benign, tumour transplantation in *Drosophila* adults induce peripheral wasting of adipose, muscle and gonadal tissues via ImpL2 (Figueroa-Clarevega and Bilder, 2015).

Impact: Because bacterial products, in addition to live bacteria, may promote ISC proliferation (Apidianakis et al., 2009), orally administered bacteria or chemicals can be used to boost the activation and expansion of wild type and tumorous ISCs. For example, Notch-defective ISCs require stress-induced divisions for tumour initiation (Patel et al., 2015). Moreover, the hindgut, as opposed to the midgut epithelium, is amenable to tumour cell invasion and dissemination, suggesting a metastasis-prone signalling network different from that of the midgut. Therefore, the aforementioned studies provide the framework for studying the interplay between intestinal microbes and the host genetic background in intestinal tumorigenesis, metastasis and cancer cachexia.

12. DIETARY CONTROL OF ISCs

Adult SCs are influenced by multiple dietary factors. For example, nutrient-dependent pathways acting within the *Drosophila* ovary control the number and mitosis of germline SCs (Armstrong et al., 2014). Similarly, when food is abundant, the ISC niche produces the insulin-like peptide, Dilp3, leading to higher ISC division rates and predominance of symmetric SC division fates. The net result is an increase in total intestinal cells, which is reversed upon withdrawal of food (O'Brien et al., 2011). Interestingly, EEs serve as an important link between diet and expression of *dilp3* in the SC niche to stimulate ISC proliferation and tissue growth (Amcheslavsky et al., 2014). These results corroborate a study published in 2010, suggesting that flies feeding on a protein-poor diet exhibit reduced germline and intestinal SC proliferation and number. The effect is reversible upon refeeding, indicating that the remaining SCs are competent to respond quickly to changes in nutritional status (McLeod et al., 2010).

Similarly, nutrient deprivation and reduced insulin signalling delay EC differentiation prolonging the contact between ISCs and EBs. Disrupting the ISC-EB contact alleviates defects in ISC proliferation, when insulin signalling is reduced or when nutrients are scarce (Choi et al., 2011). In this study ISC proliferation, but not ISC numbers, appear affected by nutrition. This could be due to technical differences among the aforementioned studies, for example, different methods to enumerate progenitors via Delta protein vs *esg*-positive cells or overlooking the fact that ISC divisions contribute to ISC maintenance. Alternatively, differences in the fly strains genetic backgrounds, diet regimes used and age of flies may account for the apparent discrepancy.

In 2011 *Acetobacter pomorum*, a commensal bacterium of *Drosophila*, was found to modulate insulin signalling via acetic acid production and concomitantly developmental rate, body size, energy metabolism and ISC activity. In the absence of this microbe acetic acid may instead be added in fly food to restore insulin signalling and homeostasis (Shin et al., 2011). Another homeostatic factor is *indy*, the expression of which changes in the midgut upon ageing and nutrition. Indy functions to increase mitochondrial biogenesis, reduce reactive oxygen species and preserve ISC homeostasis via the mitochondrial regulator dPGC-1/Spargel (Rogers and Rogina, 2015).

More recently, dietary L-glutamate was found to stimulate ISC division and gut growth via the metabotropic glutamate receptor, mGluR, which modulates ISC cytosolic Ca^{2+} oscillations to sustain high cytosolic Ca^{2+} concentrations (Deng et al., 2015). High concentrations of Ca^{2+} in the cytosol induce ISC proliferation through Calcineurin and the transcription factor Crtc. Strikingly, not only dietary L-glutamate, but many other stimuli, such as initial response to infection, ageing and oncogene expression, correlate with high cytosolic Ca^{2+} concentrations, suggesting that many ISC proliferating signals go through this novel pathway (Deng et al., 2015).

Finally, Lin-28 is identified as an intrinsic factor highly enriched in ISCs that boosts insulin signalling cell autonomously and promotes ISC symmetric division in response to nutrients. In *lin-28* null mutants ISCs do not expand, have reduced rates of symmetric division and reduced insulin signalling under rich nutrient conditions. Moreover Lin-28 binds to InR mRNAs, suggesting a mechanism as to how it promotes insulin signalling and ISC divisions (Chen et al., 2015).

Impact: The effect of nutrient-rich vs nutrient-poor diet on ISCs was anticipated. The effect of nutrition on insulin signalling, which in turn affects ISCs could also be expected. Nevertheless, the aforementioned studies provide mechanistic aspects on the role of the ISC niche, namely, the visceral muscle and the EEs, the intestinal microbiota and *lin-28* in insulin signalling modulation in response to nutrition. In addition, the role of glutamate and Ca^{2+} pathway provides a novel perspective in nutrition signalling and the integration of signalling of disparate pathways in ISC-mediated homeostasis and disease.

13. MOLECULARLY DEFINED MIDGUT COMPARTMENTALIZATION AND ISCs

The study of the anatomy and cellular architecture of the *Drosophila* intestine reveals two main features, plasticity and segmentation (Lemaitre and Miguel-Aliaga, 2013). Partitioning is an important feature of the digestive tract because it optimizes digestion, allowing the intake and processing of food, absorption of nutrients and elimination of solid waste (Buchon et al., 2013b). The foregut is of ectodermal origin and includes the pharynx, the oesophagus and the crop, which is used to store food. At the foregut–midgut junction resides a specialized structure, the cardia, which acts as a clamp responsible for the regulation of food passage. The midgut derives from the endoderm and it is the main area of digestion and absorption of nutrients. Finally, the posterior intestine, the hindgut, is of ectodermal origin and is the main area of reabsorption of water and the concentration of faeces before elimination (Cognigni et al., 2011).

Interestingly, within the midgut, which is the most studied intestinal part of the adult fly, there are three main areas with distinct morphological and functional features: the anterior midgut, the middle midgut that encompasses a low pH region known as the area of iron/copper cells and the posterior midgut (Li et al., 2013a). The posterior midgut differs from the anterior immunologically, because the downstream targets of the Imd/NF-kB pathway, the antimicrobial peptides, are suppressed by Caudal in the posterior, while other factors act predominantly in the anterior to suppress this pathway. Strikingly, in 2013 evidence indicated that the adult *Drosophila* midgut could be divided into six main regions and 14 subregions with distinct morphological, histological and genetic properties (Buchon et al., 2013b; Marianes and Spradling, 2013). Histologically the midgut can be separated in 9 sequential subregions. R0 corresponds to the cardia or proventriculus, a pear-shaped structure that contains both ectodermally derived foregut cells and endodermally derived midgut cells (King, 1988; Singh et al., 2011). R1a, which is composed of flat ECs, is additionally characterized by the large lumen and a multilayered peritrophic membrane. R1b is characterized by a highly folded lumen and contains long ECs with extended membrane infoldings (labyrinth). R2 is composed of columnar ECs with an apical extrusion containing lipid vesicles. R3ab is composed mainly of copper cells and interstitial cells. Copper cells have a basally located nucleus and a deeply invaginated apical membrane covered with long microvilli.

Interstitial cells have a more apically localized nucleus, short microvilli and a broad apical surface restricting that of adjacent copper cells (Shanbhag and Tripathi, 2009). R3c is composed of large flat cells (Strand and Micchelli, 2011). R4a, which is characterized by a highly folded lumen, is composed of ECs organized in a monolayer, sporadically in a multilayer and even forming villi reminiscent of those of the mammalian small intestine. R4bc is composed of ECs with small apical protrusions. R5 contains fewer cells with smooth brush border.

Not surprisingly, genetic disruption of such an elaborate compartmentalization leads to loss of intestinal homeostasis characterized by the increased ISC proliferation (Buchon et al., 2013b; Marianes and Spradling, 2013).

Impact: Due to the distinct features of each part of the midgut, it is now necessary to study cellular and molecular parameters in specific parts of the midgut. Future studies should also compare findings among midgut parts and how the whole intestine integrates such differences to optimize its function. For example, the anterior, the middle and the posterior midgut confer different predisposition for dysplasia induced by *Ras* oncogene and *Apc* mutant clones (Martorell et al., 2014b).

14. DRUG TESTING AND *DROSOPHILA* ISCs

A major problem of chemotherapy is the resistance of cancer cells to the treatment due to preexisting mutations or mutations that accumulate over time (Foo and Michor, 2014) and the presence of cancer stem cells that regenerate the tumour (Foo and Michor, 2014). Drug discovery usually relies on cell lines or primary cells (Winquist et al., 2010), which do not fully reflect human biology because they do not recapitulate the tumour microenvironment (Lander et al., 2012). It is thus necessary to utilize new approaches involving tumour microenvironment, such as model organisms or organoids (Markstein, 2013). In 2014 an array of chemotherapy and other drugs were tested for their effect on tumours emanating from midgut ISCs expressing the human oncogene *Raf* (Markstein et al., 2014). The authors identified two categories of effective drugs: those that reduced tumours without any side effects, such as Rapamycin, Thiotepa, Floxuridine, Topotecan, Methotrexate and Gemcitabine; and drugs, such as Bortezomib, Daunorubicine, Mitomycin C, Paclitaxel, Vinblastine and D-actinomycin, that while effective against tumours, induce damage/stress to the ECs, JAK/STAT pathway activation and concomitant ISC-mediated midgut regeneration (Markstein et al., 2014). The latter class of drugs is more likely to affect tumour microenvironment or tumours per se potentially contributing to tumour recurrence following chemotherapy (Markstein et al., 2014).

Another important field of *Drosophila* ISC research focuses on the relationship between ageing and the appearance of cancer. Lifespan can be extended pharmacologically or by genetic manipulation, targeting, for example, the InR or the TOR pathway (Ewald et al., 2015; Nuzzo et al., 2015; Slack et al., 2015). Dietary Rapamycin can inhibit the TOR pathway providing a means to alter the ageing process and potentially cancer immergence (Bjedov et al., 2010). In a recent study the effect of Rapamycin was assessed in *Drosophila* gut homeostasis during ageing (Fan et al., 2015). Old flies develop intestinal dysplasia and die faster due to microbiota deregulation (Guo et al., 2014). Interestingly, Rapamycin slows down the intestinal deterioration by limiting the ISC proliferation rate and by delaying the microbial expansion and intestinal barrier dysfunction during ageing. At the molecular level, Rapamycin inhibits TOR to activate autophagy in the ageing guts (Fan et al., 2015). In addition, Rapamycin reduces the Imd-Rel-mediated immune response in the ageing gut by inducing the negative regulators of this pathway, namely, Caudal, USP36 and PGRP-SC. Similarly, expression of *duox,* which induces ROS production, is reduced in the presence of Rapamycin, limiting the levels of ROS and concomitant intestinal dysplasia (Fan et al., 2015).

Metformin was also specifically studied in 2013 and 2015 in relation to ISCs of *Drosophila* (Na et al., 2013, 2015). Metformin is implicated in cancer due its antiproliferative effects (Alimova et al., 2009; Ashinuma et al., 2012; Cantrell et al., 2010) via the inhibition of the TOR pathway (Na et al., 2015). The Yoo group published a series of reports on Metformin. In 2012 it was reported that foci of γH2AvD, the *Drosophila* orthologue of the mammalian DNA double strand break marker γH2AX, and the oxidative DNA damage marker 8-oxo-dG accumulates with age and in response to oxidative stress in ISCs (Park et al., 2012). In 2013 the group reported that Metformin inhibits age-related accumulation of DNA damage in *Drosophila* ISCs and concomitant intestinal hyperplasia (Na et al., 2013). In 2014 they reported that centrosome amplification is increased in ISCs upon ageing or oxidative stress (Park et al., 2014). And in 2015 Metformin was shown to prevent age- and oxidative stress-induced centrosome amplification in *Drosophila* midgut ISCs via the downregulation of the AKT/TOR signalling pathway.

Impact: A key question in drug discovery using flies is whether chemical screens in *Drosophila* can identify compounds that are effective in mammals. While this is a recurrent and reasonable question because the *Drosophila* drug discovery platforms are in their infancy, the opposite appears to be true: compounds originally identified in mammalian cells are increasingly shown to have the same molecular mechanism of action in *Drosophila* (Fernandez-Hernandez et al., 2016). Due to the easy and sophisticated genetic manipulation of ISCs and their microenvironment and the relatively easy delivery of orally administered drugs to the *Drosophila* intestinal epithelium (Tzelepis et al., 2013), more and more studies are likely to take advantage of the *Drosophila* intestinal epithelium as a model for low- or large-scale drug screening.

15. HOW TO KILL ISCs

Jiang and colleagues first noted in 2009 that ISCs could be ablated only partially upon expression of *rpr*, *hid* or *p53* for 20 days using the *esg-Gal4* driver (Jiang et al., 2009), suggesting that ISCs are quite resistant to apoptosis. An alternative way to deplete ISCs is to induce their differentiation via Notch signalling overactivation, for example, by expressing a constitutively active Notch intracellular domain in the midgut progenitors (Micchelli and Perrimon, 2006; Ohlstein and Spradling, 2006). A more efficient but artificial way to kill all progenitor cells is to express a short hairpin RNA against *prickle* (line# HMS00408) for 7 days esg-Gal4 at 29°C (Lu and Li, 2015).

In 2015 a mechanistic study corroborated previous studies about adult *Drosophila* midgut ISCs being resistant to various cellular damages. Strikingly though, Ras oncogene-expressing ISCs, unlike wild-type ISCs, are easily eliminated by apoptosis. In addition, tumour ISC proliferation is inhibited by autophagy, which acts independently from caspase-dependent apoptosis. Instead, inhibition of tumour-ISCs by autophagy is likely through the sequestration and degradation of mitochondria (Ma et al., 2016).

More recently necrosis was found to kill wild-type and tumourous ISCs. COPI–Arf1 complex knockdown attenuates the lipolysis pathway and selectively kills wild-type and *Ras1* oncogene-expressing ISCs through necrosis, sparing all differentiated cells (Singh et al., 2016). Interestingly, neighbouring ECs engulf dying ISCs by inducing a *draper-myoblast city-Rac1-JNK*-dependent autophagy pathway (Singh et al., 2016).

Impact: While killing ISCs provides a means to study their regulation and impact on tissue homeostasis, killing tumourous ISCs is more relevant to translational research. Mammalian tumour cells can be insensitive to apoptosis, and this is a hallmark of cancer. However, the above studies indicate that tumour SCs can be very effectively eliminated by the induction of autophagy or by induction of the lipolysis pathway. Since many other signals might be able to kill tumour ISCs, it will be interesting to find those that selectively kill cancer-like SCs exhibiting some or all the hallmarks of cancer, including insensitivity to apoptosis.

16. SEXUAL IDENTITY OF ISCs

Do *Drosophila* SCs have a sexual identity? Do male and female SCs respond differently to infection or stress? Is there a need to customize treatments for males vs females regarding intestinal inflammation and cancer? Two recent studies on *Drosophila* ISCs deal with these questions (Hudry et al., 2016; Regan et al., 2016). Using a combination of genetic techniques, Hudry et al. (2016) recorded and altered the expression of gender-specific genes (*Sex lethal, transformer* and others) in the ISCs and progenitors of the *Drosophila* midgut changing the rate of ISC proliferation in a gender-specific manner, that is, following the principle that "female" SCs proliferate more (Hudry et al., 2016). Regan et al. (2016) achieved "feminization" of males by misexpressing the female-specific spliceform of *transformer* (traF) in midgut EBs and ECs cells (Regan et al., 2016). Interestingly, midgut feminization and concomitant tissue enlargement is reversible upon restoration of expression of key genes (Hudry et al., 2016). Moreover, there is a trade-off between ISC-mediated regeneration and midgut pathology in old flies. In females, the midgut epithelium undergoes major deterioration upon ageing, propelled by ISC mitosis, compared to males that exhibit lower ISC mitosis. Nevertheless, males are prone to leakiness of the gut and succumb easier to intestinal stress and infection (Regan et al., 2016).

Impact: Until now sexual differences in animal organs were attributed to different circulating hormones shaping the body during development and adulthood. Contrary to this notion, the aforementioned studies attribute differences in midgut size and ISC proliferation in sexual identity genes. Curiously enough sex genes were altered in ISCs and EBs in one study (Hudry et al., 2016), but in EBs and mature ECs in the other (Regan et al., 2016). Nevertheless, both studies identify sex genes acting as regulators of ISC proliferation, and raise the possibility that similar or other genes confer sex-biased characteristics in a tissue-intrinsic manner in other fly organs and other organisms.

17. CONCLUSIONS

ISCs are found in all deuterostomes, such as echinoderms (e.g. starfish) and chordates (e.g. vertebrates), but not in other phyla of the protostomes except arthropods (e.g. insects). Thus, the ISCs of *Drosophila* and those of humans must be the result of convergent evolution. Accordingly, similarities are expected in terms of the effect of ageing and environmental factors, such as diet, on ISCs, but also differences at the molecular and cellular level between flies and humans. For example, the Notch, Wg/Wnt and Hh pathways signal among the ISC lineage cells, but not always in a similar way between flies and mammals. Studies on the development of the *Drosophila* GI tract can also be very informative, for example, the differences between the adult mesoderm-derived midgut and ectoderm-derived hindgut in ISC quiescence and tumour cell invasion and dissemination. The wealth of conserved signalling pathways and secreted ligands, the conserved mechanism of action of medical drugs in flies, the role of regenerative inflammation in host defence and the role of oncogenes and tumour suppressors in ISC-driven tumorigenesis are some of the examples that validate the use of *Drosophila* in biomedical research. The studies so far reveal many novel aspects of ISC function, such as the cell intrinsic sexual identity, the importance of gut compartmentalization, signals controlling ISC death, survival, proliferation and differentiation, signalling integration and the role of ISCs in healthy lifespan. Future studies are expected to further explore basic parameters of ISC function and the use of *Drosophila* ISCs in translational research.

REFERENCES

Aghajanian, P., Takashima, S., Paul, M., Younossi-Hartenstein, A., Hartenstein, V., 2016. Metamorphosis of the *Drosophila* visceral musculature and its role in intestinal morphogenesis and stem cell formation. Dev. Biol. 420, 43–59.

Albrecht, S.C., Barata, A.G., Grosshans, J., Teleman, A.A., Dick, T.P., 2011. In vivo mapping of hydrogen peroxide and oxidized glutathione reveals chemical and regional specificity of redox homeostasis. Cell Metab. 14, 819–829.

Alimova, I.N., Liu, B., Fan, Z., Edgerton, S.M., Dillon, T., Lind, S.E., Thor, A.D., 2009. Metformin inhibits breast cancer cell growth, colony formation and induces cell cycle arrest in vitro. Cell Cycle 8, 909–915.

Amcheslavsky, A., Jiang, J., Ip, Y.T., 2009. Tissue damage-induced intestinal stem cell division in *Drosophila*. Cell Stem Cell 4, 49–61.

Amcheslavsky, A., Ito, N., Jiang, J., Ip, Y.T., 2011. Tuberous sclerosis complex and Myc coordinate the growth and division of *Drosophila* intestinal stem cells. J. Cell Biol. 193, 695–710.

Amcheslavsky, A., Song, W., Li, Q., Nie, Y., Bragatto, I., Ferrandon, D., Perrimon, N., Ip, Y.T., 2014. Enteroendocrine cells support intestinal stem-cell-mediated homeostasis in *Drosophila*. Cell Rep. 9, 32–39.

Apidianakis, Y., Pitsouli, C., Perrimon, N., Rahme, L., 2009. Synergy between bacterial infection and genetic predisposition in intestinal dysplasia. Proc. Natl. Acad. Sci. U.S.A. 106, 20883–20888.

Armstrong, A.R., Laws, K.M., Drummond-Barbosa, D., 2014. Adipocyte amino acid sensing controls adult germline stem cell number via the amino acid response pathway and independently of target of rapamycin signaling in *Drosophila*. Development 141, 4479–4488.

Ashinuma, H., Takiguchi, Y., Kitazono, S., Kitazono-Saitoh, M., Kitamura, A., Chiba, T., Tada, Y., Kurosu, K., Sakaida, E., Sekine, I., Tanabe, N., Iwama, A., Yokosuka, O., Tatsumi, K., 2012. Antiproliferative action of metformin in human lung cancer cell lines. Oncol. Rep. 28, 8–14.

Ayyaz, A., Li, H., Jasper, H., 2015. Haemocytes control stem cell activity in the *Drosophila* intestine. Nat. Cell Biol. 17, 736–748.

Baldwin, K.M., Hakim, R.S., 1991. Growth and differentiation of the larval midgut epithelium during molting in the moth, *Manduca sexta*. Tissue Cell 23, 411–422.

Bangi, E., Pitsouli, C., Rahme, L.G., Cagan, R., Apidianakis, Y., 2012. Immune response to bacteria induces dissemination of Ras-activated *Drosophila* hindgut cells. EMBO Rep. 13, 569–576.

Bardin, A.J., Perdigoto, C.N., Southall, T.D., Brand, A.H., Schweisguth, F., 2010. Transcriptional control of stem cell maintenance in the *Drosophila* intestine. Development 137, 705–714.

Baumann, O., 2001. Posterior midgut epithelial cells differ in their organization of the membrane skeleton from other *Drosophila* epithelia. Exp. Cell Res. 270, 176–187.

Beehler-Evans, R., Micchelli, C.A., 2015. Generation of enteroendocrine cell diversity in midgut stem cell lineages. Development 142 (4), 654–664.

Belenkaya, T.Y., Wu, Y., Tang, X., Zhou, B., Cheng, L., Sharma, Y.V., Yan, D., Selva, E.M., Lin, X., 2008. The retromer complex influences Wnt secretion by recycling wntless from endosomes to the trans-Golgi network. Dev. Cell 14, 120–131.

Bely, A.E., 2014. Early events in annelid regeneration: a cellular perspective. Integr. Comp. Biol. 54, 688–699.

Biteau, B., Jasper, H., 2011. EGF signaling regulates the proliferation of intestinal stem cells in *Drosophila*. Development 138, 1045–1055.

Biteau, B., Jasper, H., 2014. Slit/Robo signaling regulates cell fate decisions in the intestinal stem cell lineage of *Drosophila*. Cell Rep. 7, 1867–1875.

Biteau, B., Hochmuth, C.E., Jasper, H., 2008. Jnk activity in somatic stem cells causes loss of tissue homeostasis in the aging *Drosophila* gut. Cell Stem Cell 3, 442–455.

Biteau, B., Karpac, J., Supoyo, S., Degennaro, M., Lehmann, R., Jasper, H., 2010. Lifespan extension by preserving proliferative homeostasis in *Drosophila*. PLoS Genet. 6, e1001159.

Bizzozero, G., 1893. Ueber die schlauchförmigen Drüsen des Magendarmkanals und die Beziehungen ihres Epithels zu dem Oberflächenepithel der Schleimhaut Dritte Mittheilung. Arch. Mikrosk. Anat. 42, 82–152.

Bjedov, I., Toivonen, J.M., Kerr, F., Slack, C., Jacobson, J., Foley, A., Partridge, L., 2010. Mechanisms of life span extension by rapamycin in the fruit fly *Drosophila melanogaster*. Cell Metab. 11, 35–46.

Bozcuk, A.N., 1972. DNA synthesis in the absence of somatic cell division associated with ageing in *Drosophila subobscura*. Exp. Gerontol. 7, 147–150.

Buchon, N., Broderick, N.A., Chakrabarti, S., Lemaitre, B., 2009a. Invasive and indigenous microbiota impact intestinal stem cell activity through multiple pathways in *Drosophila*. Genes Dev. 23, 2333–2344.

Buchon, N., Broderick, N.A., Poidevin, M., Pradervand, S., Lemaitre, B., 2009b. *Drosophila* intestinal response to bacterial infection: activation of host defense and stem cell proliferation. Cell Host Microbe 5, 200–211.
Buchon, N., Broderick, N.A., Kuraishi, T., Lemaitre, B., 2010. *Drosophila* EGFR pathway coordinates stem cell proliferation and gut remodeling following infection. BMC Biol. 8, 152.
Buchon, N., Broderick, N.A., Lemaitre, B., 2013a. Gut homeostasis in a microbial world: insights from *Drosophila melanogaster*. Nat. Rev. Microbiol. 11, 615–626.
Buchon, N., Osman, D., David, F.P., Fang, H.Y., Boquete, J.P., Deplancke, B., Lemaitre, B., 2013b. Morphological and molecular characterization of adult midgut compartmentalization in *Drosophila*. Cell Rep. 3, 1725–1738.
Cantrell, L.A., Zhou, C., Mendivil, A., Malloy, K.M., Gehrig, P.A., Bae-Jump, V.L., 2010. Metformin is a potent inhibitor of endometrial cancer cell proliferation—implications for a novel treatment strategy. Gynecol. Oncol. 116, 92–98.
Chakrabarti, S., Liehl, P., Buchon, N., Lemaitre, B., 2012. Infection-induced host translational blockage inhibits immune responses and epithelial renewal in the *Drosophila* gut. Cell Host Microbe 12, 60–70.
Chakrabarti, S., Dudzic, J.P., Li, X., Collas, E.J., Boquete, J.P., Lemaitre, B., 2016. Remote control of intestinal stem cell activity by haemocytes in *Drosophila*. PLoS Genet. 12, e1006089.
Chen, H., Zheng, X., Zheng, Y., 2014. Age-associated loss of lamin-B leads to systemic inflammation and gut hyperplasia. Cell 159, 829–843.
Chen, C.H., Luhur, A., Sokol, N., 2015. Lin-28 promotes symmetric stem cell division and drives adaptive growth in the adult *Drosophila* intestine. Development 142, 3478–3487.
Chen, H.J., Wei, Z., Sun, J., Bhattacharya, A., Savage, D.J., Serda, R., Mackeyev, Y., Curley, S.A., Bu, P., Wang, L., Chen, S., Cohen-Gould, L., Huang, E., Shen, X., Lipkin, S.M., Copeland, N.G., Jenkins, N.A., Shuler, M.L., 2016. A recellularized human colon model identifies cancer driver genes. Nat. Biotechnol. 34 (8), 845.
Cheng, H., Leblond, C.P., 1974. Origin, differentiation and renewal of the four main epithelial cell types in the mouse small intestine. V. Unitarian theory of the origin of the four epithelial cell types. Am. J. Anat. 141, 537–561.
Choi, N.H., Kim, J.G., Yang, D.J., Kim, Y.S., Yoo, M.A., 2008. Age-related changes in *Drosophila* midgut are associated with PVF2, a PDGF/VEGF-like growth factor. Aging Cell 7, 318–334.
Choi, N.H., Lucchetta, E., Ohlstein, B., 2011. Nonautonomous regulation of *Drosophila* midgut stem cell proliferation by the insulin-signaling pathway. Proc. Natl. Acad. Sci. U.S.A. 108, 18702–18707.
Christofi, T., Apidianakis, Y., 2013. Ras-oncogenic *Drosophila* hindgut but not midgut cells use an inflammation-like program to disseminate to distant sites. Gut Microbes 4, 54–59.
Christofi, T., Apidianakis, Y., 2015. Parallels between mammals and flies in inflammatory bowel disease. In: Vaiserman, A.M., Moskalev, A.A., Pasyukova, E.G. (Eds.), Life Extension: Lessons from *Drosophila*. Springer International Publishing, Switzerland.
Clevers, H., 2013. The intestinal crypt, a prototype stem cell compartment. Cell 154, 274–284.
Cognigni, P., Bailey, A.P., Miguel-Aliaga, I., 2011. Enteric neurons and systemic signals couple nutritional and reproductive status with intestinal homeostasis. Cell Metab. 13, 92–104.
Cordero, J.B., Sansom, O.J., 2012. Wnt signalling and its role in stem cell-driven intestinal regeneration and hyperplasia. Acta Physiol. (Oxf.) 204, 137–143.
Cordero, J.B., Stefanatos, R.K., Myant, K., Vidal, M., Sansom, O.J., 2012a. Non-autonomous crosstalk between the Jak/Stat and Egfr pathways mediates Apc1-driven intestinal stem cell hyperplasia in the *Drosophila* adult midgut. Development 139, 4524–4535.

Cordero, J.B., Stefanatos, R.K., Scopelliti, A., Vidal, M., Sansom, O.J., 2012b. Inducible progenitor-derived Wingless regulates adult midgut regeneration in *Drosophila*. EMBO J. 31, 3901–3917.
Cronin, S.J., Nehme, N.T., Limmer, S., Liegeois, S., Pospisilik, J.A., Schramek, D., Leibbrandt, A., Simoes Rde, M., Gruber, S., Puc, U., Ebersberger, I., Zoranovic, T., Neely, G.G., von Haeseler, A., Ferrandon, D., Penninger, J.M., 2009. Genome-wide RNAi screen identifies genes involved in intestinal pathogenic bacterial infection. Science 325, 340–343.
Crosnier, C., Stamataki, D., Lewis, J., 2006. Organizing cell renewal in the intestine: stem cells, signals and combinatorial control. Nat. Rev. Genet. 7, 349–359.
Day, M., Powning, R., 1949. A study of the processes of digestion in certain insects. Aust. J. Biol. Sci. 2, 175–215.
De Navascues, J., Perdigoto, C.N., Bian, Y., Schneider, M.H., Bardin, A.J., Martinez-Arias, A., Simons, B.D., 2012. *Drosophila* midgut homeostasis involves neutral competition between symmetrically dividing intestinal stem cells. EMBO J. 31, 2473–2485.
Demerec, M., 1950. Biology of Drosophila. Hafner Pub. Co., New York; London.
Deng, H., Gerencser, A.A., Jasper, H., 2015. Signal integration by Ca(2+) regulates intestinal stem-cell activity. Nature 528, 212–217.
Ewald, C.Y., Landis, J.N., Porter Abate, J., Murphy, C.T., Blackwell, T.K., 2015. Dauer-independent insulin/IGF-1-signalling implicates collagen remodelling in longevity. Nature 519, 97–101.
Fan, X., Liang, Q., Lian, T., Wu, Q., Gaur, U., Li, D., Yang, D., Mao, X., Jin, Z., Li, Y., Yang, M., 2015. Rapamycin preserves gut homeostasis during *Drosophila* aging. Oncotarget 6, 35274–35283.
Fernandez-Hernandez, I., Scheenaard, E., Pollarolo, G., Gonzalez, C., 2016. The translational relevance of *Drosophila* in drug discovery. EMBO Rep. 17, 471–472.
Figueroa-Clarevega, A., Bilder, D., 2015. Malignant *Drosophila* tumors interrupt insulin signaling to induce cachexia-like wasting. Dev. Cell 33, 47–55.
Foo, J., Michor, F., 2014. Evolution of acquired resistance to anti-cancer therapy. J. Theor. Biol. 355, 10–20.
Fox, D.T., Spradling, A.C., 2009. The *Drosophila* hindgut lacks constitutively active adult stem cells but proliferates in response to tissue damage. Cell Stem Cell 5, 290–297.
Fre, S., Huyghe, M., Mourikis, P., Robine, S., Louvard, D., Artavanis-Tsakonas, S., 2005. Notch signals control the fate of immature progenitor cells in the intestine. Nature 435, 964–968.
Gonzalez, C., 2013. *Drosophila melanogaster*: a model and a tool to investigate malignancy and identify new therapeutics. Nat. Rev. Cancer 13, 172–183.
Guo, Z., Driver, I., Ohlstein, B., 2013. Injury-induced BMP signaling negatively regulates *Drosophila* midgut homeostasis. J. Cell Biol. 201, 945–961.
Guo, L., Karpac, J., Tran, S.L., Jasper, H., 2014. PGRP-SC2 promotes gut immune homeostasis to limit commensal dysbiosis and extend lifespan. Cell 156, 109–122.
Guo, Z., Lucchetta, E., Rafel, N., Ohlstein, B., 2016. Maintenance of the adult *Drosophila* intestine: all roads lead to homeostasis. Curr. Opin. Genet. Dev. 40, 81–86.
Hartenstein, V., 1993. Atlas of Drosophila Development. Cold Spring Harbor Laboratory Press, Cold Spring Harbor, NY.
Hartenstein, V., Jan, Y.N., 1992. Studying *Drosophila* embryogenesis with P-lacZ enhancer trap lines. Rouxs Arch. Dev. Biol. 201, 194–220.
Hudry, B., Khadayate, S., Miguel-Aliaga, I., 2016. The sexual identity of adult intestinal stem cells controls organ size and plasticity. Nature 530, 344–348.
Hur, J.H., Bahadorani, S., Graniel, J., Koehler, C.L., Ulgherait, M., Rera, M., Jones, D.L., Walker, D.W., 2013. Increased longevity mediated by yeast NDI1 expression in *Drosophila* intestinal stem and progenitor cells. Aging (Albany NY) 5, 662–681.

Jiang, H., Edgar, B.A., 2009. EGFR signaling regulates the proliferation of *Drosophila* adult midgut progenitors. Development 136, 483–493.
Jiang, H., Edgar, B.A., 2011. Intestinal stem cells in the adult *Drosophila* midgut. Exp. Cell Res. 317, 2780–2788.
Jiang, H., Patel, P.H., Kohlmaier, A., Grenley, M.O., McEwen, D.G., Edgar, B.A., 2009. Cytokine/Jak/Stat signaling mediates regeneration and homeostasis in the *Drosophila* midgut. Cell 137, 1343–1355.
Jiang, H., Grenley, M.O., Bravo, M.J., Blumhagen, R.Z., Edgar, B.A., 2011. EGFR/Ras/MAPK signaling mediates adult midgut epithelial homeostasis and regeneration in *Drosophila*. Cell Stem Cell 8, 84–95.
Jones, R.M., Luo, L., Ardita, C.S., Richardson, A.N., Kwon, Y.M., Mercante, J.W., Alam, A., Gates, C.L., Wu, H., Swanson, P.A., Lambeth, J.D., Denning, P.W., Neish, A.S., 2013. Symbiotic lactobacilli stimulate gut epithelial proliferation via Nox-mediated generation of reactive oxygen species. EMBO J. 32, 3017–3028.
Karin, M., Clevers, H., 2016. Reparative inflammation takes charge of tissue regeneration. Nature 529, 307–315.
Karpowicz, P., Perez, J., Perrimon, N., 2010. The Hippo tumor suppressor pathway regulates intestinal stem cell regeneration. Development 137, 4135–4145.
King, D.G., 1988. Cellular organization and peritrophic membrane formation in the cardia (proventriculus) of *Drosophila melanogaster*. J. Morphol. 196 (3), 253–282.
Klapper, R., 2000. The longitudinal visceral musculature of *Drosophila melanogaster* persists through metamorphosis. Mech. Dev. 95 (1–2), 47–54.
Kocks, C., Cho, J.H., Nehme, N., Ulvila, J., Pearson, A.M., Meister, M., Strom, C., Conto, S.L., Hetru, C., Stuart, L.M., Stehle, T., Hoffmann, J.A., Reichhart, J.M., Ferrandon, D., Ramet, M., Ezekowitz, R.A., 2005. Eater, a transmembrane protein mediating phagocytosis of bacterial pathogens in *Drosophila*. Cell 123, 335–346.
Kux, K., Pitsouli, C., 2014. Tissue communication in regenerative inflammatory signaling: lessons from the fly gut. Front. Cell. Infect. Microbiol. 4, 49.
Kwon, H.J., Waghmare, I., Verghese, S., Singh, A., Kango-Singh, M., 2015. *Drosophila* C-terminal Src kinase regulates growth via the Hippo signaling pathway. Dev. Biol. 397, 67–76.
Lander, A.D., Kimble, J., Clevers, H., Fuchs, E., Montarras, D., Buckingham, M., Calof, A.L., Trumpp, A., Oskarsson, T., 2012. What does the concept of the stem cell niche really mean today? BMC Biol. 10, 19.
Lee, W.C., Beebe, K., Sudmeier, L., Micchelli, C.A., 2009. Adenomatous polyposis coli regulates *Drosophila* intestinal stem cell proliferation. Development 136, 2255–2264.
Lemaitre, B., Hoffmann, J., 2007. The host defense of *Drosophila melanogaster*. Annu. Rev. Immunol. 25, 697–743.
Lemaitre, B., Miguel-Aliaga, I., 2013. The digestive tract of *Drosophila melanogaster*. Annu. Rev. Genet. 47, 377–404.
Li, H., Qi, Y., Jasper, H., 2013a. Dpp signaling determines regional stem cell identity in the regenerating adult *Drosophila* gastrointestinal tract. Cell Rep. 4, 10–18.
Li, Z., Zhang, Y., Han, L., Shi, L., Lin, X., 2013b. Trachea-derived dpp controls adult midgut homeostasis in *Drosophila*. Dev. Cell 24, 133–143.
Li, Z., Guo, Y., Han, L., Zhang, Y., Shi, L., Huang, X., Lin, X., 2014a. Debra-mediated Ci degradation controls tissue homeostasis in *Drosophila* adult midgut. Stem Cell Rep. 2, 135–144.
Li, Z., Liu, S., Cai, Y., 2014b. Differential Notch activity is required for homeostasis of malpighian tubules in adult *Drosophila*. J. Genet. Genomics 41, 649–652.
Li, Z., Liu, S., Cai, Y., 2015. EGFR/MAPK signaling regulates the proliferation of *Drosophila* renal and nephric stem cells. J. Genet. Genomics 42, 9–20.

Li, H., Qi, Y., Jasper, H., 2016. Preventing age-related decline of gut compartmentalization limits microbiota dysbiosis and extends lifespan. Cell Host Microbe 19, 240–253.

Limmer, S., Haller, S., Drenkard, E., Lee, J., Yu, S., Kocks, C., Ausubel, F.M., Ferrandon, D., 2011. *Pseudomonas aeruginosa* RhlR is required to neutralize the cellular immune response in a *Drosophila melanogaster* oral infection model. Proc. Natl. Acad. Sci. U.S.A. 108, 17378–17383.

Lin, G., Xu, N., Xi, R., 2008. Paracrine Wingless signalling controls self-renewal of *Drosophila* intestinal stem cells. Nature 455, 1119–1123.

Linneweber, G.A., Jacobson, J., Busch, K.E., Hudry, B., Christov, C.P., Dormann, D., Yuan, M., Otani, T., Knust, E., De Bono, M., Miguel-Aliaga, I., 2014. Neuronal control of metabolism through nutrient-dependent modulation of tracheal branching. Cell 156, 69–83.

Lu, Y., Li, Z., 2015. No intestinal stem cell regeneration after complete progenitor ablation in *Drosophila* adult midgut. J. Genet. Genomics 42, 83–86.

Ma, M., Zhao, H., Binari, R., Perrimon, N., Li, Z., 2016. Wildtype adult stem cells, unlike tumor cells, are resistant to cellular damages in *Drosophila*. Dev. Biol. 411, 207–216.

Marianes, A., Spradling, A.C., 2013. Physiological and stem cell compartmentalization within the *Drosophila* midgut. Elife 2. e00886.

Markstein, M., 2013. Modeling colorectal cancer as a 3-dimensional disease in a dish: the case for drug screening using organoids, zebrafish, and fruit flies. Drug Discov. Today Technol. 10, e73–e81.

Markstein, M., Dettorre, S., Cho, J., Neumuller, R.A., Craig-Muller, S., Perrimon, N., 2014. Systematic screen of chemotherapeutics in *Drosophila* stem cell tumors. Proc. Natl. Acad. Sci. U.S.A. 111, 4530–4535.

Martorell, O., Barriga, F.M., Merlos-Suarez, A., Stephan-Otto Attolini, C., Casanova, J., Batlle, E., Sancho, E., Casali, A., 2014a. Iro/IRX transcription factors negatively regulate Dpp/TGF-beta pathway activity during intestinal tumorigenesis. EMBO Rep. 15, 1210–1218.

Martorell, O., Merlos-Suarez, A., Campbell, K., Barriga, F.M., Christov, C.P., Miguel-Aliaga, I., Batlle, E., Casanova, J., Casali, A., 2014b. Conserved mechanisms of tumorigenesis in the *Drosophila* adult midgut. PLoS One 9, e88413.

Mashanov, V.S., Dolmatov, I.Y., Heinzeller, T., 2005. Transdifferentiation in holothurian gut regeneration. Biol. Bull. 209, 184–193.

Mashanov, V.S., Zueva, O., Garcia-Arraras, J.E., 2014a. Postembryonic organogenesis of the digestive tube: why does it occur in worms and sea cucumbers but fail in humans? Curr. Top. Dev. Biol. 108, 185–216.

Mashanov, V.S., Zueva, O.R., GarcÍA-ArrarÁS, J.E., 2014b. Transcriptomic changes during regeneration of the central nervous system in an echinoderm. BMC Genomics 15, 357.

Mashanov, V.S., Zueva, O.R., García-Arrarás, J.E., 2015. Expression of pluripotency factors in echinoderm regeneration. Cell Tissue Res. 359, 521–536.

Mathur, D., Bost, A., Driver, I., Ohlstein, B., 2010. A transient niche regulates the specification of *Drosophila* intestinal stem cells. Science 327, 210–213.

McLeod, C.J., Wang, L., Wong, C., Jones, D.L., 2010. Stem cell dynamics in response to nutrient availability. Curr. Biol. 20, 2100–2105.

Micchelli, C.A., 2012. The origin of intestinal stem cells in *Drosophila*. Dev. Dyn. 241, 85–91.

Micchelli, C.A., Perrimon, N., 2006. Evidence that stem cells reside in the adult *Drosophila* midgut epithelium. Nature 439, 475–479.

Micchelli, C.A., Sudmeier, L., Perrimon, N., Tang, S., Beehler-Evans, R., 2011. Identification of adult midgut precursors in *Drosophila*. Gene Expr. Patterns 11, 12–21.

Miller, A., 1950. The internal anatomy and histology of the imago of *Drosophila melanogaster*. In: Demerec, M. (Ed.), Biology of Drosophila. Cold Spring Harbor Laboratory Press, Cold Spring Harbor, NY, pp. 420–534.

Moreira, C.G., Regan, J.C., Zaidman-Remy, A., Jacinto, A., Prag, S., 2011. *Drosophila* hemocyte migration: an in vivo assay for directional cell migration. Methods Mol. Biol. 769, 249–260.

Na, H.J., Park, J.S., Pyo, J.H., Lee, S.H., Jeon, H.J., Kim, Y.S., Yoo, M.A., 2013. Mechanism of metformin: inhibition of DNA damage and proliferative activity in *Drosophila* midgut stem cell. Mech. Ageing Dev. 134, 381–390.

Na, H.J., Park, J.S., Pyo, J.H., Jeon, H.J., Kim, Y.S., Arking, R., Yoo, M.A., 2015. Metformin inhibits age-related centrosome amplification in *Drosophila* midgut stem cells through AKT/TOR pathway. Mech. Ageing Dev. 149, 8–18.

Nehme, N.T., Liegeois, S., Kele, B., Giammarinaro, P., Pradel, E., Hoffmann, J.A., Ewbank, J.J., Ferrandon, D., 2007. A model of bacterial intestinal infections in *Drosophila melanogaster*. PLoS Pathog. 3, e173.

Nuzzo, D., Picone, P., Baldassano, S., Caruana, L., Messina, E., Marino Gammazza, A., Cappello, F., Mule, F., Di Carlo, M., 2015. Insulin resistance as common molecular denominator linking obesity to Alzheimer's disease. Curr. Alzheimer Res. 12, 723–735.

O'Brien, L.E., Soliman, S.S., Li, X., Bilder, D., 2011. Altered modes of stem cell division drive adaptive intestinal growth. Cell 147, 603–614.

Ohlstein, B., Spradling, A., 2006. The adult *Drosophila* posterior midgut is maintained by pluripotent stem cells. Nature 439, 470–474.

Ohlstein, B., Spradling, A., 2007. Multipotent *Drosophila* intestinal stem cells specify daughter cell fates by differential notch signaling. Science 315, 988–992.

Osman, D., Buchon, N., Chakrabarti, S., Huang, Y.T., Su, W.C., Poidevin, M., Tsai, Y.C., Lemaitre, B., 2012. Autocrine and paracrine unpaired signaling regulate intestinal stem cell maintenance and division. J. Cell Sci. 125, 5944–5949.

Panayidou, S., Apidianakis, Y., 2013. Regenerative inflammation: lessons from *Drosophila* intestinal epithelium in health and disease. Pathogens 2, 209–231.

Panayidou, S., Ioannidou, E., Apidianakis, Y., 2014. Human pathogenic bacteria, fungi, and viruses in *Drosophila*: disease modeling, lessons, and shortcomings. Virulence 5, 253–269.

Pappenheim, A., 1896. Ueber Entwickelung und Ausbildung der Erythroblasten. Arch. Pathol. Anat. Physiol. Klin. Med. 145, 587–643.

Park, J.S., Kim, Y.S., Yoo, M.A., 2009. The role of p38b Mapk in age-related modulation of intestinal stem cell proliferation and differentiation in *Drosophila*. Aging (Albany NY) 1, 637–651.

Park, J.S., Lee, S.H., Na, H.J., Pyo, J.H., Kim, Y.S., Yoo, M.A., 2012. Age- and oxidative stress-induced DNA damage in *Drosophila* intestinal stem cells as marked by Gamma-H2AX. Exp. Gerontol. 47 (5), 401.

Park, J.S., Pyo, J.H., Na, H.J., Jeon, H.J., Kim, Y.S., Arking, R., Yoo, M.A., 2014. Increased centrosome amplification in aged stem cells of the *Drosophila* midgut. Biochem. Biophys. Res. Commun. 450, 961–965.

Park, J.S., Na, H.J., Pyo, J.H., Jeon, H.J., Kim, Y.S., Yoo, M.A., 2015. Requirement of ATR for maintenance of intestinal stem cells in aging *Drosophila*. Aging (Albany NY) 7, 307–318.

Patel, P.H., Dutta, D., Edgar, B.A., 2015. Niche appropriation by *Drosophila* intestinal stem cell tumours. Nat. Cell Biol. 17, 1182–1192.

Perdigoto, C.N., Schweisguth, F., Bardin, A.J., 2011. Distinct levels of Notch activity for commitment and terminal differentiation of stem cells in the adult fly intestine. Development 138, 4585–4595.

Pinto, D., Gregorieff, A., Begthel, H., Clevers, H., 2003. Canonical Wnt signals are essential for homeostasis of the intestinal epithelium. Genes Dev. 17, 1709–1713.

Pitsouli, C., Perrimon, N., 2008. Developmental biology: our fly cousins' gut. Nature 454, 592–593.

Potten, C.S., Al-Barwari, S.E., Searle, J., 1978. Differential radiation response amongst proliferating epithelial cells. Cell Tissue Kinet. 11, 149–160.
Ramalho-Santos, M., Willenbring, H., 2007. On the origin of the term "stem cell" Cell Stem Cell 1, 35–38.
Regan, J.C., Khericha, M., Dobson, A.J., Bolukbasi, E., Rattanavirotkul, N., Partridge, L., 2016. Sex difference in pathology of the ageing gut mediates the greater response of female lifespan to dietary restriction. Elife 5, e10956.
Ren, F., Wang, B., Yue, T., Yun, E.Y., Ip, Y.T., Jiang, J., 2010. Hippo signaling regulates *Drosophila* intestine stem cell proliferation through multiple pathways. Proc. Natl. Acad. Sci. U.S.A. 107, 21064–21069.
Ren, F., Shi, Q., Chen, Y., Jiang, A., Ip, Y.T., Jiang, H., Jiang, J., 2013. *Drosophila* Myc integrates multiple signaling pathways to regulate intestinal stem cell proliferation during midgut regeneration. Cell Res. 23, 1133–1146.
Rera, M., Bahadorani, S., Cho, J., Koehler, C.L., Ulgherait, M., Hur, J.H., Ansari, W.S., Lo Jr., T., Jones, D.L., Walker, D.W., 2011. Modulation of longevity and tissue homeostasis by the *Drosophila* PGC-1 homolog. Cell Metab. 14, 623–634.
Rera, M., Clark, R.I., Walker, D.W., 2012. Intestinal barrier dysfunction links metabolic and inflammatory markers of aging to death in *Drosophila*. Proc. Natl. Acad. Sci. U.S.A. 109, 21528–21533.
Rogers, R.P., Rogina, B., 2015. The role of INDY in metabolism, health and longevity. Front. Genet. 6, 204.
Salomon, R.N., Jackson, F.R., 2008. Tumors of testis and midgut in aging flies. Fly (Austin) 2, 265–268.
Scopelliti, A., Cordero, J.B., Diao, F., Strathdee, K., White, B.H., Sansom, O.J., Vidal, M., 2014. Local control of intestinal stem cell homeostasis by enteroendocrine cells in the adult *Drosophila* midgut. Curr. Biol. 24, 1199–1211.
Shanbhag, S., Tripathi, S., 2009. Epithelial ultrastructure and cellular mechanisms of acid and base transport in the *Drosophila* midgut. J. Exp. Biol. 212 (Pt 11), 1731–1744.
Shaw, R.L., Kohlmaier, A., Polesello, C., Veelken, C., Edgar, B.A., Tapon, N., 2010. The Hippo pathway regulates intestinal stem cell proliferation during *Drosophila* adult midgut regeneration. Development 137, 4147–4158.
Shin, S.C., Kim, S.H., You, H., Kim, B., Kim, A.C., Lee, K.A., Yoon, J.H., Ryu, J.H., Lee, W.J., 2011. *Drosophila* microbiome modulates host developmental and metabolic homeostasis via insulin signaling. Science 334, 670–674.
Simon, T.C., Gordon, J.I., 1995. Intestinal epithelial cell differentiation: new insights from mice, flies and nematodes. Curr. Opin. Genet. Dev. 5, 577–586.
Singh, S.R., Liu, W., Hou, S.X., 2007. The adult *Drosophila* malpighian tubules are maintained by multipotent stem cells. Cell Stem Cell 1, 191–203.
Singh, S.R., Zeng, X., Zheng, Z., Hou, S.X., 2011. The adult *Drosophila* gastric and stomach organs are maintained by a multipotent stem cell pool at the foregut/midgut junction in the cardia (proventriculus). Cell Cycle 10, 1109–1120.
Singh, S.R., Zeng, X., Zhao, J., Liu, Y., Hou, G., Liu, H., Hou, S.X., 2016. The lipolysis pathway sustains normal and transformed stem cells in adult *Drosophila*. Nature 538, 109–113.
Siudeja, K., Nassari, S., Gervais, L., Skorski, P., Lameiras, S., Stolfa, D., Zande, M., Bernard, V., Rio Frio, T., Bardin, A.J., 2015. Frequent somatic mutation in adult intestinal stem cells drives neoplasia and genetic mosaicism during aging. Cell Stem Cell 17, 663–674.
Slack, C., Alic, N., Foley, A., Cabecinha, M., Hoddinott, M.P., Partridge, L., 2015. The Ras-Erk-ETS-signaling pathway is a drug target for longevity. Cell 162, 72–83.
Snodgrass, R.E., 1935. Principles of Insect Morphology. McGraw-Hill Book Company, Inc., New York, London.

Spradling, A., Drummond-Barbosa, D., Kai, T., 2001. Stem cells find their niche. Nature 414, 98–104.
Staley, B.K., Irvine, K.D., 2012. Hippo signaling in *Drosophila*: recent advances and insights. Dev. Dyn. 241, 3–15.
Stevens, C.E., Leblond, C.P., 1947. Rate of renewal of the cells of the intestinal epithelium in the rat. Anat. Rec. 97, 373.
Strand, M., Micchelli, C.A., 2011. Quiescent gastric stem cells maintain the adult *Drosophila* stomach. Proc. Natl. Acad. Sci. U.S.A. 108, 17696–17701.
Strand, M., Micchelli, C.A., 2013. Regional control of *Drosophila* gut stem cell proliferation: EGF establishes GSSC proliferative set point & controls emergence from quiescence. PLoS One 8, e80608.
Strasburger, M., 1932. Bau, Funktion und Variabilität des Darmtraktus von *Drosophila melanogaster* (Meigen). Z. Wiss. Zool. 140, 539–649.
Takashima, S., Hartenstein, V., 2012. Genetic control of intestinal stem cell specification and development: a comparative view. Stem Cell Rev. 8, 597–608.
Takashima, S., Mkrtchyan, M., Younossi-Hartenstein, A., Merriam, J.R., Hartenstein, V., 2008. The behaviour of *Drosophila* adult hindgut stem cells is controlled by Wnt and Hh signalling. Nature 454, 651–655.
Takashima, S., Adams, K.L., Ortiz, P.A., Ying, C.T., Moridzadeh, R., Younossi-Hartenstein, A., Hartenstein, V., 2011. Development of the *Drosophila* entero-endocrine lineage and its specification by the Notch signaling pathway. Dev. Biol. 353, 161–172.
Takashima, S., Gold, D., Hartenstein, V., 2013a. Stem cells and lineages of the intestine: a developmental and evolutionary perspective. Dev. Genes Evol. 223, 85–102.
Takashima, S., Paul, M., Aghajanian, P., Younossi-Hartenstein, A., Hartenstein, V., 2013b. Migration of *Drosophila* intestinal stem cells across organ boundaries. Development 140, 1903–1911.
Takashima, S., Aghajanian, P., Younossi-Hartenstein, A., Hartenstein, V., 2016. Origin and dynamic lineage characteristics of the developing *Drosophila* midgut stem cells. Dev. Biol. 416, 347–360.
Tepass, U., Hartenstein, V., 1994. Epithelium formation in the *Drosophila* midgut depends on the interaction of endoderm and mesoderm. Development 120, 579–590.
Tepass, U., Hartenstein, V., 1995. Neurogenic and proneural genes control cell fate specification in the *Drosophila* endoderm. Development 121, 393–405.
Tian, A., Jiang, J., 2014. Intestinal epithelium-derived BMP controls stem cell self-renewal in *Drosophila* adult midgut. Elife 3, e01857.
Tian, A., Shi, Q., Jiang, A., Li, S., Wang, B., Jiang, J., 2015. Injury-stimulated Hedgehog signaling promotes regenerative proliferation of *Drosophila* intestinal stem cells. J. Cell Biol. 208, 807–819.
Tiozzo, S., Brown, F.D., De Tomaso, A.W., 2008. Regeneration and stem cells in ascidians. In: Bosch, T.C.G. (Ed.), Stem Cells: From Hydra to Man. Springer Netherlands, Dordrecht.
Tzelepis, I., Kapsetaki, S.E., Panayidou, S., Apidianakis, Y., 2013. *Drosophila melanogaster*: a first step and a stepping-stone to anti-infectives. Curr. Opin. Pharmacol. 13 (5), 763–768.
van Der Flier, L.G., van Gijn, M.E., Hatzis, P., Kujala, P., Haegebarth, A., Stange, D.E., Begthel, H., van Den Born, M., Guryev, V., Oving, I., van Es, J.H., Barker, N., Peters, P.J., van De Wetering, M., Clevers, H., 2009. Transcription factor *achaete scute-like 2* controls intestinal stem cell fate. Cell 136 (5), 903–912.
Veenstra, J.A., Ida, T., 2014. More Drosophila enteroendocrine peptides: Orcokinin B and the CCHamides 1 and 2. Cell Tissue Res. 357, 607–621.
Veenstra, J.A., Agricola, H.J., Sellami, A., 2008. Regulatory peptides in fruit fly midgut. Cell Tissue Res. 334, 499–516.

Vogt, G., 2012. Hidden treasures in stem cells of indeterminately growing bilaterian invertebrates. Stem Cell Rev. 8, 305–317.

Wang, Z., Hang, S., Purdy, A.E., Watnick, P.I., 2013. Mutations in the IMD pathway and mustard counter *Vibrio cholerae* suppression of intestinal stem cell division in *Drosophila*. MBio 4. e00337–13.

Wigglesworth, V.B., 1965. The Principles of Insect Physiology. Methuen & Co, London.

Winquist, R.J., Furey, B.F., Boucher, D.M., 2010. Cancer stem cells as the relevant biomass for drug discovery. Curr. Opin. Pharmacol. 10, 385–390.

Xu, N., Wang, S.Q., Tan, D., Gao, Y., Lin, G., Xi, R., 2011. EGFR, Wingless and JAK/STAT signaling cooperatively maintain *Drosophila* intestinal stem cells. Dev. Biol. 354, 31–43.

Zeng, X., Hou, S.X., 2015. Enteroendocrine cells are generated from stem cells through a distinct progenitor in the adult *Drosophila* posterior midgut. Development 142, 644–653.

Zhou, J., Florescu, S., Boettcher, A.L., Luo, L., Dutta, D., Kerr, G., Cai, Y., Edgar, B.A., Boutros, M., 2015. Dpp/Gbb signaling is required for normal intestinal regeneration during infection. Dev. Biol. 399, 189–203.

CHAPTER SIX

Insect Symbiosis and Immunity: The Bean Bug–*Burkholderia* Interaction as a Case Study

Jiyeun K. Kim*, Bok L. Lee†
*Kosin University College of Medicine, Busan, South Korea
†Global Research Laboratory of Insect Symbiosis, College of Pharmacy, Pusan National University, Busan, South Korea

Contents

1. Introduction 180
   1.1 Insect–Bacteria Symbiosis 180
2. The Bean Bug–*Burkholderia* Symbiosis 183
3. The Effect of the *Burkholderia* Symbiont on the Immunity of Bean Bug 185
4. The Effect of the Immunity of Bean Bug on *Burkholderia* Symbionts 187
   4.1 Molecular Changes of *Burkholderia* Symbiont 187
   4.2 The Susceptibility of the *Burkholderia* Symbiont to the Immunity of Bean Bug 188
   4.3 The Suppression of a Bean Bug's Immunity in the Symbiotic Organ 190
   4.4 Regulation of the *Burkholderia* Symbiont Population by the Immunity of Bean Bug 190
5. Conclusions 193
Acknowledgements 193
References 194

Abstract

Most insects are symbiotically associated with bacteria. In an insect–bacteria symbiosis, understanding the mechanisms of bacterial persistence in the presence of host immunity is an important question to answer. Recently, bean bugs possessing *Burkholderia* symbiont in their midgut region have been recognized as a useful insect model to study symbiosis mechanisms at the molecular level. In the bean bug symbiosis, the *Burkholderia* symbionts positively affect the immunity of bean bugs, and the immunity of bean bugs regulates the population of the symbionts. The symbiotic association with the host induces drastic changes in the cell envelope of the *Burkholderia* symbionts, which make the *Burkholderia* symbionts become highly susceptible to the host's immunity. However, the bean bug suppresses the immune responses of the symbiotic midgut region to support the survival of the immune-susceptible *Burkholderia* symbionts. The bean bug–*Burkholderia* studies demonstrate the intricate interplay between symbiosis and immunity.

Advances in Insect Physiology, Volume 52
ISSN 0065-2806
http://dx.doi.org/10.1016/bs.aiip.2016.11.003

1. INTRODUCTION

Symbiosis with bacteria is universally present in animal and plant hosts. Contrary to pathogenic bacteria, persistent infection by symbiotic bacteria is supported by their hosts. One of the important questions in symbiosis is how the symbionts harmoniously associate with their hosts in the presence of host immunity. To answer this question, it is crucial to have experimental symbiosis models that enable us to investigate the molecular crosstalk between hosts and bacteria.

1.1 Insect–Bacteria Symbiosis

Insects are the most diverse group of animals (Grimaldi and Engel, 2005), and the majority of insects possess symbiotic bacteria in their tissues and inside of their cells (Buchner, 1965). Thus, insects have been the most studied animal group in the symbiosis field. The symbionts in insects exert various biological effects on their hosts. Some symbionts provide essential nutrients (Moran et al., 2008). Some symbionts provide a defence ability against natural enemies, such as fungi and bacterial infections, or an adaptive ability to specific ecological conditions (Oliver et al., 2010). Other symbionts regulate the reproduction of insects (Werren et al., 2008).

Many studies on insect symbiosis have focused on the maternally transmitted obligate symbionts, which are referred to as primary symbionts. These obligate symbionts are known to associate with the host over a long period of evolutionary time and become vital to a host's life. The adaptive evolution of symbionts inside the host is manifested in the dramatic reduction of the bacterial genome size, which results in the loss of free-living ability (McCutcheon and Moran, 2012; Moran, 2003). In addition to obligate symbionts, insects also harbour facultative symbionts. The facultative symbionts are mostly vertically transmitted, but some can be horizontally transmitted. While obligate symbionts are always beneficial and essential to the host's vitality, facultative symbionts are either beneficial or harmful and are not essential to the hosts. Because of the recent association with the host, the genomes of facultative symbionts are not reduced, and they still possess a free-living ability.

Several insect–bacteria symbiosis models have been studied to understand the symbiotic interactions between the host and the symbiont.

1.1.1 *Pea Aphid*–Buchnera

In insect symbiosis, pea aphid, *Acyrthosiphon pisum*, is the best studied insect (Baumann, 2005). Pea aphid possesses *Buchnera aphidicola* as an obligate symbiont. The pea aphid–*Buchnera* symbiosis is a nutritional symbiosis because aphids with an unbalanced diet of plant phloem sap are nutritionally dependent on their *Buchnera* symbiont for acquiring essential amino acids (Douglas, 1998). *Buchnera* is located within specialized aphid cells, known as bacteriocytes, which are assembled in an organ-like structure called bacteriome. The start of their symbiotic association is estimated to be 150 million years ago, and the long-term coevolution results in the loss of autonomy in both the host and the symbiont. The genome of *Buchnera* is greatly reduced in size, missing many essential genes for free-living life, but a large portion of its genome is dedicated to the synthesis of the essential amino acids required by host, clearly demonstrating intimate adaptation to host (Shigenobu and Wilson, 2011; Shigenobu et al., 2000).

1.1.2 *Pea Aphid–Facultative Symbionts*

Pea aphids also possess facultative symbionts, including *Hamiltonella defensa*, *Serratia symbiotica* and *Regiella insecticola* (Oliver et al., 2010). Because the aphid can survive without these facultative symbionts, the infection with facultative symbionts has been controlled to investigate the effects of these symbionts to pea aphids. The microinjection of the bacteria to uninfected aphids or the treatment of infected aphids with antibiotics has been performed (Koga et al., 2003). Experimental studies comparing aphid lines with different facultative symbionts showed that particular facultative symbionts play a role in defence against parasitoid wasps and fungi, in heat tolerance and in aphid adaptation to host plants (Oliver et al., 2010).

1.1.3 *Tsetse Fly*–Wigglesworthia

The tsetse fly, genus *Glossina*, is the sole vector of Human African Trypanosomiasis also known as sleeping sickness caused by the protozoan *Trypanosoma brucei*. Similar to *Buchnera* in aphids, tsetse flies possess *Wigglesworthia* as an obligate symbiont (Aksoy, 1995). *Wigglesworthia* provides essential vitamin metabolites to the tsetse fly host whose diet is only a vertebrate blood. *Wigglesworthia* symbionts are localized within the bacteriocytes, and their genome has undergone dramatic size reduction due to the coevolution with the host (Akman et al., 2002).

1.1.4 Tsetse Fly–Sodalis

Tsetse flies possess *Sodalis glossinidius* as a facultative symbiont (Wang et al., 2013). The vertically transmitted *Sodalis* can colonize different organs both intra- and extracellularly. Even though *Sodalis* has lost genes involved in energy metabolism, it retains many of the functions associated with free-living ability (Akman et al., 2001). Due to the success of in vitro cultivation of *Sodalis*, genetically manipulated *Sodalis* expressing foreign products is subjected to paratransgenic approach, attempting to eliminate pathogens from vectors through transgenic symbionts (Sassera et al., 2013).

1.1.5 Tsetse Fly–Wolbachia

Wolbachia has infected a wide range of insects and is a facultative symbiont of tsetse flies. *Wolbachia* localized in the germ line tissues in tsetse flies has the ability to manipulate host reproduction through cytoplasmic incompatibility, that is few offspring between infected males and uninfected females, and male-killing, that is no male offspring from infected females (Werren et al., 2008). The reproductive manipulation by *Wolbachia* has been proposed as an environmentally friendly way to control disease vectors including mosquitoes and tsetse flies (Doudoumis et al., 2013; Ricci et al., 2012).

1.1.6 Fruit Fly–Facultative Symbionts

Wolbachia and *Spiroplasma* are two known vertically transmitted facultative symbionts in *Drosophila* (Mateos et al., 2006). Not only their effects on the reproductive manipulation, *Wolbachia* and *Spiroplasma* play defensive roles against broad enemies, such as viruses, nematodes, pathogenic bacteria and parasitoid wasps (Hamilton and Perlman, 2013).

Many of these models use physiological approaches to examine the effect of symbionts on the host's biology (Douglas, 1998) and investigate genome of the symbionts to analyse the bacterial outcomes of symbiosis (McCutcheon and Moran, 2012; Tamas et al., 2002; Toft and Andersson, 2010). Compared to the extensive knowledge regarding the physiological and genomic phenomena of the insect–bacteria symbiosis models, the molecular and biochemical characteristics of the crosstalk between hosts and symbionts have been somewhat less investigated due to the difficulty in isolating and culturing symbiotic bacteria outside of the insect hosts (Baumann and Moran, 1997; Pontes and Dale, 2006). Obligate symbionts, such as *Buchnera* in aphids and *Wigglesworthia* in tsetse flies, are incapable of independent living and, thus, are uncultivable (McCutcheon and Moran, 2012; Moran, 2003). Some facultative symbionts, such as *Wolbachia* in various insects and *Sodalis* in tsetse flies, are cultivable outside of their host insects, but culturing these symbionts is generally not easy. In many cases, the culturing of symbionts requires complex culture media containing either mammalian sera or live insect cells and is prone to contamination. The symbionts tend to grow very slowly and form only a few colonies on agar plates (Pontes and Dale, 2006).

In other noninsect symbiosis models, such as squid–*Vibrio*, legume–*Rhizobium* and nematode–*Photorhabdus* symbioses, the symbiotic bacteria are easily cultivable and genetically manipulable. Therefore, they are suitable models for studying the symbiotic crosstalk at the molecular level (Goodrich-Blair and Clarke, 2007; Nyholm and McFall-Ngai, 2004). The symbionts in these model systems are acquired by their hosts from the environment during each generation. Such horizontal transmission is not common in insect symbiosis. Insects normally transmit their symbiont from mother to offspring. The tight evolutionary association of obligate symbionts to their insect hosts is ensured by transovarial transmission, where the symbionts are passed to the next generation at the stages of oogenesis or embryogenesis (Braendle et al., 2003; Mira and Moran, 2002). Other vertical transmissions are accomplished through the superficial bacterial contamination of the eggs, probing of the faeces or deposition of bacterium-containing capsules with eggs or jelly-like secretions (Salem et al., 2015).

In 2005, Kikuchi and his colleagues determined that bean bugs, *Riptortus pedestris*, possess beneficial symbionts that are not vertically transmitted but are rather acquired from the environment during every generation (Kikuchi et al., 2005). Because these symbionts are cultivable and manipulable, the bean bug has been recognized as a good experimental symbiosis model. Using molecular and biochemical approaches, several novel symbiotic factors of bacteria have been elucidated (Kim and Lee, 2015) and some host mechanisms for managing their symbiont have been proposed. In this chapter, we outline the recent findings of the interactions between symbionts and host immunity in the bean bug symbiosis model.

2. THE BEAN BUG–*BURKHOLDERIA* SYMBIOSIS

Members of the insect suborder Heteroptera are known as true bugs or stinkbugs and comprise more than 40,000 insect species with incomplete metamorphosis (Weirauch and Schuh, 2011). Many plant-sucking Heteroptera species possess a specialized symbiotic organ in their posterior region of the midgut. The symbiotic midgut region among stinkbug taxa exhibits different morphologies, including tubular outgrowths, small crypts arranged in two rows, small crypts arranged in four rows and a pair of large flat assemblages of crypts (Kikuchi et al., 2011). The lumen of the midgut crypts is usually occupied by a single bacterial species, and elimination of the symbiont causes a negative effect on the host development and fitness, which indicates the beneficial role of the gut symbionts (Bistolas et al., 2014; Kikuchi et al., 2009).

The bean bug, *R. pedestris*, is a member of the family Alydidae in the superfamily Coreoidea which is under infraorder Pentatomomorpha and suborder Heteroptera (Fig. 1A). It is a notorious pest of leguminous crops in East Asia. The midgut of bean bugs is divided into morphologically distinct regions called M1, M2, M3, M4B and M4, where M4 is the posterior symbiotic midgut region (Fig. 1B). The bean bug M4 symbiotic organ has two rows of numerous crypts (Fig. 1C) whose lumens are densely populated with betaproteobacterial symbionts of the genus *Burkholderia* (Fig. 1D). Unlike many other insects that acquire their symbionts vertically from the mother, bean bugs orally acquire *Burkholderia* from the rhizosphere environment during their early nymphal stages (Kikuchi et al., 2007).

Fig. 1 The bean bug–*Burkholderia* symbiosis. (A) An adult bean bug. (B) The dissected midgut of a bean bug with morphologically distinct regions M1, M2, M3, M4B and M4. (C) A closer look at the M4 region with numerous crypts aligned in *two rows*. (D) A transmission electron microscopic image of the cross-sectioned crypts. Cs indicate crypt epithelia, whereas Bs show crypt cavities filled with *Burkholderia* cells. The *yellow lines* highlight boundaries of the crypt epithelia and the crypt cavities. *Reprinted from Kim, J.K., Won, Y.J., Nikoh, N., Nakayama, H., Han, S.H., Kikuchi, Y., Rhee, Y.H., Park, H.Y., Kwon, J.Y., Kurokawa, K., Dohmae, N., Fukatsu, T., Lee, B.L., 2013b. Polyester synthesis genes associated with stress resistance are involved in an insect–bacterium symbiosis. Proc. Natl. Acad. Sci. U.S.A. 110 (26), E2381–E2389.*

The genus *Burkholderia* is a soil bacterium. Even after becoming gut symbionts of bean bugs, *Burkholderia* symbionts retain their free-living ability and are easily cultured on standard bacterial media. Furthermore, they can be genetically manipulated using their genome information (Shibata et al., 2013). In the laboratory, we infect the second instar bean bugs with cultured *Burkholderia* cells suspended in drinking water to establish symbiosis. To further develop comparative insect models, symbiotic, aposymbiotic and mutant-infected insect lines are generated by providing a drinking solution with wild-type *Burkholderia*, no *Burkholderia* and mutant *Burkholderia*, respectively. After oral uptake, the *Burkholderia* cells monospecifically colonize and prosper in the M4 symbiotic organ (Kikuchi et al., 2005) and reached nearly 10^8 cells exclusively in the M4 midgut region. These large numbers of *Burkholderia* symbionts can be easily isolated from the M4 midgut region (Kikuchi and Yumoto, 2013; Kim et al., 2014a) and directly compared with the cultured *Burkholderia* cells to find the molecular mechanisms of their host adaptation. All of these aspects make the bean bug–*Burkholderia* symbiosis model an attractive experimental model to study molecular crosstalk between the insects and the symbionts.

3. THE EFFECT OF THE *BURKHOLDERIA* SYMBIONT ON THE IMMUNITY OF BEAN BUG

Bean bugs are hemimetabolous insects. While the immunity of holometabolous insects has been comprehensively investigated (Buchon et al., 2014; Iwanaga and Lee, 2005; Lemaitre and Hoffmann, 2007), the understanding of the immunity of hemimetabolous insects is limited to the pea aphid *A. pisum* (Gerardo et al., 2010) and brown planthopper *Nilaparvata lugens* (Bao et al., 2013) through their genomic and transcriptomic analyses. The immunity of the bean bug is not well known; however, recently three antimicrobial peptides (AMPs) of bean bugs were purified and identified as follows: riptocin, which is a pyrrhocoricin-like proline-rich AMP, rip-defensin and rip-thanatins (Kim et al., 2015b). The identification of the AMPs of bean bugs supported us to investigate the relationship between *Burkholderia* symbionts and bean bug immunity at the molecular level.

Certain insect symbiosis models have previously shown the association of symbionts and the host's innate immunity. Tsetse flies (*Glossina* spp.) possess commensal *Sodalis* and obligate *Wigglesworthia* as their major gut symbionts. The tsetse flies without *Wigglesworthia* exhibit an irregular expression of humoural and epithelial immunity-related genes and decreased numbers of phagocytic haemocytes (Weiss et al., 2011; Weiss et al., 2012). In fruit flies (*Drosophila melanogaster*) and mosquitoes (*Aedes aegypti*), *Wolbachia* symbiont increased the host survival against pathogen infections (Hedges et al., 2008; Moreira et al., 2009; Walker et al., 2011). Similarly, in the bean bug symbiosis model, we reported the notion that *Burkholderia* symbionts also positively affect host immunity.

Upon bacterial challenges, the symbiotic bean bugs exhibited significantly better survival than the aposymbiotic (possessing no *Burkholderia* symbionts) bean bugs (Kim et al., 2015a) (Fig. 2). Even after the cellular immune responses were inhibited, the symbiotic insects still exhibited significantly better survival than the aposymbiotic insects, which indicated the existence of stronger humoural immune responses in symbiotic insects. To address the molecular basis of the strong humoural immune defence of the symbiotic insects, the expressions of the bean bug AMPs in the fat body were compared between the symbiotic and aposymbiotic insects. At the basal condition, without a bacterial injection, the expression levels of riptocin, rip-defensin and rip-thanatin were similar in both insects. However, the expression of AMPs was significantly increased more in the symbiotic insects than in the aposymbiotic insects in response to the bacterial septic infection (Fig. 3). These results showed that *Burkholderia* symbionts enhanced the systemic immunity of host bean bugs (Kim et al., 2015a). The molecular mechanisms of how *Burkholderia* symbionts affect the host systemic immunity, through general health improvement by nutritional support or specific immune enhancement by symbiont-derived signals, are not known and should be pursued further.

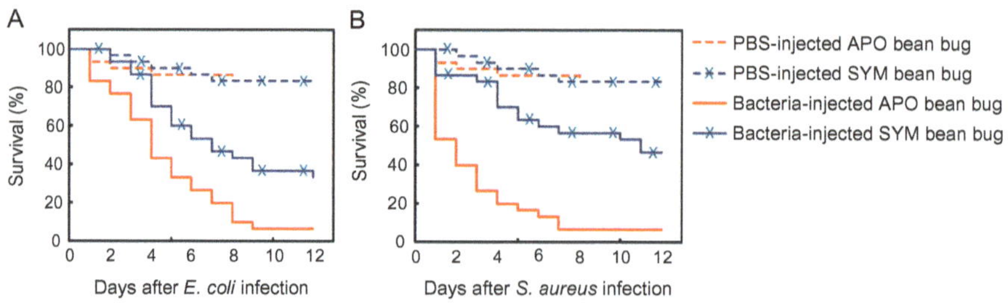

Fig. 2 The bean bug survival rate after bacterial challenges. The symbiotic (SYM) insects (*blue line*) and the aposymbiotic (APO) insects (*red line*) were injected with *E. coli* cells (A) or *S. aureus* cells (B) and monitored for their survival. For control insect groups, the same volume of buffer was injected (*dotted lines*). For each sample group, 30 insects were monitored.

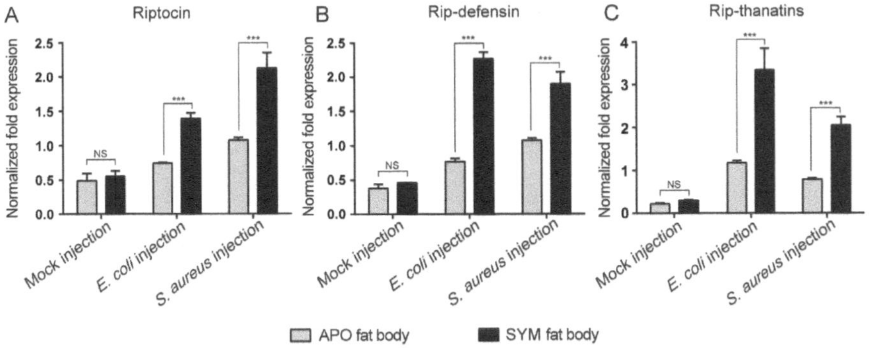

Fig. 3 Expression levels of AMPs. The fat bodies of the symbiotic and aposymbiotic insects were collected to measure the expression of riptocin (A), rip-defensin (B) and rip-thanatins (C) at 17 h postinjection with buffer, *E. coli* cells or *S. aureus* cells. The expression level was normalized by the comparative C_T ($\Delta\Delta C_T$) method using *EF1α* as a reference gene. The columns and bars indicate the means and SDs ($n=3$). Statistically significant differences between the symbiotic and aposymbiotic insects were analysed by a two-way ANOVA with Sidak's correction (*NS*, not significant; ***, $P<0.0001$).

4. THE EFFECT OF THE IMMUNITY OF BEAN BUG ON *BURKHOLDERIA* SYMBIONTS

4.1 Molecular Changes of *Burkholderia* Symbiont

Similar to a symbiont's capability to change the host's biology, the biology of a symbiont must be altered by the symbiotic association with the host. To elucidate the symbiotic alteration of the *Burkholderia* symbionts, the intact symbionts were isolated from the M4 midgut regions of bean bugs and compared with the cultured *Burkholderia* cells. The cell envelope of the symbionts was the target of our investigation because it is located at the frontline of interacting with the host. The cell envelope of Gram-negative *Burkholderia* consists of inner and outer membranes separated by peptidoglycan-containing periplasmic space (Costerton et al., 1974; Silhavy et al., 2010). Lipopolysaccharide (LPS) is located at the outer part of the outer membrane and is composed of lipid A that is embedded in the outer membrane and core-oligosaccharide that connects lipid A and the O-antigen with the oligosaccharides repeats (Raetz and Whitfield, 2002). Interestingly, when the LPS of the symbiotic *Burkholderia* cells is compared with that of the cultured cells, the O-antigen polysaccharide was lost in the symbiotic cells (Kim et al., 2015b) (Fig. 4A). By analyzing

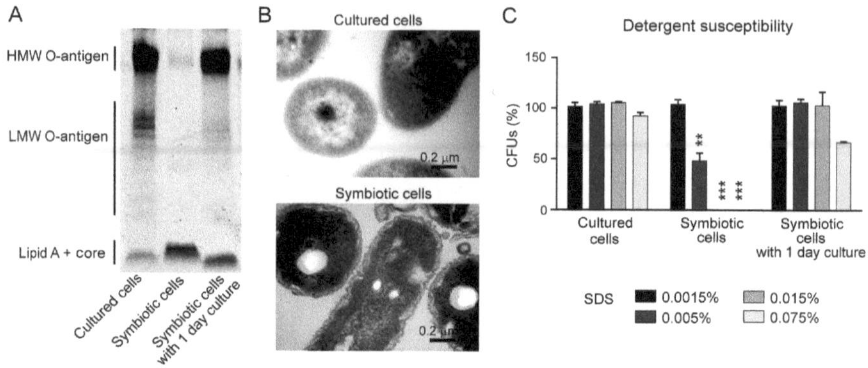

Fig. 4 Changes in the cell envelopes of the *Burkholderia* symbionts. (A) LPS patterns on SDS-PAGE gels. The LPS of the cultured, symbiotic and symbiotic *Burkholderia* cells that have been cultured in vitro for 1 day is resolved on SDS-PAGE gels and visualized by Pro-Q emerald staining. (B) The TEM images of the cultured and symbiotic *Burkholderia* cells. (C) SDS susceptibility of the cultured, symbiotic and symbiotic *Burkholderia* cells that have been cultured in vitro for 1 day. The cells were subjected to CFU assays after incubation with different SDS concentrations. The means and SDs are shown ($n=3$). *Asterisks* indicate statistically significant differences for the SDS-untreated CFUs (unpaired *t*-test: **, $P<0.005$; ***, $P<0.0001$).

the matrix-assisted laser desorption ionization-mass spectrometric data and the nuclear magnetic resonance spectroscopic data, the LPS structure of the symbiotic cells was determined to have a full length of lipid A and core-oligosaccharide but not the O-antigen. We also observed the differences in cell membranes between the symbiotic and cultured cells in the transmission electron microscopic images (Fig. 4B) and further examined the difference in detergent susceptibility (Fig. 4C). The symbiotic *Burkholderia* cells were more susceptible to the detergent than the cultured cells, which suggests the compromised integrity of the cell membrane in the symbiotic cells (Kim et al., 2015b). Both the O-antigen and the cell membrane integrity are restored by culturing the symbiotic cells, which indicates that the cell envelope changes in the *Burkholderia* cells are symbiosis-derived characteristics.

4.2 The Susceptibility of the *Burkholderia* Symbiont to the Immunity of Bean Bug

The drastic cell envelope changes of *Burkholderia* symbionts consequently altered the bacterial response to the host immunity. While the cultured *Burkholderia* cells were resistant to the antimicrobial activity of the haemolymph of bean bugs, reflecting *Burkholderia*'s intrinsic resistance to antimicrobials

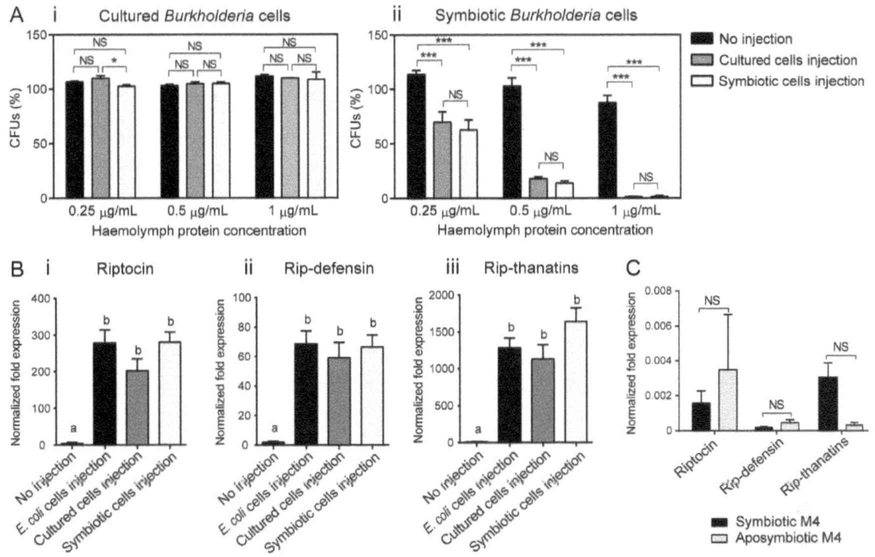

Fig. 5 (A) The antimicrobial activity of the *Burkholderia*-injected haemolymph. The non-injected, the cultured *Burkholderia*-injected and the symbiotic *Burkholderia*-injected haemolymphs were compared for their antimicrobial activities on the cultured (i) and symbiotic *Burkholderia* (ii) cells. The means and SDs are shown ($n=3$). Statistically significant differences were analysed by a two-way ANOVA with Tukey's correction (*, $P<0.05$; ***, $P<0.0001$; *NS*, not significant). (B) The comparison of the expression level of the bean bug AMPs after bacterial septic infection. The expression of riptocin (i), rip-defensin (ii) and rip-thanatins (iii) was measured in the fat body of adult insects after injection of the *E. coli*, cultured *Burkholderia* or symbiotic *Burkholderia* cells. The expression level was normalized by the C_T ($\Delta\Delta C_T$) method using *EF1α* as a reference gene. Columns and bars indicate means and SEM ($n=9$). Different letters (a, b) indicate statistically significant differences (one-way ANOVA with Tukey's correction: $P\leq 0.0001$). (C) The comparison of the expression level of the AMPs in M4 midgut between the symbiotic insect and the aposymbiotic insect. *NS* means not significant (unpaired *t*-test).

(Loutet and Valvano, 2011), the symbiotic *Burkholderia* cells were highly susceptible to the haemolymph (Fig. 5A). When the bean bug AMPs purified from haemolymph were used to examine the susceptibilities of the *Burkholderia* cells, the symbiotic cells were much more susceptible to riptocin and rip-defensin than the cultured cells (Kim et al., 2015b). Ultimately, we performed an in vivo bacterial clearance experiment that represented both the humoural and cellular innate immune responses of the host. Upon systemic injection of the symbiotic cells to the bean bugs, these cells were cleared much faster than the cultured cells, which showed that the *Burkholderia* symbionts are highly susceptible to the host immune responses (Kim et al., 2015b).

4.3 The Suppression of a Bean Bug's Immunity in the Symbiotic Organ

Considering that members of the genus *Burkholderia* are known to be resistant to antimicrobial agents (Mahenthiralingam et al., 2005) and that the cultured *Burkholderia* cells are highly resistant to insect immune responses (Kim et al., 2015b), it is clear that the symbiosis with bean bugs induces dramatic changes in the *Burkholderia* cells and results in the loss of intrinsic resistance against the host immunity.

Our findings raised a question regarding how the symbionts that are highly susceptible to the host immune responses can survive inside the host. We first tested whether the *Burkholderia* symbiont induces an immune response in the bean bug. The immunogenicity of the symbiotic *Burkholderia* cells was accessed by measuring the antimicrobial activity of haemolymph after injecting *Burkholderia* cells into the bean bug. Upon systemic injection, the symbiotic *Burkholderia* cells induced a similar level of antimicrobial activity as the cultured *Burkholderia* cells (Fig. 5A). The AMP expression levels in the fat body were similar among the *Escherichia coli*-injected, cultured *Burkholderia*-injected and symbiotic *Burkholderia*-injected insects, which showed the immunogenicity of the symbiotic *Burkholderia* cells (Fig. 5B). However, when we examined the immune responses in the M4 midgut region, the AMP expression levels of the symbiotic insects were not significantly different from those of the aposymbiotic insects, indicating that the *Burkholderia* symbionts do not induce immune responses in the M4 region (Fig. 5C). The AMP expression levels in the M4 region were significantly lower than the basal expression level of the fat body (Kim et al., 2015b). Therefore, the suppressed immunity of the symbiotic M4 region enables the immune-susceptible *Burkholderia* symbionts to persist in the bean bug host.

4.4 Regulation of the *Burkholderia* Symbiont Population by the Immunity of Bean Bug

In the view of symbionts, the cell envelope changes that increase the susceptibility to the host's immune responses seem to be a disadvantage for their survival inside the host. However, from the perspective of the host, the immune-susceptible symbionts could be an advantage for them to easily manage their symbiont population. In insect symbiosis, it is very important to maintain an optimal population of symbionts because the level of the symbiont population affects the host fitness, the fidelity of vertical transmission and the intensity of the reproductive aberrations (Koga et al., 2003;

McGraw et al., 2002; Mouton et al., 2004; Sakurai et al., 2005). Therefore, insects have several mechanisms for controlling the population of their symbionts (Brennan et al., 2008; Login et al., 2011; Nakabachi et al., 2005; Nishikori et al., 2009). Our previous studies also show an intimate relationship between the population of the *Burkholderia* symbionts and the development and fitness of bean bugs (Kim et al., 2013b, 2014b,c), and recently, two mechanisms to control the population of *Burkholderia* symbionts in the bean bug were proposed (Kim et al., 2013a, 2014a).

To understand how the symbiont population is regulated, the population size of the *Burkholderia* symbionts in the M4 region was monitored over the course of bean bug development (Kim et al., 2014a). The size of the symbiont population steadily increased during the nymphal stages. However, pattern of transient decrease of the symbiont population prior to the moulting period was observed in each instar stage. Especially in the fifth instar stage, the symbiont population abruptly decreased to approximately one-eighth of the size just before the moulting period (Fig. 6A). The lysate of the M4 region at the premoult stage exhibited antimicrobial activity against the symbiotic *Burkholderia* cells but not the cultured cells. The antimicrobial activity was due to the increased expression of c-type lysozyme and riptocin (Kim et al., 2014a). These results suggest that the *Burkholderia* symbiont population can be efficiently controlled by regulation of the antimicrobial protein expression in the symbiotic organ.

Another mechanism of regulating the symbiont population in bean bugs is related to the M4B region of the midgut that is adjacent to the M4 region. Initially, the M4B region was considered to be a symbiotic organ due to the nucleic acid detection of *Burkholderia* symbionts. However, we surprisingly found that the *Burkholderia* cells in the M4B region were not alive (Kim et al., 2013a). The M4B lysate exhibited strong antimicrobial activity against the symbiotic *Burkholderia* cells (Fig. 6B). Because the M4B region of aposymbiotic insect had little antimicrobial activity, the antimicrobial activity of the M4B region seems to be induced by the presence of *Burkholderia* symbionts (Kim et al., 2013a) (Fig. 6C). One of the components responsible for the antimicrobial activity of the M4B region was cathepsin-L-like protease (Byeon et al., 2015). The expression sequence tag data showed that cathepsin-L-like protease genes were highly and preferentially expressed in the M4B region of the symbiotic insects (Futahashi et al., 2013), which supported the fact that the cathepsin-L-like proteases play a role in killing and digesting the *Burkholderia* symbionts in the M4B region. The direct connection of the M4B region to the symbiont-propagating M4 region seems to

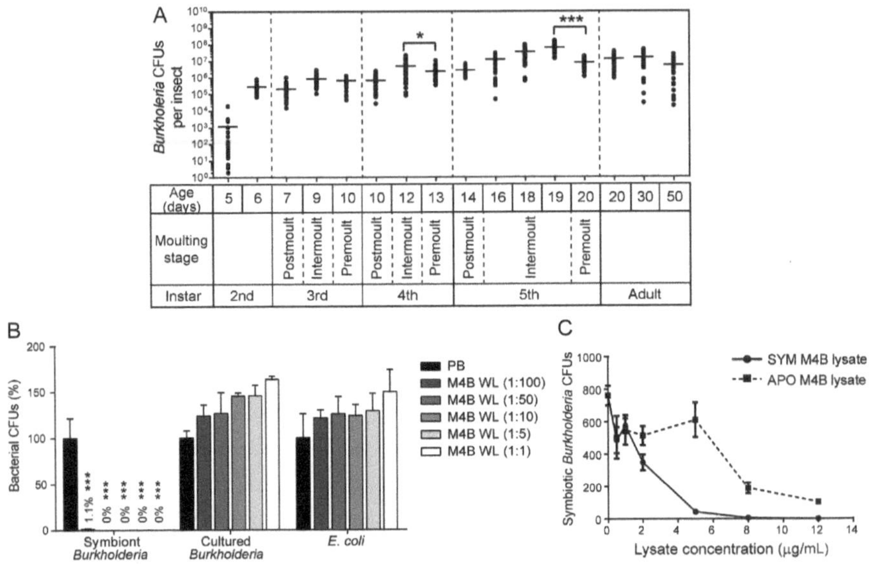

Fig. 6 (A) The population sizes of the *Burkholderia* symbiont during insect growth. Age indicates days after hatching. $N=40$; $*P<0.05$, $***P<0.0001$, unpaired *t*-test. (B) The specific antimicrobial activity of M4B against the symbiotic *Burkholderia*. Cell suspensions of the symbiotic *Burkholderia*, cultured *Burkholderia* and *E. coli* cells were subjected to CFU assay after incubating with M4B lysate samples. *M4B WL*, whole lysate of M4B midgut region; *PB*, phosphate buffer without lysate. As for concentration of M4 lysates, M4B WL (1:1) is equivalent to 50 dissected M4B samples per mL, and M4B WL (1:50) is equivalent to a dissected M4B sample per mL. The means and SDs are shown ($n=3$). *Asterisks* indicate statistically significant differences (unpaired *t*-test: *, $P<0.05$; **, $P<0.01$; ***, $P<0.005$). (C) The comparison of the M4B antimicrobial activity between symbiotic (SYM) insects and aposymbiotic (APO) insects. The symbiotic *Burkholderia* cells isolated from M4 region were incubated with different concentrations of SYM M4B lysates or APO M4B lysates, and subjected to CFU assay. The means and standard errors are plotted ($n=3$).

be appropriate for the host to use the M4B region for controlling the symbiont population.

Both controlling mechanisms would only be effective if the *Burkholderia* symbionts were susceptible to the host's immune responses. Contrary to the susceptibility of the symbiotic *Burkholderia* cells, the cultured *Burkholderia* cells exhibited resistance to the M4 lysate of the premoulting stage, the M4B lysate and the cathepsin-L-like protease (Byeon et al., 2015; Kim et al., 2013a). Combined, the data collected in our studies demonstrate the effectiveness of the host's immunity in controlling the population of their *Burkholderia* symbionts through symbiosis-derived bacterial changes.

5. CONCLUSIONS

The symbiosis of an insect is intimately interconnected with its immunity. In bean bug symbiosis, *Burkholderia* symbionts enhance the host's systemic immunity, and bean bug's immune responses tightly regulate the population of the symbionts in the symbiotic organ. For the delicate correlation between symbiosis and immunity, the *Burkholderia* cells undergo drastic cell envelope changes and, thus, become highly susceptible to the host's immunity. At the same time, the bean bug suppresses the immune responses of the symbiotic midgut region to support the survival of *Burkholderia* symbionts.

To understand the complete picture of the symbiosis and immunity interconnection, it is necessary to elucidate the biological mechanisms of cell envelope changes of the *Burkholderia* symbionts. In several symbiosis models, modification of the bacterial cell envelope by the host's molecules has been reported. The *Drosophila* model showed that peptidoglycans of gut symbionts are hydrolysed by the host's peptidoglycan recognition protein-LB (PGRP-LB) harbouring amidase activity to prevent the overactivation of the innate immune responses (Bischoff et al., 2006; Paredes et al., 2011). Similarly, LPS from commensal bacteria is dephosphorylated by the host's intestinal alkaline phosphatase in the zebrafish gut model (Bates et al., 2007). These mechanisms are important in the modulation of gut homeostasis. Therefore, identification of the host molecules involved in the alteration of the cell envelopes of *Burkholderia* symbionts would provide important information for understanding the homeostatic mechanism of gut symbiosis.

Bean bug immunity itself is a very valuable subject to pursue in understanding a hemimetabolous insect's immunity. The genomic and transcriptomic analyses of the bean bug will provide information regarding the presence and expression of immune-related genes. With information of the bean bug's immunity, a comprehensive understanding of the molecular crosstalk between the host insects and symbionts will be achievable.

ACKNOWLEDGEMENTS

This study was supported by the Global Research Laboratory Program (Grant No. 2011-0021535) and Basic Science Research Program (Grant No. 2014R1A1A4A01007507) of the National Research Foundation of Korea.

REFERENCES

Akman, L., Rio, R.V., Beard, C.B., Aksoy, S., 2001. Genome size determination and coding capacity of *Sodalis glossinidius*, an enteric symbiont of tsetse flies, as revealed by hybridization to *Escherichia coli* gene arrays. J. Bacteriol. 183, 4517–4525.

Akman, L., Yamashita, A., Watanabe, H., Oshima, K., Shiba, T., Hattori, M., Aksoy, S., 2002. Genome sequence of the endocellular obligate symbiont of tsetse flies, *Wigglesworthia glossinidia*. Nat. Genet. 32, 402–407.

Aksoy, S., 1995. *Wigglesworthia* gen. nov. and *Wigglesworthia glossinidia* sp. nov., taxa consisting of the mycetocyte-associated, primary endosymbionts of tsetse flies. Int. J. Syst. Bacteriol. 45, 848–851.

Bao, Y.Y., Qu, L.Y., Zhao, D., Chen, L.B., Jin, H.Y., Xu, L.M., Cheng, J.A., Zhang, C.X., 2013. The genome- and transcriptome-wide analysis of innate immunity in the brown planthopper, *Nilaparvata lugens*. BMC Genomics 14, 160.

Bates, J.M., Akerlund, J., Mittge, E., Guillemin, K., 2007. Intestinal alkaline phosphatase detoxifies lipopolysaccharide and prevents inflammation in zebrafish in response to the gut microbiota. Cell Host Microbe 2, 371–382.

Baumann, P., 2005. Biology bacteriocyte-associated endosymbionts of plant sap-sucking insects. Annu. Rev. Microbiol. 59, 155–189.

Baumann, P., Moran, N.A., 1997. Non-cultivable microorganisms from symbiotic associations of insects and other hosts. Antonie Van Leeuwenhoek 72, 39–48.

Bischoff, V., Vignal, C., Duvic, B., Boneca, I.G., Hoffmann, J.A., Royet, J., 2006. Down-regulation of the *Drosophila* immune response by peptidoglycan-recognition proteins SC1 and SC2. PLoS Pathog. 2, e14.

Bistolas, K.S., Sakamoto, R.I., Fernandes, J.A., Goffredi, S.K., 2014. Symbiont polyphyly, co-evolution, and necessity in pentatomid stinkbugs from Costa Rica. Front. Microbiol. 5, 349.

Braendle, C., Miura, T., Bickel, R., Shingleton, A.W., Kambhampati, S., Stern, D.L., 2003. Developmental origin and evolution of bacteriocytes in the aphid-*Buchnera* symbiosis. PLoS Biol. 1, 70–76.

Brennan, L.J., Keddie, B.A., Braig, H.R., Harris, H.L., 2008. The endosymbiont *Wolbachia pipientis* induces the expression of host antioxidant. PLoS One 3, e2083.

Buchner, P., 1965. Endosymbiosis of Animals With Plant Microorganisms. Interscience Publishers, New York.

Buchon, N., Silverman, N., Cherry, S., 2014. Immunity in *Drosophila melanogaster*—from microbial recognition to whole-organism physiology. Nat. Rev. Immunol. 14, 796–810.

Byeon, J.H., Seo, E.S., Lee, J.B., Lee, M.J., Kim, J.K., Yoo, J.W., Jung, Y., Lee, B.L., 2015. A specific cathepsin-L-like protease purified from an insect midgut shows antibacterial activity against gut symbiotic bacteria. Dev. Comp. Immunol. 53, 79–84.

Costerton, J.W., Ingram, J.M., Cheng, K.J., 1974. Structure and function of the cell envelope of gram-negative bacteria. Bacteriol. Rev. 38, 87–110.

Doudoumis, V., Alam, U., Aksoy, E., Abd-Alla, A.M., Tsiamis, G., Brelsfoard, C., Aksoy, S., Bourtzis, K., 2013. Tsetse-*Wolbachia* symbiosis: comes of age and has great potential for pest and disease control. J. Invertebr. Pathol. 112 (Suppl.), S94–S103.

Douglas, A.E., 1998. Nutritional interactions in insect-microbial symbioses: aphids and their symbiotic bacteria *Buchnera*. Annu. Rev. Entomol. 43, 17–37.

Futahashi, R., Tanaka, K., Tanahashi, M., Nikoh, N., Kikuchi, Y., Lee, B.L., Fukatsu, T., 2013. Gene expression in gut symbiotic organ of stinkbug affected by extracellular bacterial symbiont. PLoS One 8, e64557.

Gerardo, N.M., Altincicek, B., Anselme, C., Atamian, H., Barribeau, S.M., de Vos, M., Duncan, E.J., Evans, J.D., Gabaldon, T., Ghanim, M., Heddi, A., Kaloshian, I.,

Latorre, A., Moya, A., Nakabachi, A., Parker, B.J., Perez-Brocal, V., Pignatelli, M., Rahbe, Y., Ramsey, J.S., Spragg, C.J., Tamames, J., Tamarit, D., Tamborindeguy, C., Vincent-Monegat, C., Vilcinskas, A., 2010. Immunity and other defenses in pea aphids, Acyrthosiphon pisum. Genome Biol. 11, R21.

Goodrich-Blair, H., Clarke, D.J., 2007. Mutualism and pathogenesis in *Xenorhabdus* and *Photorhabdus*: two roads to the same destination. Mol. Microbiol. 64, 260–268.

Grimaldi, D., Engel, M.S., 2005. Evolution of the Insects. Cambridge University Press, New York.

Hamilton, P.T., Perlman, S.J., 2013. Host defense via symbiosis in *Drosophila*. PLoS Pathog. 9, e1003808.

Hedges, L.M., Brownlie, J.C., O'Neill, S.L., Johnson, K.N., 2008. *Wolbachia* and virus protection in insects. Science 322, 702.

Iwanaga, S., Lee, B.L., 2005. Recent advances in the innate immunity of invertebrate animals. J. Biochem. Mol. Biol. 38, 128–150.

Kikuchi, Y., Yumoto, I., 2013. Efficient colonization of the bean bug *Riptortus pedestris* by an environmentally transmitted *Burkholderia* symbiont. Appl. Environ. Microbiol. 79, 2088–2091.

Kikuchi, Y., Meng, X.Y., Fukatsu, T., 2005. Gut symbiotic bacteria of the genus *Burkholderia* in the broad-headed bugs *Riptortus clavatus* and *Leptocorisa chinensis* (Heteroptera: Alydidae). Appl. Environ. Microbiol. 71, 4035–4043.

Kikuchi, Y., Hosokawa, T., Fukatsu, T., 2007. Insect-microbe mutualism without vertical transmission: a stinkbug acquires a beneficial gut symbiont from the environment every generation. Appl. Environ. Microbiol. 73, 4308–4316.

Kikuchi, Y., Hosokawa, T., Nikoh, N., Meng, X.Y., Kamagata, Y., Fukatsu, T., 2009. Host-symbiont co-speciation and reductive genome evolution in gut symbiotic bacteria of acanthosomatid stinkbugs. BMC Biol. 7, 2.

Kikuchi, Y., Hosokawa, T., Fukatsu, T., 2011. An ancient but promiscuous host-symbiont association between *Burkholderia* gut symbionts and their heteropteran hosts. ISME J. 5, 446–460.

Kim, J.K., Lee, B.L., 2015. Symbiotic factors in *Burkholderia* essential for establishing an association with the bean bug, *Riptortus pedestris*. Arch. Insect Biochem. Physiol. 88, 4–17.

Kim, J.K., Kim, N.H., Jang, H.A., Kikuchi, Y., Kim, C.H., Fukatsu, T., Lee, B.L., 2013a. Specific midgut region controlling the symbiont population in an insect-microbe gut symbiotic association. Appl. Environ. Microbiol. 79, 7229–7233.

Kim, J.K., Won, Y.J., Nikoh, N., Nakayama, H., Han, S.H., Kikuchi, Y., Rhee, Y.H., Park, H.Y., Kwon, J.Y., Kurokawa, K., Dohmae, N., Fukatsu, T., Lee, B.L., 2013b. Polyester synthesis genes associated with stress resistance are involved in an insect-bacterium symbiosis. Proc. Natl. Acad. Sci. U.S.A. 110 (26), E2381–E2389.

Kim, J.K., Han, S.H., Kim, C.H., Jo, Y.H., Futahashi, R., Kikuchi, Y., Fukatsu, T., Lee, B.L., 2014a. Molting-associated suppression of symbiont population and up-regulation of antimicrobial activity in the midgut symbiotic organ of the *Riptortus-Burkholderia* symbiosis. Dev. Comp. Immunol. 43, 10–14.

Kim, J.K., Jang, H.A., Won, Y.J., Kikuchi, Y., Heum Han, S., Kim, C.H., Nikoh, N., Fukatsu, T., Lee, B.L., 2014b. Purine biosynthesis-deficient *Burkholderia* mutants are incapable of symbiotic accommodation in the stinkbug. ISME J. 8, 552–563.

Kim, J.K., Kwon, J.Y., Kim, S.K., Han, S.H., Won, Y.J., Lee, J.H., Kim, C.H., Fukatsu, T., Lee, B.L., 2014c. Purine biosynthesis, biofilm formation, and persistence of an insect-microbe gut symbiosis. Appl. Environ. Microbiol. 80, 4374–4382.

Kim, J.K., Lee, J.B., Huh, Y.R., Jang, H.A., Kim, C.H., Yoo, J.W., Lee, B.L., 2015a. *Burkholderia* gut symbionts enhance the innate immunity of host *Riptortus pedestris*. Dev. Comp. Immunol. 53, 265–269.

Kim, J.K., Son, D.W., Kim, C.H., Cho, J.H., Marchetti, R., Silipo, A., Sturiale, L., Park, H.Y., Huh, Y.R., Nakayama, H., Fukatsu, T., Molinaro, A., Lee, B.L., 2015b. Insect gut symbiont susceptibility to host antimicrobial peptides caused by alteration of the bacterial cell envelope. J. Biol. Chem. 290, 21042–21053.

Koga, R., Tsuchida, T., Fukatsu, T., 2003. Changing partners in an obligate symbiosis: a facultative endosymbiont can compensate for loss of the essential endosymbiont *Buchnera* in an aphid. Proc. Biol. Sci. 270, 2543–2550.

Lemaitre, B., Hoffmann, J., 2007. The host defense of *Drosophila melanogaster*. Annu. Rev. Immunol. 25, 697–743.

Login, F.H., Balmand, S., Vallier, A., Vincent-Monegat, C., Vigneron, A., Weiss-Gayet, M., Rochat, D., Heddi, A., 2011. Antimicrobial peptides keep insect endosymbionts under control. Science 334, 362–365.

Loutet, S.A., Valvano, M.A., 2011. Extreme antimicrobial peptide and polymyxin B resistance in the genus *Burkholderia*. Front. Microbiol. 2, 159.

Mahenthiralingam, E., Urban, T.A., Goldberg, J.B., 2005. The multifarious, multireplicon *Burkholderia cepacia* complex. Nat. Rev. Microbiol. 3, 144–156.

Mateos, M., Castrezana, S.J., Nankivell, B.J., Estes, A.M., Markow, T.A., Moran, N.A., 2006. Heritable endosymbionts of *Drosophila*. Genetics 174, 363–376.

McCutcheon, J.P., Moran, N.A., 2012. Extreme genome reduction in symbiotic bacteria. Nat. Rev. Microbiol. 10, 13–26.

McGraw, E.A., Merritt, D.J., Droller, J.N., O'Neill, S.L., 2002. *Wolbachia* density and virulence attenuation after transfer into a novel host. Proc. Natl. Acad. Sci. U.S.A. 99, 2918–2923.

Mira, A., Moran, N.A., 2002. Estimating population size and transmission bottlenecks in maternally transmitted endosymbiotic bacteria. Microb. Ecol. 44, 137–143.

Moran, N.A., 2003. Tracing the evolution of gene loss in obligate bacterial symbionts. Curr. Opin. Microbiol. 6, 512–518.

Moran, N.A., McCutcheon, J.P., Nakabachi, A., 2008. Genomics and evolution of heritable bacterial symbionts. Annu. Rev. Genet. 42, 165–190.

Moreira, L.A., Iturbe-Ormaetxe, I., Jeffery, J.A., Lu, G., Pyke, A.T., Hedges, L.M., Rocha, B.C., Hall-Mendelin, S., Day, A., Riegler, M., Hugo, L.E., Johnson, K.N., Kay, B.H., McGraw, E.A., van den Hurk, A.F., Ryan, P.A., O'Neill, S.L., 2009. A *Wolbachia* symbiont in *Aedes aegypti* limits infection with dengue, Chikungunya, and Plasmodium. Cell 139, 1268–1278.

Mouton, L., Dedeine, F., Henri, H., Bouletreau, M., Profizi, N., Vavre, F., 2004. Virulence, multiple infections and regulation of symbiotic population in the *Wolbachia-Asobara tabida* symbiosis. Genetics 168, 181–189.

Nakabachi, A., Shigenobu, S., Sakazume, N., Shiraki, T., Hayashizaki, Y., Carninci, P., Ishikawa, H., Kudo, T., Fukatsu, T., 2005. Transcriptome analysis of the aphid bacteriocyte, the symbiotic host cell that harbors an endocellular mutualistic bacterium, *Buchnera*. Proc. Natl. Acad. Sci. U.S.A. 102, 5477–5482.

Nishikori, K., Morioka, K., Kubo, T., Morioka, M., 2009. Age- and morph-dependent activation of the lysosomal system and *Buchnera* degradation in aphid endosymbiosis. J. Insect Physiol. 55, 351–357.

Nyholm, S.V., McFall-Ngai, M.J., 2004. The winnowing: establishing the squid-vibrio symbiosis. Nat. Rev. Microbiol. 2, 632–642.

Oliver, K.M., Degnan, P.H., Burke, G.R., Moran, N.A., 2010. Facultative symbionts in aphids and the horizontal transfer of ecologically important traits. Annu. Rev. Entomol. 55, 247–266.

Paredes, J.C., Welchman, D.P., Poidevin, M., Lemaitre, B., 2011. Negative regulation by amidase PGRPs shapes the *Drosophila* antibacterial response and protects the fly from innocuous infection. Immunity 35, 770–779.

Pontes, M.H., Dale, C., 2006. Culture and manipulation of insect facultative symbionts. Trends Microbiol. 14, 406–412.
Raetz, C.R., Whitfield, C., 2002. Lipopolysaccharide endotoxins. Annu. Rev. Biochem. 71, 635–700.
Ricci, I., Valzano, M., Ulissi, U., Epis, S., Cappelli, A., Favia, G., 2012. Symbiotic control of mosquito borne disease. Pathog. Glob. Health 106, 380–385.
Sakurai, M., Koga, R., Tsuchida, T., Meng, X.Y., Fukatsu, T., 2005. *Rickettsia* symbiont in the pea aphid *Acyrthosiphon pisum*: novel cellular tropism, effect on host fitness, and interaction with the essential symbiont *Buchnera*. Appl. Environ. Microbiol. 71, 4069–4075.
Salem, H., Florez, L., Gerardo, N., Kaltenpoth, M., 2015. An out-of-body experience: the extracellular dimension for the transmission of mutualistic bacteria in insects. Proc. Biol. Sci. 282, 20142957.
Sassera, D., Epis, S., Pajoro, M., Bandi, C., 2013. Microbial symbiosis and the control of vector-borne pathogens in tsetse flies, human lice, and triatomine bugs. Pathog. Glob. Health 107, 285–292.
Shibata, T.F., Maeda, T., Nikoh, N., Yamaguchi, K., Oshima, K., Hattori, M., Nishiyama, T., Hasebe, M., Fukatsu, T., Kikuchi, Y., Shigenobu, S., 2013. Complete genome sequence of *Burkholderia* sp. strain RPE64, bacterial symbiont of the bean bug *Riptortus pedestris*. Genome Announc. 1, e00441-13.
Shigenobu, S., Wilson, A.C., 2011. Genomic revelations of a mutualism: the pea aphid and its obligate bacterial symbiont. Cell. Mol. Life Sci. 68, 1297–1309.
Shigenobu, S., Watanabe, H., Hattori, M., Sakaki, Y., Ishikawa, H., 2000. Genome sequence of the endocellular bacterial symbiont of aphids *Buchnera* sp. APS. Nature 407, 81–86.
Silhavy, T.J., Kahne, D., Walker, S., 2010. The bacterial cell envelope. Cold Spring Harb. Perspect. Biol. 2, a000414.
Tamas, I., Klasson, L., Canback, B., Naslund, A.K., Eriksson, A.S., Wernegreen, J.J., Sandstrom, J.P., Moran, N.A., Andersson, S.G.E., 2002. 50 million years of genomic stasis in endosymbiotic bacteria. Science 296, 2376–2379.
Toft, C., Andersson, S.G.E., 2010. Evolutionary microbial genomics: insights into bacterial host adaptation. Nat. Rev. Genet. 11, 465–475.
Walker, T., Johnson, P.H., Moreira, L.A., Iturbe-Ormaetxe, I., Frentiu, F.D., McMeniman, C.J., Leong, Y.S., Dong, Y., Axford, J., Kriesner, P., Lloyd, A.L., Ritchie, S.A., O'Neill, S.L., Hoffmann, A.A., 2011. The wMel *Wolbachia* strain blocks dengue and invades caged *Aedes aegypti* populations. Nature 476, 450–453.
Wang, J., Weiss, B.L., Aksoy, S., 2013. Tsetse fly microbiota: form and function. Front. Cell. Infect. Microbiol. 3, 69.
Weirauch, C., Schuh, R.T., 2011. Systematics and evolution of Heteroptera: 25 years of progress. Annu. Rev. Entomol. 56, 487–510.
Weiss, B.L., Wang, J., Aksoy, S., 2011. Tsetse immune system maturation requires the presence of obligate symbionts in larvae. PLoS Biol. 9, e1000619.
Weiss, B.L., Maltz, M., Aksoy, S., 2012. Obligate symbionts activate immune system development in the tsetse fly. J. Immunol. 188, 3395–3403.
Werren, J.H., Baldo, L., Clark, M.E., 2008. *Wolbachia*: master manipulators of invertebrate biology. Nat. Rev. Microbiol. 6, 741–751.

CHAPTER SEVEN

Exploiting Innate Immunity for Biological Pest Control

Fei Liu*,†, Wuren Huang‡, Kai Wu§, Zhongying Qiu*, Yuan Huang*, Erjun Ling‡,1

*College of Life Sciences, Shaanxi Normal University, Xi'an, Shaanxi, China
†College of Life Sciences and Food Engineering, Shaanxi Xueqian Normal University, Xi'an, Shaanxi, China
‡Key Laboratory of Insect Developmental and Evolutionary Biology, Institute of Plant Physiology and Ecology, Shanghai Institutes for Biological Sciences, Chinese Academy of Sciences, Shanghai, China
§School of Life Science, Shangrao Normal University, Shangrao, Jiangxi, China
[1]Corresponding author: e-mail addresses: ejling@sibs.ac.cn; erjunling@sippe.ac.cn

Contents

1. Biological Pest Control 200
2. Insect Innate Immunity 201
3. Agents for Biological Pest Control and Their Relationship With Host Immune Responses 205
   3.1 Entomopathogenic Bacteria 205
   3.2 Entomopathogenic Viruses 207
   3.3 Entomopathogenic Fungi 208
   3.4 Parasitoids 210
4. The DUOX Pathway in the Gut Is Crucial in Protecting Host Development 212
5. Additional Factors Involved in Gut Protection 213
6. Immunodeficiency in Insects 214
7. Prospects of Increasing Host Susceptibility to Pathogens Through Immune Suppression 215
8. Outlook 218
Acknowledgements 219
References 219

Abstract

Insects are the most abundant animals on earth, and many are agriculture pests. Currently, we have to resort to chemical pesticides for suppressing pest populations, which lead to environmental pollution, health risk and ecological imbalance. The living environments of most insects are full of various species of pathogens that can enter insects via the mouths, tracheae and wounds. In this review, we summarize on the insect immune responses against pathogens and correspondingly the pathogenic suppression on the host innate immunity. Many individuals with mutations in important immune-related genes are genetically manipulated in laboratories to show low rates of survival even when kept in conventionally reared environment, which indicates that

Advances in Insect Physiology, Volume 52
ISSN 0065-2806
http://dx.doi.org/10.1016/bs.aiip.2017.02.001

inhibition on the innate immunity could be a potential mode of biological pest control. With the knowledge accumulated on insect immunity, it is time for us to consider developing novel methods for biological pest control by suppressing immune activity.

Insects are the most abundant animals on earth. Insects such as silkworms and bees are economically important and are vital to agricultural production (Morse and Calderone, 1999; Xia et al., 2004). Many insects are pests that cause significant damage to agriculture and threaten human health (Chowanski et al., 2016). So far, we still have to depend on chemical insecticides for pest control. However, application of chemical pesticides causes many environmental, ecological and health-related problems (Damalas and Eleftherohorinos, 2011). Therefore, it is urgent to seek new strategies for biological pest control to replace the application of chemical pesticides. Many microorganisms (bacteria, fungi and viruses), parasitoids and nematodes have been applied as biopesticides for biological pest control (Glare et al., 2012; Lacey et al., 2015). However, after entering the bodies of insects via wounds, the tracheae or the guts, these microorganisms and parasitoids are detected and defended against by the innate immune system of the insect (reviewed in Kounatidis and Ligoxygakis, 2012; Lemaitre and Hoffmann, 2007). Unlike mammalians, insects have no adaptive immune system (see Gillespie et al., 1997; Jiang et al., 2010; Kanost et al., 2004; Lavine and Strand, 2002; Strand, 2008; Tanaka and Yamakawa, 2011). However, their innate immune system is sensitive and effective in defending against most entomopathogenic microorganisms and parasitoids (Lemaitre and Hoffmann, 2007). Based on the accumulated knowledge about immune responses in insects, there is now a very clear understanding of immune pathways leading to pathogen clearance and their underlying mechanisms. The question therefore arises as to whether we can utilize this knowledge for biological pest control. In this review, we summarize the research concerning the immune responses found in insects against agents such as microorganisms and parasitoids that may be utilized for biological pest control. Evidence regarding the suppression of the host innate immunity by pathogens and the existence of immune-deficient insects produced in laboratories indicate that it may be possible to regulate the innate immunity of insects for biological pest control in the future.

1. BIOLOGICAL PEST CONTROL

Pests cause huge losses in agricultural and forest crops, and some of them are even vectors for animal diseases. So far, chemical pesticides are still the main choice for pest control. The mass production and widespread use of synthetic pesticides have effectively controlled the hazards caused by many pests and made tremendous contributions to agricultural production (Damalas and Eleftherohorinos, 2011). However, the inappropriate use of synthetic insecticides also leads to the development of resistance in pests, and the insecticide residues in the soil, water and food threaten human health (Damalas and Eleftherohorinos, 2011; Kogan, 1998). The application of chemical pesticides also disturbs the balance of ecosystems and reduces biological diversity, which results in a series of ecological, environmental and social issues (Damalas and Eleftherohorinos, 2011). These problems have directed the development of biological control strategies that result in less environmental pollution and reduce health risks.

Biological control agents include beneficial organisms (bacteria, viruses, fungi, parasitoids, protozoa and nematodes) and/or their metabolites (hormones or extractives), which are used to suppress the pest population (Glare et al., 2012; Lacey et al., 2001). The majority of commercially available microbial products are based on the species *Bacillus thuringiensis* (Bt) (Gonzalez et al., 2016; Lacey et al., 2015), which is a Gram-positive soil bacterium that can produce crystal (Cry) and cytolytic (Cyt) toxins that are normally referred to as Bt proteins (Bravo et al., 2007). These toxins have been widely applied for controlling pests of cotton and other crops that have been modified via transgenesis to express Bt proteins (Lacey et al., 2015). However, many pests have already shown an increased resistance to Bt toxins (Lacey et al., 2015), which leaves us with a serious problem. Therefore, it is necessary to understand host immune responses against each microorganism or parasitoid and to develop strategies to suppress host immunity activity if we hope to enhance the virulence of biopesticides.

2. INSECT INNATE IMMUNITY

As described below, insect host defence has several layers.

(1) *Physical barriers:* Physical barriers include the cuticle, the peritrophic membrane of the gut and the tracheal system (Christophides et al., 2004; Hillyer, 2016). They keep pathogens from entering the body cavity and reduce the probability of infection. (2) *Cellular immunity:* Cellular immunity leads to encapsulation, phagocytosis and nodulation of pathogens. It is primarily mediated by circulating haemocytes—the insect equivalent of the mammalian phagocytes (Gonzalez et al., 2013; Lavine and Strand, 2002; Strand, 2008). (3) *Humoral immunity:* The hallmark of humoral immunity

is the synthesis and production by the fat body of a potent mix of antimicrobial peptides. In addition, a proteolyic cascade leads to the production of phenoloxidase (PO) from prophenoloxidase (PPO), a fundamental element for melanization of pathogens too big (or too numerous) for phagocytosis (Gillespie et al., 1997; Jiang et al., 2010; Kanost et al., 2004; Lavine and Strand, 2002; Liu et al., 2009; Strand, 2008; Tanaka and Yamakawa, 2011). The production of bactericidal reactive oxygen species (ROS) in haemocytes and the midgut is also an important part of innate immunity (Buchon et al., 2014; Kim and Lee, 2014; Lee et al., 2015a). Viruses are eliminated through RNAi, apoptosis and autophagy in infected cells (Buchon et al., 2014; Clem, 2005; Nakamoto et al., 2012; Wu et al., 2016). The innate immune responses described above take place in the haemocoel and midgut of the host upon the detection of the invading pathogens.

Pathogen-associated molecular patterns (PAMPs) are highly conserved motifs, which are present in pathogenic microorganisms but absent in insects (Buchon et al., 2014; Hillyer, 2016; Lemaitre and Hoffmann, 2007). PAMPs include lipoteichoic acid (LTA) from Gram-positive bacteria, lipopolysaccharide (LPS) from Gram-negative bacteria, peptidoglycan (PGN) from Gram-positive and Gram-negative bacteria, and ß-1,3-glucan from fungi (Cerenius et al., 2008; Hillyer, 2016; Palmer and Jiggins, 2015). The activation of both humoral and cellular immune responses depends on the recognition of pathogens, which is mediated through a specific interaction between host-derived pattern-recognition receptors (PRRs) and PAMPs on the surface of the microbes (Hillyer, 2016; Hughes, 2012). Several PRRs have been identified. Their functions in regulating insect innate immunity have also been carefully studied, and they mainly consist of peptidoglycan-recognition proteins (PGRPs), Gram-negative binding proteins (GNBPs), C-type lectins, thioester-containing proteins, fibrinogen-related proteins (FREPs), galectins and immunoglobulin domain proteins (Palmer and Jiggins, 2015; Zhang et al., 2015).

Two classical humoral signalling pathways that produce antimicrobial peptides have been studied in *Drosophila melanogaster* and other insects: the Toll pathway and the immune deficiency (Imd) pathway (Hoffmann, 2003; Lemaitre and Hoffmann, 2007; Wang and Ligoxygakis, 2006). The Toll signalling pathway is mainly activated by Gram-positive bacteria and fungi, and the Imd pathway is mainly activated by Gram-negative bacteria (Buchon et al., 2014; Hoffmann, 2003; Lemaitre and Hoffmann, 2007). Under the regulation of these two pathways, antimicrobial peptides are produced and released into the haemolymph, where they kill the invading bacteria and fungi (Hoffmann, 2003; Lemaitre and Hoffmann, 2007). In addition to the Toll and Imd pathways, there are other pathways that participate in immune responses in insects, such as the JAK/STAT, JNK, p38 and insulin pathways (Hillyer, 2016; Wu et al., 2016).

Cellular immunity is primarily mediated by circulating haemocytes (Strand, 2008). However, except from *Drosophila* and mosquitoes and compared with humoral immunity, we have very limited information on the interactions between haemocyte PRRs and PAMPs derived from foreign bodies and how these interactions activate haemocyte-mediated responses. Insects have several types of haemocytes. In the larvae of *Bombyx mori*, there are four types of differentiated circulating haemocytes, which all develop from prohaemocytes: granulocytes, oenocytoids, plasmatocytes and spherulocytes (spherule cells) (Liu et al., 2013; Strand, 2008). However, in *D. melanogaster*, there are three types of differentiated haemocytes, which also develop from prohaemocytes: plasmatocytes, crystal cells and lamellocytes (Liu et al., 2013; Strand, 2008). Insect haemocytes are mainly involved in the phagocytosis of invading small microbes and encapsulating large parasitoids and nematodes that cannot be phagocytosed (Strand, 2008). However, each type of haemocytes also has its specific functions. In *Drosophila*, the primary function of plasmatocytes is to phagocytose microbial invaders and dead cells (Buchon et al., 2014; Liu et al., 2013; Tang, 2009). Crystal cells produce PPO (Buchon et al., 2014; Liu et al., 2013; Tang, 2009). The main function of lamellocytes is to encapsulate parasitoids and large pathogens (Tang, 2009). Haemocytes can also produce antimicrobial peptides to induce humoral immune responses (Strand, 2008). Upon the invasion of pathogens, the number of haemocytes may fluctuate although the mechanism for this is unclear (Strand, 2008).

PPO is an important innate immune protein that mainly exists in haemolymph (Ashida and Brey, 1998; Kanost et al., 2004; Lu et al., 2014). This protein is also detected in the foregut and hindgut of insects, where it has different functions (Shao et al., 2012; Wu et al., 2015). PPO is involved in both cellular and humoral immunity (Lemaitre and Hoffmann, 2007; Lu et al., 2014). Upon detection, PPO binds to invading pathogens via some other proteins and induces melanization around the pathogens to help kill them in the haemocoel (Ashida and Brey, 1998; Kanost et al., 2004; Lu et al., 2014), which is a part of humoral immunity. When haemocyte-based encapsulation occurs around large parasitoids, melanization is also induced (Strand, 2008), which is a part of cellular immunity. PPO must be cleaved at a conserved sequence to form activated phenoloxidase (PO), which is regulated by a cascade composed of serine proteases and serine proteases inhibitors (serpins) (Kanost et al., 2004; Lu et al., 2014). PO catalyses the oxidation of phenols to quinones, which subsequently polymerize into melanin (Ashida and Brey, 1998; Kanost et al., 2004; Lu et al., 2014). In the genomes of most insect species, there are only one to three *PPO* genes. In mosquitoes, there are 9 *PPO* genes in *Anopheles gambiae* and 10 in *Aedes aegypti* (Waterhouse et al., 2007; Zou et al., 2010). In *Drosophila*, there are three PPO genes. *PPO1* and *PPO2* are primarily expressed in crystal cells, while *PPO3* is mainly expressed in lamellocytes that are differentiated from prohaemocytes after parasitoid wasp infections (Wertheim et al., 2005). When *PPO1* and *PPO2* are deleted in *Drosophila*, the mutants show decreased immune activity against many pathogens (Binggeli et al., 2014), which indicates that PPO is a very important immune protein.

The insect digestive tract is composed of foregut, midgut and hindgut (Lehane and Billingsley, 1996; Mistry et al., 2016; Wu et al., 2016). Since there are microbes on the foods that insects ingest, to avoid intestinal infections, the gut epithelium produces antimicrobial peptides through the Imd pathway and produces ROS via dual oxidase (DUOX) and NADPH oxidase (Nox) to destroy pathogens in the midgut (Buchon et al., 2014; Engel and Moran, 2013; Kim and Lee, 2014). In phytophagous insects, the hindgut epithelium can produce PPO to kill microorganisms in the faeces via melanization, thereby reducing the bacterial flora in the environment (Shao et al., 2012). In the foregut, toxic plant phenolics are metabolized and detoxified by PPO (Wu et al., 2015). These mechanisms demonstrate that the digestive tract of insects has a complete protection system.

In addition to their protective function, many immunity proteins also help regulate development and various physiological functions in the host. Toll is a key gene for the production of Drosomycin and other antimicrobial peptides that act to defend against Gram-positive bacteria and fungi (Lemaitre and Hoffmann, 2007). During the progress of embryo development, Toll also determines the formation of the embryonic dorsal–ventral pattern (Gerttula et al., 1988). PPO is an important immune protein for inducing melanization around invading pathogens (Lemaitre and Hoffmann, 2007; Lu et al., 2014). However, this protein is also closely involved with wound healing (Ashida and Brey, 1998; Kanost et al., 2004; Lu et al., 2014; Tang, 2009). Therefore, the immune systems are very important to insects, not only because of their role in immune protection but also because of their involvement in development.

3. AGENTS FOR BIOLOGICAL PEST CONTROL AND THEIR RELATIONSHIP WITH HOST IMMUNE RESPONSES

For most insects, including pests, there are many bacteria, fungi and viruses in their living environment (Kounatidis and Ligoxygakis, 2012). These microorganisms or parasitoids can enter the insect bodies through the guts, tracheae or wounds (Kounatidis and Ligoxygakis, 2012). Owing to the protection provided by their simple but efficient innate immune system, insects can successfully develop. However, if the immune system is impaired, is it still possible for insects, including pests, to grow and develop normally? To answer this question, it is necessary to know how pathogens and hosts interact. As biopesticides, bacteria and viruses can infect the insects orally, and fungi and parasitoids may break through the integument of the host and reach the haemocoel. The immune responses against bacteria and viruses are also discussed in other sections in detail.

3.1 Entomopathogenic Bacteria

Normally, it is not easy for bacteria to break through the integument and reach the haemocoel unless there are wounds in the integument. Nevertheless, predators can inflect such wounds that make systemic infection a real possibility in some environments. In addition, food ingested by insects may contain many bacteria. Therefore, the oral route is also an important pathway for microbes to enter the bodies of insects. The immune responses in the gut against bacteria have been extensively studied. There are two main strategies that insects utilize to defend against bacteria in food: the production of antimicrobial peptides and ROS in the gut (Engel and Moran, 2013; Kim and Lee, 2014). In insect midguts, PGRP-LE and PGRP-LC bind to PGN, activating the Imd pathway to produce antimicrobial peptides (reviewed in Kim and Lee, 2014; Mistry et al., 2016). When the bacteria *Erwinia carotovora* was used to orally infect adult *Drosophila*, the Imd, JAK/STAT and EGFR pathways were activated to produce antimicrobial peptides to kill the pathogens and repair the epithelial damage in the gut (Buchon et al., 2009). Compared with wild-type flies, when *E. carotovora* was used to infect mutants with defective Imd components, bacteria persisted in higher numbers (Mistry et al., 2016). In addition to the production of antimicrobial peptides, upon oral infection, the transcription of DUOX was also induced (Buchon et al., 2014). DUOX can produce ROS, which is an important response in eliminating pathogenic bacteria from the midgut (Kim and Lee, 2014;

Lee et al., 2015a; Mistry et al., 2016). In insect midguts, the activation of DUOX is not PGN dependent (Kim and Lee, 2014). Pathogenic bacteria release uracil into the midgut of the host, which can be detected and bound by G protein-coupled receptors (GPCRs), activating DUOX through the $G\alpha q$-PLCβ-Ca^{2+} pathway and regulating the production of ROS (Kim and Lee, 2014; Lee et al., 2015a). In *B. mori* larvae, the midgut expresses DUOX upon *Escherichia coli* and nucleopolyhedrovirus (NPV) oral infections (Hu et al., 2013). When DUOX was knocked down, the number of *E. coli* surviving on the peritrophic membrane increased (Hu et al., 2013). In insects, when the production of ROS is excessive, they can also cause oxidative damage to the guts. In *B. mori*, peroxiredoxins (Prxs), a type of antioxidant enzyme, balance ROS levels in the guts to avoid the damage caused by excessive levels of ROS (Zhang and Lu, 2015). In *Drosophila*, immune-regulated catalase (IRC) degrades H_2O_2 to avoid oxidative stress as a result of excessive ROS levels (Ha et al., 2005b). Some bacteria can produce insecticidal toxins that damage the midgut after being ingested (Crava et al., 2015; Ferré and Rie, 2002). Bt toxins, such as Cry and Cyt, secreted by *B. thuringiensis* can directly cause serious damage to insect midguts (Bravo et al., 2007). Some toxins may interrupt host gene transcription and translation. When adult *Drosophila* were orally infected by *Pseudomonas entomophila*, protein translation in midgut cells was significantly blocked as a result of pore-forming toxins and ROS production (Chakrabarti et al., 2012). In *Drosophila*, *E. carotovora* can also damage the guts, which induces stem cell proliferation and epithelial renewal via the JAK/STAT and EGFR pathways (Buchon et al., 2009). Oral bacterial infections induce immune responses in the midgut, and epithelial renewal is necessary to allow the host to recover from damage in the midgut caused by various factors.

Measurements of transcriptional changes in the midguts of Lepidoptera orally infected with different types of bacteria show that many immunity-related genes are also changed (Wu et al., 2016). When *B. mori* larvae were fed with *Bacillus bombysepticus*, antimicrobial peptides in the midgut, such as attacin, lebocin, enbocin, gloverin and moricin, were all upregulated (Huang et al., 2009). After feeding *B. mori* larvae with *Pseudomonas aeruginosa*, immune-related genes, such as PGRP-L1, the serine protease precursor gene and 30 kP protease A were significantly upregulated (Zhu and Lu, 2013). In the gypsy moth, *Lymantria dispar*, gut immune responses were found to occur upon oral infection with *B. thuringiensis* (Sparks et al., 2013). Obviously, insect hosts have extensive responses to oral bacterial infections to ensure normal development.

3.2 Entomopathogenic Viruses

Entomopathogenic viruses are an important type of biopesticide. Viruses that have been applied for biological pest control include the NPV, granuloviruses (GVs), cytoplasmic polyhedrosis viruses (CPVs) and densonucleosis viruses (DNVs) (Lacey et al., 2015). Most viruses invade insects via the oral route. Some viruses proliferate in the epithelial cells of the midgut, and the others transfer into the haemocoel cavity to infect different tissues after breaking through the midgut (Cheng and Lynn, 2009). When viruses reach the midgut, changes in gene transcription are induced in many species of insects (Gao et al., 2014; Kolliopoulou et al., 2015; Wu et al., 2013). Some important genes have been identified that affect viral infections. In *B. mori*, BmLipase-1, BmNox and BmSerine protease-2 are closely related to antiviral activity (Cheng et al., 2014; Nakazawa et al., 2004). BmLipase-1 expression was specifically induced in midgut epithelial cells when *B. mori* larvae were infected with BmNPV (Hu et al., 2015; Ponnuvel et al., 2003). Highlighting its importance in acting against viruses, the overexpression of BmLipase-1 in *B. mori* after transgenesis resulted in the transgenic larvae showing strong antiviral activities (Jiang et al., 2012a, 2013). It is possible that additional antiviral genes could be identified after analysing the changes in the transcriptome and/or proteome in the midguts of insects infected by different viruses. Further study on the functions of these genes may reveal important genes that can regulate antiviral activity.

Peritrophic membranes protect insects from the mechanical damage caused by foods and keep pathogens from contacting the cells of the midgut and inducing infections (Lehane, 1997; Wu et al., 2016). The integrity of peritrophic membranes is important to protect hosts from viral infections. *Autographa californica* multicapsid nucleopolyhedrovirus (AcMNPV) secretes chitinase, degrading the chitin of the peritrophic membrane and producing perforations, which enhances viral infections (Rao et al., 2004). Members of the species *Anticarsia gemmatalis* that are sensitive to its nucleopolyhedrovirus (AgMNPV) show fragile peritrophic membranes, and AgMNPV can easily breakthrough and increase the occurrence of infections (Levy et al., 2011). This illustrates how the structure of the peritrophic membrane is another important factor in protecting insects from viral infections.

In insects, NF-κB-dependent pathways can be triggered in response to viral infections via unknown mechanisms, and it is still unclear whether the processes lead to the production of effector proteins (Buchon et al., 2014; Costa et al., 2009; Zambon et al., 2005). However, insects have evolved other immune responses to defend against viral infections

(Buchon et al., 2014). In resistant species of *Cydia pomonella*, *C. pomonella* granulovirus (CpGV) DNA replication is blocked in the midgut, as demonstrated via quantitative PCR assays (Asser-Kaiser et al., 2011). However, in susceptible insects, CpGV DNA duplicate normally. In many insects, virus-infected cells are induced into apoptosis or autophagy to prevent viral duplication and the infection of neighbouring cells (Nakamoto et al., 2012; Wu et al., 2016). In *Drosophila*, Toll-7 expression is induced upon infection by the vesicular stomatitis virus (VSV), and Toll-7 interacts with VSV on the plasma membrane to activate antiviral autophagy (Nakamoto et al., 2012).

3.3 Entomopathogenic Fungi

Entomopathogenic fungi are the most abundant type of microorganisms that infects insects. Approximately, 60% of insect diseases are caused by pathogenic fungi (Faria and Wraight, 2007). As the natural pathogens of a variety of insects, entomopathogenic fungi can be environment friendly alternatives to chemical insecticides for biological pest control. Mycopesticides are defined as products based on living fungal propagules intended to control pests through inundative or inoculative applications (Faria and Wraight, 2007). *Beauveria bassiana* is the most widely used fungus for controlling agricultural and forestry pests. Over the years, nearly 40 types of agricultural and forestry pests have been effectively controlled by *B. bassiana* in China, including *Dendrolimus kikuchii*, *Empoasca pirisuga*, *Pyrausta nubilalis* and *Carposina nipponensis* (Xie et al., 2012). Fungal spores of *B. bassiana* grown on cooked rice have been used to suppress the population of the coffee berry borer, *Hypothenemus hampei*, in Colombia, and the pathogenic effect of this fungus against *H. hampei* was found to be over 92.5% (Posada-Flórez, 2008). Effective oil formulations containing *Metarhizium anisopliae* spores have been shown to kill 70%–90% of treated locusts within 14–20 days in Africa, Australia and Brazil (Lomer et al., 2001). Entomopathogenic fungi have also been demonstrated to be a potential biocontrol agent against adult *Culicoides* (Ansari et al., 2011). The application of 'dry' conidia on surfaces where the midges tend to rest causes a reduction in their survival and effectively reduces disease transmission (Ansari et al., 2011).

As a type of ecological pesticide, entomopathogenic fungi have been widely applied and play an important role in biological control. Just like other microorganisms, the immune responses that fungi have to face upon invading the insect limit their application. However, there are an increasing number of studies showing that entomopathogenic fungi can reach the bodies of insects and regulate host immunity through extracellular toxins or proteins. Bassiacridin, a protein secreted by *B. bassiana*, is toxic to *Locusta migratoria*. Injections of bassiacridin resulted in a mortality rate up to 50%, and melanization occurred in the tracheae, air sacs and fat bodies of the insects (Quesada-Moraga and Vey, 2004). Destruxins, secondary metabolites of *M. anisopliae* (Han et al., 2013), are important antiimmunity agents (Vilcinskas et al., 1997; Wang et al., 2012). Destruxins alter the morphology and cytoskeleton of Lepidopteran plasmatocytes, thus inhibiting the processes involved in phagocytosis, including cell attachment and spreading (Vilcinskas et al., 1997; Wang et al., 2012). Destruxin E treatments resulted in cytochemical changes in granulocytes, suggesting that changes in nonself-recognition and cellular defence occur in *Galleria mellonella* (Vey et al., 2002). The modulation of cellular immune responses in hosts is the key function of destruxins during fungal infections (Kershaw et al., 1999).

Insect humoral immunity can also be modulated by destruxin A. When *B. mori* larvae were injected with destruxin A, immunity-related proteins, including PPO1, PPO2, serine proteinase-like protein, antitrypsin isoform 3, p50 protein and calreticulin precursor became unregulated or downregulated (Fan et al., 2014). Injections of destruxin A reduced the expression of various antimicrobial peptides in *Drosophila* (Pal et al., 2007). The applications of destruxin A upregulated numerous genes, such as PGRP, scavenger receptor, lectin, most serpins, spätzle 6 precursor and spätzle 6, whereas cactus and dorsal interacting protein were downregulated in the larvae of *Plutella xylostella*, indicating the suppression of the Toll pathway (Han et al., 2013). However, destruxin A caused significant reductions of serpin-2, 4 and 5 in *P. xylostella* larvae (Han et al., 2014). Therefore, toxins produced by fungi may influence host humoral immunity by regulating the transcription of immunity-related genes.

In addition to toxins derived from entomopathogenic fungi can interfere with the cellular and humoral immunity of the host, some entomopathogenic fungi have evolved strategies to evade the recognition of immune system by changing the structure of their cell surfaces (Wang and St Leger, 2006). Host haemocytes can recognize the conidia but not the hyphal bodies (Wang and St Leger, 2006). After invasion, *M. anisopliae* expresses a *Metarhizium* collagen-like protein that is coated on the cell surface, which is helpful in evading immune recognition, phagocytosis and encapsulation by host haemocytes, allowing the type of fungus to efficiently kill the hosts (Wang and St Leger, 2006). Widely used as biopesticides, entomopathogenic fungi at different developmental stages have many novel strategies to escape or suppress host immune responses, which may provide new ideas for the development of these organisms for use in biological pest control.

3.4 Parasitoids

Parasitoids are another popular organism for use in biological pest control. The interactions between parasitoids and hosts are very complicated. Insects can recognize and suppress the development of parasitoid, and in turn, parasitoids have to overcome the immune barrier of the host for their normal growth (Asgari, 2006; Strand, 2008). After being parasitized, the cellular and humoral immunities of the host are activated to defend against parasitoids (Strand, 2008; Tang, 2009). Meanwhile, many genes are transcribed after parasitization. When *Drosophila* larvae were parasitized by *Asobara tabida*, some genes involved in the JAK/STAT pathway, melanization and the Toll and Imd pathways were found to be significantly upregulated to protect the host from the parasitoid attack (Wertheim et al., 2005). Conversely, parasitic wasps use venom, polydnaviruses (PDVs), teratocytes, ovarian proteins (OP), virus-like particles (VLPs), virus-like filaments (VLFs), teratocyte secretary proteins (TSPs) or other factors to suppress host immune responses to ensure successful parasitism (Asgari, 2006).

As a result of long-term coevolution, parasitic wasps produce some anti-defence strategies to overcome host immune responses. Venom is utilized by most parasitic wasps to regulate the immune response, metabolism and behaviour of the host, to cause the atrophy of internal tissues and secretory organs, and even to deter metamorphosis of the hosts (Asgari and Rivers, 2011; Moreau and Guillot, 2005). Analysing the components of the venom of parasitic wasps may provide clues for understanding their strategy to evade host immunity. In the venom of *Aphidius ervi*, the key component has been identified as γ-glutamyl transpeptidase, and the functions of this compound are being studied (Nguyen et al., 2013). PDVs are a special type of virus that accesses the body of the host along with the eggs of the parasitic wasp and proliferates in the ovaries of the parasitoids (Webb and Strand, 2005). The most significant impacts that PDVs have on the host are that they can suppress the immune response and regulate development by reducing food ingestion, delaying growth and interrupting endocrine function and nutrient metabolism (Webb and Strand, 2005). Teratocytes are a special type of cell that supplies nutrients to parasitic wasp larvae (Dahlman, 1990). Meanwhile, these cells suppress host immune function (Dahlman, 1990; Dahlman and Vinson, 1993).

Encapsulation is an important cellular defensive response against invading parasitoids (Strand, 2008). The process of encapsulation requires the recognition and aggregation of haemocytes around the invading parasitoid, resulting in the formation of a multilayer capsule primarily composed of different types of haemocytes (Strand, 2008). Nevertheless, parasitoids have also developed many mechanisms to suppress encapsulation by interfering with the recognition, adherence and spreading of haemocytes (Stettler et al., 1998; Zhang et al., 2006). In some parasitoids, their eggs and/or larvae have evolved special surface molecules that are not recognized by the host and evade haemocyte binding (Corley and Strand, 2003). During oviposition, some wasps inject parasitic factors into the hosts to interfere with the immune system (Zhang et al., 2006). Both *Leptopilina boulardi* and *Leptopilina heterotoma* can successfully parasitize *Drosophila* by injecting venom loaded with VLPs into hosts to prevent encapsulation and to avoid the killing of their eggs (Schlenke et al., 2007). The venom of *Pteromalus puparum* can also suppress haemocyte-mediated encapsulation and phagocytosis in hosts (Yu et al., 2007; Zhang et al., 2006, 2011). The immune suppressive factors secreted in the venoms of *L. boulardi* and other parasitoids can alter the number and composition of haemocytes through cell lysis and apoptosis (Schlenke et al., 2007; Stettler et al., 1998). Surprisingly, to partially escape encapsulation, *Asobara* wasps lay eggs that adhere to the fat body and other internal organs in which incomplete capsules are formed and thus the parasitoid egg can still hatch in the host (Wertheim et al., 2005).

During melanization, many toxic molecules are produced, and melanization can accelerate to kill pathogens to defend against infection. Melanization is crucial to deter the development of parasitoids in hosts. Serine proteases and their inhibitors (serpins) have been shown to be involved in the activation of PPO to induce melanization (Ashida and Brey, 1998; Kanost et al., 2004; Lu et al., 2014). Invading parasitoids can regulate the transcription of genes involved in melanization. When *Bemisia tabaci* were parasitized by *Eretmocerus mundus*, *serpin A3K* and other melanization-related genes were found to be changed according to microarray analyses, suggesting that parasitization in *B. tabaci* involves the regulation of the melanization cascade (Mahadav et al., 2008). Components of *Cotesia rubecula* venom have been found to inhibit melanization in *Pieris rapae* by interfering with the PPO activation cascade of the host (Asgari et al., 2003). Genes involved in the PPO activation cascade system in *P. rapae* were downregulated by *P. puparum* venom, and melanization was likely suppressed by the infecting endoparasitoids

(Fang et al., 2010). Clearly, the PPO activation cascade in insects is also a target that is suppressed by some parasitoids.

In summary, for insects that survive in the wild, the invasion of microorganisms and parasitoids into their bodies via wounds, the tracheae and the guts are unavoidable, and insects utilize cellular and humoral immunity to defend against these pathogens. However, as a result of coevolution, many pathogens have also evolved novel strategies to overcome the innate immunity of the host, which is likely good news for the development of new methods for biological pest control.

4. THE DUOX PATHWAY IN THE GUT IS CRUCIAL IN PROTECTING HOST DEVELOPMENT

The mucosal epithelia come into direct contact with a large number of microorganisms, including both symbionts and pathogens (Engel and Moran, 2013; Lemaitre and Miguel-Aliaga, 2013). Hosts must be tolerant of symbionts and simultaneously effectively defend against pathogens (Bae et al., 2010). *Drosophila* is a model species for studies on gut immunity. According to the recent research on gut immunity in *Drosophila* and other insects, there are two local immune responses that happen in the intestinal epithelial cell layer: the production of bactericidal ROS and antimicrobial peptides (Engel and Moran, 2013).

The mechanism for producing antimicrobial peptides in the midgut is not exactly the same as in the haemocoel. In the haemocoel, the Toll and Imd pathways are involved in the production of antimicrobial peptides. However, in the *Drosophila* midgut, the Imd pathway is the main mechanism leading to the production of antimicrobial peptides after oral bacterial infections (Engel and Moran, 2013). Bacterial PGN binds to PGRP receptors to activate the Imd pathway, resulting in the activation of the transcriptional factor Relish and thus inducing the expression of several antimicrobial peptides (Engel and Moran, 2013). Genetic studies on *Drosophila* show that the level of ROS is enhanced through the activity of DUOX after oral bacterial infections (Lee et al., 2015a,b). When DUOX expression was silenced in adult flies, the mortality rate increased significantly after the intake of microbially contaminated food (Ha et al., 2005a). The uracil produced by pathogenic bacteria acts as a ligand to modulate DUOX-dependent gut immunity in *Drosophila*, which is under the control of the Hedgehog pathway (Lee et al., 2013, 2015a,b). Flies with impaired Hedgehog signalling had no ability to form Cad99C-dependent endosomes, and DUOX activity was decreased, which was lethal after intestinal infections (Lee et al., 2015a). DUOX activation is under the control of a GPCR-associated pathway, and the mutation of genes involved in this pathway leads to a decrease in the production of ROS, which is lethal to insects following oral infections (Bae et al., 2010; Kim and Lee, 2014). Commensal bacteria do not produce uracil, or they produce it at very low levels. The DUOX system can discriminate between symbionts and pathogens on that basis (Kim and Lee, 2014). Therefore, *Drosophila* DUOX is crucial in maintaining gut antimicrobial activities (Bae et al., 2010; Ha et al., 2005a).

5. ADDITIONAL FACTORS INVOLVED IN GUT PROTECTION

Beyond DUOX, there are several factors involved in midgut immunity largely identified in studies of *Drosophila* host defence. An example is *p38*, one of the subfamilies of mitogen-activated protein (MAP) kinases (Zarubin and Han, 2005), and it can be activated by cellular stresses, such as infection, and is involved in *Drosophila* host defence (Chen et al., 2010). Compared with wild-type flies, *p38a* mutants and *p38b* mutants were found to be more sensitive to infections. The *p38b;p38a* double-mutant flies were even sensitive to the naturally existing microbes in their food, which were able to cause melanization in the hindgut and larval-stage lethality (Chen et al., 2010). The peritrophic membrane serves as an important barrier for the immune system in the gut. The main components of the peritrophic membrane are chitin polymers and glycoproteins (Mistry et al., 2016). The *Drosocrystallin* (*Dcy*) gene is important in forming the peritrophic membrane in adult flies (Kuraishi et al., 2011). The loss of *Dcy* function results in a decrease in gut width and an increase in the permeability of the peritrophic membrane (Kuraishi et al., 2011). *Dcy*-deficient flies show an increased sensitivity to oral infections by *P. entomophila* and *Serratia marcescens*, and they were found to die faster than wild-type flies even after the ingestion of an extract of *P. entomophila* (Kuraishi et al., 2011). Additional studies indicate that calcofluor and chitinase breakdown the structure of the peritrophic membrane, which can also facilitate pathogenic infections in insects (Jiang et al., 2012b). Thus, the integrity of the peritrophic membrane plays an important role in protecting against intestinal pathogens in *Drosophila* (Kuraishi et al., 2011).

In the wild, it is unavoidable that insects will ingest microorganisms with their food everyday. The PGN on the cell wall of pathogenic bacteria and the uracil secreted by those bacteria are important components for the host to detect and defend against pathogenic bacteria (Buchon et al., 2014; Lee et al., 2013; Lemaitre and Hoffmann, 2007). Under the protection of ROS, antimicrobial peptides and peritrophic membranes, hosts can ingest foods and absorb the nutrients required for normal development. Any defect in the midgut protection system, especially DUOX pathway, can be lethal to the insects. Therefore, insect midgut immunity, and especially the DUOX pathway, is an ideal target for biological pest control.

6. IMMUNODEFICIENCY IN INSECTS

The immune system is important for protecting invertebrates and vertebrates from infections of invading bacteria, fungi, viruses and parasitoids. Immunodeficiencies can have dramatic consequences in humans (reviewed in Champi, 2002; Lloyd, 1996; Notarangelo, 2010; Rosen et al., 1995). As a concept, however, they provide an enticing hint to entomologists, suggesting that we may be able to control pests effectively if we could also induce immunodeficiencies in these insects.

Insects have an astonishing ability to proliferate. Among such a huge population, it is unavoidable that some individuals will have certain genes that have been mutated as a result of solar radiation, insecticides, herbicides, fertilizers and environment pollution from industry and transportation. In Britain, the peppered moth, *Biston betularia*, became mutated into a black (carbonaria) form during the industrial revolution, which was attributed to interaction with bird predation and coal pollution (Cook, 2003). A recent work demonstrates that the *cortex* gene mutated due to the insertion of a transportable element into its first intron, which led to the black form of the moth (Van't Hof et al., 2016). Similarly, among the huge populations of pest insects, there might be some individuals with mutations in immunity-related genes that end up being immunodeficient. However, under the influence of natural selection, insects with decreased immune function may not survive to pass their genes on to the next generation. This is likely the reason why we have not observed immunodeficient insects in the field.

In laboratories, there are many species of insects, especially *Drosophila*, that are immunodeficient. These immunodeficient insects have been produced via chemical mutagenesis or genetic manipulation. *Drosophila* pests are often observed on fresh soft fruit (*Drosophila suzukii*), as well as fermenting and rotting fruits and vegetables in family kitchens, restaurants, fruit markets and fields. In addition, *Drosophila* has been the model of choice for studying innate immune defences (Dionne and Schneider, 2008; Hoffmann, 2003; Lemaitre and Hoffmann, 2007). These studies identified and biochemically characterized antimicrobial peptides produced by the insect that can kill pathogens (Hoffmann, 2003; Lemaitre and Hoffmann, 2007). The production of antimicrobial peptides is under the control of the Toll and Imd pathways (Hoffmann, 2003; Lemaitre and Hoffmann, 2007). In *Drosophila* mutants with altered genes associated with the Toll and Imd pathways that have been produced in laboratories, bacterial growth has been shown to increase significantly compared with growth in wild-type flies. Survival probabilities for different Toll pathway mutants are dramatically reduced upon fungal infection (Lemaitre et al., 1996). Dif and Dorsal are transactivators in the Toll pathway (Lemaitre and Hoffmann, 2007). In Dif and Dorsal loss-of-functions mutants, there are very few blood cells in the haemocoel, and these mutants easily become infected by opportunistic bacteria and die at the larval stage (Matova and Anderson, 2006). *Imd* mutants show a lower resistance to Gram-negative bacteria and have a lower survival ratio than wild-type flies after systematic infections (Hoffmann, 2003). This is also the reason why the mutation was referred to as *immune deficiency* (*Imd*) (Lemaitre et al., 1995). Melanization is also induced around the sites of infection. In *Drosophila*, when two *PPO* genes were deleted, the mutants were found to be significantly more susceptible to many bacteria and fungi (Binggeli et al., 2014), although melanization is not essential for defence against bacteria in mosquitoes (Schnitger et al., 2007). It is hard to obtain immunodeficient insects in the field, probably due to natural selection. Laboratory studies have demonstrated that insects do become immunodeficient when certain important immunity-related genes are mutated or inhibited. Individual insects with an impaired immune system are easily infected, either systemically and/or locally, which are exciting news for the development of biological pest control. Therefore, it is at least theoretically feasible to envisage utilizing the innate immune system of insects as a component in biological pest control if we can effectively block the function or transcription of key immune proteins.

7. PROSPECTS OF INCREASING HOST SUSCEPTIBILITY TO PATHOGENS THROUGH IMMUNE SUPPRESSION

Over the course of evolution, individual insects with immunodeficiencies have been presumably eliminated due to natural selection. In the interaction of hosts and pathogens, many microorganisms and parasitoids have evolved tactics to interrupt or suppress host immunity for a successful infection (Clem and Miller, 1994; Feng et al., 2015). The oosporein produced by *B. bassiana* is crucial for maintain fungal virulence as it allows the type of fungus to evade the immune response of the host (Feng et al., 2015). In AcMNPV, the expression of *p35* and *iap* can inhibit programmed cell death in infected insects, which facilitates viral infections (Clem and Miller, 1994). When *p35* was mutated in AcMNPV, the pathogenicity of this type of virus was found to be decreased (Clem and Miller, 1994). Obviously, the strategies that are used by pathogens to decrease the innate immunity of the host are valuable for use in biological pest control.

Extrinsic factors have been studied and found to affect the insect immune system. Botanical extracts, microbes, entomopathogenic nematodes and low doses of insecticides were found to increase the virulence of entomopathogenic fungi (Ansari et al., 2006, 2010; Kryukov et al., 2009; Shapiro-Ilan et al., 2004). Some of these can even impair the insect immune system (Hiromori and Nishigaki, 2001). Synergistic effects have been observed when fungal components and Bt proteins are applied for biological pest control (Gao et al., 2012; Kryukov et al., 2009; Wraight and Ramos, 2005). Bt proteins can impair the host's immune system, which enhances the efficacy of *B. bassiana* and *M. anisopliae* infections (Butt et al., 2016). During infection, entomopathogenic nematodes release symbiotic bacteria into the host (Caldas et al., 2002; Ji and Kim, 2004). Some proteases are secreted by bacterial symbiont of nematodes that degrade host antimicrobial peptides (Caldas et al., 2002; Ji and Kim, 2004). In *Xenorhabdus nematophila*, symbiotic bacteria produce benzylideneacetone, which can suppress cellular and humoral immune responses in *Spodoptera exigua* (Park and Kim, 2011). The synergistic application of *M. anisopliae* and insecticides reduces the number of granular cells and PO activity in *Anomala cuprea* larvae (Hiromori and Nishigaki, 2001). The injection of a recombinant protein from a gene in the venom of the endoparasitic wasp *Pimpla hypochondriaca* into the bodies of *Mamestra brassicae* larvae synergistically enhanced the sensitivity of the hosts to *B. bassiana* and Bt infections (Richards and Dani, 2010; Richards et al., 2011). Further research has demonstrated that toxins injected by wasps can also inhibit host cellular immunity (Richards et al., 2013). In some insects (*L. dispar*), the application of Bt was found to activate cellular immune responses (Butt et al., 2016). As a result, when the immune response was interrupted, the insecticidal efficiency of Bt was increased (Broderick et al., 2010). Some studies have found that chemical insecticides can significantly interrupt the insect immune system, which also enhances the susceptibility of the insects to microbes (James and Xu, 2012). Therefore, there is increasing evidence showing that the impairment on the insect immune system can significantly enhance the efficiency of biological control.

One very successful example of this is based on the function of GNBP-2. The termite GNBP-2 protein has β-1,3-glucanase activity, and it is important for sensing pathogenic infections and triggering host immune responses (Bulmer et al., 2009). Termites incorporate the protein into their nests, and it promptly cleaves and releases the components of invading pathogens. A small glycomimetic molecule was designed to block the function of the immune recognition protein GNBP-2, and eventually, the termites were easily killed by opportunistic pathogens (Bulmer et al., 2009). This new pest control method is nontoxic, inexpensive and effective. Based on this example, it is likely that new methods of biological pest control could be developed that act via suppressing immune activity (Bulmer et al., 2009).

Another pertinent example is *Pantoea agglomerans*, a bacterial symbiont of mosquito. When *P. agglomerans* was genetically manipulated to express four copies of the plasmodium enolase–plasminogen interaction peptide $(EPIP)_4$ or scorpine, the ratio of mosquitoes carrying parasites significantly decreased (Wang and Jacobs-Lorena, 2013), although the recombinant *P. agglomerans* did not affect mosquito longevity. These studies indicate that many pathogenic and even symbiotic bacteria can be manipulated for biological pest control.

As an immunity protein, PPO is present in haemolymph and can be activated quickly upon detecting invading pathogens. Almost all microorganisms that might be applied as potential bioinsecticides have to face the threat of PPO-induced melanization upon entering the host. When the expression of *PPO* was reduced via RNAi, some invertebrates were found to be vulnerable to exogenous pathogenic bacteria or viruses (Liu et al., 2007; Paria et al., 2013). In *Drosophila*, when *PPO1* and *PPO2* were deleted, the mutant flies were easily infected with many species of bacteria and fungi (Binggeli et al., 2014). A recent study showed that there is PPO in the foreguts of silkworms and *Drosophila* (Wu et al., 2015). Plants can produce many secondary metabolites (De Filippis, 2016). Plant phenolics are one of these metabolites and pose no health risks to humans (Dai and Mumper, 2010). However, phenolics are toxic to insects (Salminen and Lempa, 2002; Usha Rani and Pratyusha, 2014). The deletion of two *PPO* genes in *Drosophila* enhances the toxicity of plant phenolics to larvae and adults, which demonstrates that PPO in the foregut can detoxify plant phenolics (Wu et al., 2015). If an effective method to block PPO activation, especially in the foregut, is developed, the phenolics produced in plants may be lethal to pests. Therefore, PPO is also a potential target for biological pest control if we could effectively block the activation and activity of this enzyme.

8. OUTLOOK

Based on our current knowledge, we are now in a position to perform more in-depth studies on innate immunity in insects. The methods for studying insect immune responses have already been well developed (Neyen et al., 2014), and they can be used for reference in different species of insects. In addition, along with the rapid development of genome sequencing technology, scientists can screen and identify immune-related genes at the genomic level in various pests, which would greatly improve the efficiency of immune-related gene identification and promote the expansion of such researches in a variety of insects. At present, at the genomic level, the species for which all the genes involved in the immune response have been identified are *Drosophila* (Adams et al., 2000; Irving et al., 2001), mosquitoes (Arensburger et al., 2010; Holt et al., 2002; Nene et al., 2007; Waterhouse et al., 2007), *Apis mellifera* (Evans et al., 2006; The Honeybee Genome Sequencing C, 2006), *B. mori* (Tanaka et al., 2008; Xia et al., 2004), *Tribolium castaneum* (Richards et al., 2008; Zou et al., 2007) and *Manduca sexta* (Gunaratna and Jiang, 2013; Kanost et al., 2016). Among these species, *Drosophila*, mosquitoes, *B. mori* and *M. sexta* are well studied in relation to immunity. In addition, the wide use of transcriptome sequencing technologies provides a convenient way to identify and screen effector genes related to infections (Gunaratna and Jiang, 2013), allowing us to quickly target key immune defence-related genes.

Numerous proteins have been proven to be involved in the production of antimicrobial peptides and ROS (Buchon et al., 2014; Hoffmann, 2003; Kim and Lee, 2014; Lee et al., 2015b; Lemaitre and Hoffmann, 2007), and the loss of the function of those genes may induce immunodeficiencies in laboratory-reared insects. The accumulated knowledge about the insect immune system is a valuable resource especially for the development of biological pest control. However, there are still many things that we need to work on before we can utilize this knowledge for biological pest control. For example, knowing the crystal structures of immune-related proteins is very important for understanding their functions. In summary, we have already discovered many things about the innate immune response in insects, and it is now time for us to consider how innate immunity in insects can be targeted for biological pest control.

ACKNOWLEDGEMENTS

This work was supported by the National Natural Science Foundation of China (31672360, 31472043), the Postdoctoral Science Foundation of China (2014M562369) and Shaanxi Science and Technology Plan Projects (2016NY-204).

REFERENCES

Adams, M.D., Celniker, S.E., Holt, R.A., Evans, C.A., Gocayne, J.D., Amanatides, P.G., Scherer, S.E., Li, P.W., Hoskins, R.A., Galle, R.F., George, R.A., Lewis, S.E., Richards, S., Ashburner, M., Henderson, S.N., Sutton, G.G., Wortman, J.R., Yandell, M.D., Zhang, Q., Chen, L.X., Brandon, R.C., Rogers, Y.H., Blazej, R.G., Champe, M., Pfeiffer, B.D., Wan, K.H., Doyle, C., Baxter, E.G., Helt, G., Nelson, C.R., Gabor, G.L., Abril, J.F., Agbayani, A., An, H.J., Andrews-Pfannkoch, C., Baldwin, D., Ballew, R.M., Basu, A., Baxendale, J., Bayraktaroglu, L., Beasley, E.M., Beeson, K.Y., Benos, P.V., Berman, B.P., Bhandari, D., Bolshakov, S., Borkova, D., Botchan, M.R., Bouck, J., Brokstein, P., Brottier, P., Burtis, K.C., Busam, D.A., Butler, H., Cadieu, E., Center, A., Chandra, I., Cherry, J.M., Cawley, S., Dahlke, C., Davenport, L.B., Davies, P., de Pablos, B., Delcher, A., Deng, Z., Mays, A.D., Dew, I., Dietz, S.M., Dodson, K., Doup, L.E., Downes, M., Dugan-Rocha, S., Dunkov, B.C., Dunn, P., Durbin, K.J., Evangelista, C.C., Ferraz, C., Ferriera, S., Fleischmann, W., Fosler, C., Gabrielian, A.E., Garg, N.S., Gelbart, W.M., Glasser, K., Glodek, A., Gong, F., Gorrell, J.H., Gu, Z., Guan, P., Harris, M., Harris, N.L., Harvey, D., Heiman, T.J., Hernandez, J.R., Houck, J., Hostin, D., Houston, K.A., Howland, T.J., Wei, M.H., Ibegwam, C., Jalali, M., Kalush, F., Karpen, G.H., Ke, Z., Kennison, J.A., Ketchum, K.A., Kimmel, B.E., Kodira, C.D., Kraft, C., Kravitz, S., Kulp, D., Lai, Z., Lasko, P., Lei, Y., Levitsky, A.A., Li, J., Li, Z., Liang, Y., Lin, X., Liu, X., Mattei, B., McIntosh, T.C., McLeod, M.P., McPherson, D., Merkulov, G., Milshina, N.V., Mobarry, C., Morris, J., Moshrefi, A., Mount, S.M., Moy, M., Murphy, B., Murphy, L., Muzny, D.M., Nelson, D.L., Nelson, D.R., Nelson, K.A., Nixon, K., Nusskern, D.R., Pacleb, J.M., Palazzolo, M., Pittman, G.S., Pan, S., Pollard, J., Puri, V., Reese, M.G., Reinert, K., Remington, K., Saunders, R.D., Scheeler, F., Shen, H., Shue, B.C., Sidén-Kiamos, I., Simpson, M., Skupski, M.P., Smith, T., Spier, E., Spradling, A.C., Stapleton, M., Strong, R., Sun, E., Svirskas, R., Tector, C., Turner, R., Venter, E., Wang, A.H., Wang, X., Wang, Z.Y., Wassarman, D.A., Weinstock, G.M., Weissenbach, J., Williams, S.M., Woodage, T., Worley, K.C., Wu, D., Yang, S., Yao, Q.A., Ye, J., Yeh, R.F., Zaveri, J.S., Zhan, M., Zhang, G., Zhao, Q., Zheng, L., Zheng, X.H., Zhong, F.N., Zhong, W., Zhou, X., Zhu, S., Zhu, X., Smith, H.O., Gibbs, R.A., Myers, E.W., Rubin, G.M., Venter, J.C., 2000. The genome sequence of *Drosophila melanogaster*. Science 287, 2185–2195.

Ansari, M.A., Shah, F.A., Tirry, L., Moens, M., 2006. Field trials against *Hoplia philanthus* (Coleoptera: Scarabaeidae) with a combination of an entomopathogenic nematode and the fungus *Metarhizium anisopliae* CLO 53. Biol. Control 39, 453–459.

Ansari, M.A., Shah, F.A., Butt, T.M., 2010. The entomopathogenic nematode *Steinernema kraussei* and *Metarhizium anisopliae* work synergistically in controlling overwintering larvae of the black vine weevil, *Otiorhynchus sulcatus*, in strawberry growbags. Biocontrol Sci. Technol. 20, 99–105.
Ansari, M.A., Pope, E.C., Carpenter, S., Scholte, E.J., Butt, T.M., 2011. Entomopathogenic fungus as a biological control for an important vector of livestock disease: the *Culicoides* biting midge. PLoS One 6, e16108.
Arensburger, P., Megy, K., Waterhouse, R.M., Abrudan, J., Amedeo, P., Antelo, B., Bartholomay, L., Bidwell, S., Caler, E., Camara, F., Campbell, C.L., Campbell, K.S., Casola, C., Castro, M.T., Chandramouliswaran, I., Chapman, S.B., Christley, S., Costas, J., Eisenstadt, E., Feshotte, C., Fraser-Liggett, C., Guigo, R., Haas, B., Hammond, M., Hansson, B.S., Hemingway, J., Hill, S., Howarth, C., Ignell, R., Kennedy, R.C., Kodira, C.D., Lobo, N.F., Mao, C., Mayhew, G., Michel, K., Mori, A., Liu, N., Naveira, H., Nene, V., Nguyen, N., Pearson, M.D., Pritham, E.J., Puiu, D., Qi, Y., Ranson, H., Ribeiro, J.M.C., Roberston, H.M., Severson, D.W., Shumway, M., Stanke, M., Strausberg, R., Sun, C., Sutton, G., Tu, Z., Tubio, J.M.C., Unger, M.F., Vanlandingham, D.L., Vilella, A.J., White, O., White, J.R., Wondji, C.S., Wortman, J., Zdobnov, E.M., Birren, B., Christensen, B.M., Collins, F.H., Cornel, A., Dimopoulos, G., Hannick, L.I., Higgs, S., Lanzaro, G.C., Lawson, D., Lee, N.H., Muskavitch, M.A.T., Raikhel, A.S., Atkinson, P.W., 2010. Sequencing of *Culex quinquefasciatus* establishes a platform for mosquito comparative genomics. Science 330, 86–88.
Asgari, S., 2006. Venom proteins from polydnavirus-producing endoparasitoids: their role in host–parasite interactions. Arch. Insect Biochem. Physiol. 61, 146–156.
Asgari, S., Rivers, D.B., 2011. Venom proteins from endoparasitoid wasps and their role in host–parasite interactions. Annu. Rev. Entomol. 56, 313–335.
Asgari, S., Zhang, G., Zareie, R., Schmidt, O., 2003. A serine proteinase homolog venom protein from an endoparasitoid wasp inhibits melanization of the host hemolymph. Insect Biochem. Mol. Biol. 33, 1017–1024.
Ashida, M., Brey, P., 1998. Recent advances on the research of the insect prophenoloxidase cascade. In: Brey, P., Hultmark, D. (Eds.), Molecular Mechanisms of Immune Responses in Insects. Chapman & Hall, London, pp. 135–172.
Asser-Kaiser, S., Radtke, P., El-Salamouny, S., Winstanley, D., Jehle, J.A., 2011. Baculovirus resistance in codling moth (*Cydia pomonella* L.) caused by early block of virus replication. Virology 410, 360–367.
Bae, Y.S., Choi, M.K., Lee, W.J., 2010. Dual oxidase in mucosal immunity and host-microbe homeostasis. Trends Immunol. 31, 278–287.
Binggeli, O., Neyen, C., Poidevin, M., Lemaitre, B., 2014. Prophenoloxidase activation is required for survival to microbial infections in *Drosophila*. PLoS Pathog. 10, e1004067.
Bravo, A., Gill, S.S., Soberón, M., 2007. Mode of action of *Bacillus thuringiensis* Cry and Cyt toxins and their potential for insect control. Toxicon 49, 423–435.
Broderick, N.A., Raffa, K.F., Handelsman, J., 2010. Chemical modulators of the innate immune response alter gypsy moth larval susceptibility to *Bacillus thuringiensis*. BMC Microbiol. 10, 129.
Buchon, N., Broderick, N.A., Poidevin, M., Pradervand, S., Lemaitre, B., 2009. *Drosophila* intestinal response to bacterial infection: activation of host defense and stem cell proliferation. Cell Host Microbe 5, 200–211.
Buchon, N., Silverman, N., Cherry, S., 2014. Immunity in *Drosophila melanogaster*—from microbial recognition to whole-organism physiology. Nat. Rev. Immunol. 14, 796–810.
Bulmer, M.S., Bachelet, I., Raman, R., Rosengaus, R.B., Sasisekharan, R., 2009. Targeting an antimicrobial effector function in insect immunity as a pest control strategy. Proc. Natl. Acad. Sci. U.S.A. 106, 12652–12657.

Butt, T.M., Coates, C.J., Dubovskiy, I.M., Ratcliffe, N.A., 2016. Entomopathogenic fungi: new insights into host–pathogen interactions. Adv. Genet. 94, 307–364.

Caldas, C., Cherqui, A., Pereira, A., Simoes, N., 2002. Purification and characterization of an extracellular protease from *Xenorhabdus nematophila* involved in insect immunosuppression. Appl. Environ. Microbiol. 68, 1297–1304.

Cerenius, L., Lee, B.L., Söderhäll, K., 2008. The proPO-system: pros and cons for its role in invertebrate immunity. Trends Immunol. 29, 263–271.

Chakrabarti, S., Liehl, P., Buchon, N., Lemaitre, B., 2012. Infection-induced host translational blockage inhibits immune responses and epithelial renewal in the *Drosophila* gut. Cell Host Microbe 12, 60–70.

Champi, C., 2002. Primary immunodeficiency disorders in children: prompt diagnosis can lead to lifesaving treatment. J. Pediatr. Health Care 16, 16–21.

Chen, J., Xie, C., Tian, L., Hong, L., Wu, X., Han, J., 2010. Participation of the p38 pathway in *Drosophila* host defense against pathogenic bacteria and fungi. Proc. Natl. Acad. Sci. U.S.A. 107, 20774–20779.

Cheng, X.W., Lynn, D.E., 2009. Baculovirus interactions *in vitro* and *in vivo*. Adv. Appl. Microbiol. 68, 217–239.

Cheng, Y., Wang, X.Y., Du, C., Gao, J., Xu, J.P., 2014. Expression analysis of several antiviral related genes to BmNPV in different resistant strains of silkworm, *Bombyx mori*. J. Insect Sci. 14, 76.

Chowanski, S., Adamski, Z., Marciniak, P., Rosinski, G., Buyukguzel, E., Buyukguzel, K., Falabella, P., Scrano, L., Ventrella, E., Lelario, F., Bufo, S.A., 2016. A review of bioinsecticidal activity of solanaceae alkaloids. Toxins (Basel) 8, 1–28.

Christophides, G.K., Vlachou, D., Kafatos, F.C., 2004. Comparative and functional genomics of the innate immune system in the malaria vector *Anopheles gambiae*. Immunol. Rev. 198, 127–148.

Clem, R.J., 2005. The role of apoptosis in defense against baculovirus infection in insects. In: Griffin, D.E. (Ed.), Role of Apoptosis in Infection. Springer, Berlin, Heidelberg, pp. 113–129.

Clem, R.J., Miller, L.K., 1994. Control of programmed cell death by the baculovirus genes *p35* and *iap*. Mol. Cell. Biol. 14, 5212–5222.

Cook, L.M., 2003. The rise and fall of the *Carbonaria* form of the peppered moth. Q. Rev. Biol. 78, 399–417.

Corley, L.S., Strand, M.R., 2003. Evasion of encapsulation by the polyembryonic parasitoid *Copidosoma floridanum* is mediated by a polar body-derived extraembryonic membrane. J. Invertebr. Pathol. 83, 86–89.

Costa, A., Jan, E., Sarnow, P., Schneider, D., 2009. The Imd pathway is involved in antiviral immune responses in *Drosophila*. PLoS One 4, e7436.

Crava, C.M., Jakubowska, A.K., Escriche, B., Herrero, S., Bel, Y., 2015. Dissimilar regulation of antimicrobial proteins in the midgut of *Spodoptera exigua* larvae challenged with *Bacillus thuringiensis* toxins or baculovirus. PLoS One 10, e0125991.

Dahlman, D.L., 1990. Evaluation of teratocyte functions: an overview. Arch. Insect Biochem. Physiol. 13, 159–166.

Dahlman, D.H., Vinson, S.B., 1993. Teratocytes: developmental and biochemical characteristics. In: Beckage, N.E., Thompson, S.N., Federici, B.A. (Eds.), Parasites and Pathogens of Insects. Academic Press, New York, pp. 145–166.

Dai, J., Mumper, R.J., 2010. Plant phenolics: extraction, analysis and their antioxidant and anticancer properties. Molecules (Basel, Switzerland) 15, 7313–7352.

Damalas, C.A., Eleftherohorinos, I.G., 2011. Pesticide exposure, safety issues, and risk assessment indicators. Int. J. Environ. Res. Public Health 8, 1402–1419.

De Filippis, L.F., 2016. Plant secondary metabolites. In: Plant–Environment Interaction. John Wiley & Sons, Ltd, New Jersey, pp. 263–299.

Dionne, M.S., Schneider, D.S., 2008. Models of infectious diseases in the fruit fly *Drosophila melanogaster*. Dis. Model. Mech. 1, 43–49.
Engel, P., Moran, N.A., 2013. The gut microbiota of insects—diversity in structure and function. FEMS Microbiol. Rev. 37, 699–735.
Evans, J.D., Aronstein, K., Chen, Y.P., Hetru, C., Imler, J.L., Jiang, H., Kanost, M., Thompson, G.J., Zou, Z., Hultmark, D., 2006. Immune pathways and defence mechanisms in honey bees *Apis mellifera*. Insect Mol. Biol. 15, 645–656.
Fan, J., Han, P., Chen, X., Hu, Q., Ye, M., 2014. Comparative proteomic analysis of *Bombyx mori* hemocytes treated with destruxin A. Arch. Insect Biochem. Physiol. 86, 33–45.
Fang, Q., Wang, L., Zhu, J., Li, Y., Song, Q., Stanley, D.W., Akhtar, Z.-R., Ye, G., 2010. Expression of immune-response genes in lepidopteran host is suppressed by venom from an endoparasitoid, *Pteromalus puparum*. BMC Genomics 11, 1–17.
Faria, M.R.D., Wraight, S.P., 2007. Mycoinsecticides and Mycoacaricides: a comprehensive list with worldwide coverage and international classification of formulation types. Biol. Control 43, 237–256.
Feng, P., Shang, Y., Cen, K., Wang, C., 2015. Fungal biosynthesis of the bibenzoquinone oosporein to evade insect immunity. Proc. Natl. Acad. Sci. U.S.A. 112, 11365–11370.
Ferré, J., Rie, J.V., 2002. Biochemistry and genetics of insect resistance to *bacillus thuringiensis*. Annu. Rev. Entomol. 47, 501–533.
Gao, Y., Oppert, B., Lord, J.C., Liu, C., Lei, Z., 2012. *Bacillus thuringiensis* Cry3Aa toxin increases the susceptibility of *Crioceris quatuordecimpunctata* to *Beauveria bassiana* infection. J. Invertebr. Pathol. 109, 260–263.
Gao, K., Deng, X.Y., Qian, H.Y., Qin, G., Guo, X.J., 2014. Digital gene expression analysis in the midgut of 4008 silkworm strain infected with cytoplasmic polyhedrosis virus. J. Invertebr. Pathol. 115, 8–13.
Gerttula, S., Jin, Y., Anderson, K.V., 1988. Zygotic expression and activity of the *Drosophila Toll* gene, a gene required maternally for embryonic dorsal-ventral pattern formation. Genetics 119, 123–133.
Gillespie, J.P., Kanost, M.R., Trenczek, T., 1997. Biological mediators of insect immunity. Annu. Rev. Entomol. 42, 611–643.
Glare, T., Caradus, J., Gelernter, W., Jackson, T., Keyhani, N., Köhl, J., Marrone, P., Morin, L., Stewart, A., 2012. Have biopesticides come of age? Trends Biotechnol. 30, 250–258.
Gonzalez, E.A., Garg, A., Tang, J., Nazario-Toole, A.E., Wu, L.P., 2013. A glutamate-dependent redox system in blood cells is integral for phagocytosis in *Drosophila melanogaster*. Curr. Biol. 23, 2319–2324.
Gonzalez, F., Tkaczuk, C., Dinu, M.M., Fiedler, Ż., Vidal, S., Zchori-Fein, E., Messelink, G.J., 2016. New opportunities for the integration of microorganisms into biological pest control systems in greenhouse crops. J. Pest. Sci. 89, 295–311.
Gunaratna, R.T., Jiang, H., 2013. A comprehensive analysis of the *Manduca sexta* immunotranscriptome. Dev. Comp. Immunol. 39, 388–398.
Ha, E.M., Oh, C.T., Bae, Y.S., Lee, W.J., 2005a. A direct role for dual oxidase in *Drosophila* gut immunity. Science 310, 847–850.
Ha, E.M., Oh, C.T., Ryu, J.H., Bae, Y.S., Kang, S.W., Jang, I.H., Brey, P.T., Lee, W.J., 2005b. An antioxidant system required for host protection against gut infection in *Drosophila*. Dev. Cell 8, 125–132.
Han, P., Jin, F., Dong, X., Fan, J., Qiu, B., Ren, S., 2013. Transcript and protein profiling analysis of the Destruxin A-induced response in larvae of *Plutella xylostella*. PLoS One 8, e60771.
Han, P., Fan, J., Liu, Y., Cuthbertson, A.G., Yan, S., Qiu, B.L., Ren, S., 2014. RNAi-mediated knockdown of serine protease inhibitor genes increases the mortality of *Plutella xylostella* challenged by destruxin A. PLoS One 9, e97863.

Hillyer, J.F., 2016. Insect immunology and hematopoiesis. Dev. Comp. Immunol. 58, 102–118.

Hiromori, H., Nishigaki, J., 2001. Factor analysis of synergistic effect between the entomopathogenic fungus *Metarhizium anisopliae* and synthetic insecticides. Appl. Entomol. Zool. 36, 231–236.

Hoffmann, J.A., 2003. The immune response of *Drosophila*. Nature 426, 33–38.

Holt, R.A., Subramanian, G.M., Halpern, A., Sutton, G.G., Charlab, R., Nusskern, D.R., Wincker, P., Clark, A.G., Ribeiro, J.C., Wides, R., Salzberg, S.L., Loftus, B., Yandell, M., Majoros, W.H., Rusch, D.B., Lai, Z., Kraft, C.L., Abril, J.F., Anthouard, V., Arensburger, P., Atkinson, P.W., Baden, H., de Berardinis, V., Baldwin, D., Benes, V., Biedler, J., Blass, C., Bolanos, R., Boscus, D., Barnstead, M., Cai, S., Center, A., Chatuverdi, K., Christophides, G.K., Chrystal, M.A., Clamp, M., Cravchik, A., Curwen, V., Dana, A., Delcher, A., Dew, I., Evans, C.A., Flanigan, M., Grundschober-Freimoser, A., Friedli, L., Gu, Z., Guan, P., Guigo, R., Hillenmeyer, M.E., Hladun, S.L., Hogan, J.R., Hong, Y.S., Hoover, J., Jaillon, O., Ke, Z., Kodira, C., Kokoza, E., Koutsos, A., Letunic, I., Levitsky, A., Liang, Y., Lin, J.-J., Lobo, N.F., Lopez, J.R., Malek, J.A., McIntosh, T.C., Meister, S., Miller, J., Mobarry, C., Mongin, E., Murphy, S.D., O'Brochta, D.A., Pfannkoch, C., Qi, R., Regier, M.A., Remington, K., Shao, H., Sharakhova, M.V., Sitter, C.D., Shetty, J., Smith, T.J., Strong, R., Sun, J., Thomasova, D., Ton, L.Q., Topalis, P., Tu, Z., Unger, M.F., Walenz, B., Wang, A., Wang, J., Wang, M., Wang, X., Woodford, K.J., Wortman, J.R., Wu, M., Yao, A., Zdobnov, E.M., Zhang, H., Zhao, Q., Zhao, S., Zhu, S.C., Zhimulev, I., Coluzzi, M., della Torre, A., Roth, C.W., Louis, C., Kalush, F., Mural, R.J., Myers, E.W., Adams, M.D., Smith, H.O., Broder, S., Gardner, M.J., Fraser, C.M., Birney, E., Bork, P., Brey, P.T., Venter, J.C., Weissenbach, J., Kafatos, F.C., Collins, F.H., Hoffman, S.L., 2002. The genome sequence of the malaria mosquito *Anopheles gambiae*. Science 298, 129–149.

Hu, X., Yang, R., Zhang, X., Chen, L., Xiang, X., Gong, C., Wu, X., 2013. Molecular cloning and functional characterization of the dual oxidase (*BmDuox*) gene from the silkworm *Bombyx mori*. PLoS One 8, e70118.

Hu, X., Zhu, M., Wang, S., Zhu, L., Xue, R., Cao, G., Gong, C., 2015. Proteomics analysis of digestive juice from silkworm during *Bombyx mori* nucleopolyhedrovirus infection. Proteomics 15, 2691–2700.

Huang, L., Cheng, T., Xu, P., Cheng, D., Fang, T., Xia, Q., 2009. A genome-wide survey for host response of silkworm, *Bombyx mori* during pathogen *Bacillus bombysepticus* infection. PLoS One 4, e8098.

Hughes, A.L., 2012. Evolution of the betaGRP/GNBP/beta-1,3-glucanase family of insects. Immunogenetics 64, 549–558.

Irving, P., Troxler, L., Heuer, T.S., Belvin, M., Kopczynski, C., Reichhart, J.M., Hoffmann, J.A., Hetru, C., 2001. A genome-wide analysis of immune responses in *Drosophila*. Proc. Natl. Acad. Sci. U.S.A. 98, 15119–15124.

James, R.R., Xu, J., 2012. Mechanisms by which pesticides affect insect immunity. J. Invertebr. Pathol. 109, 175–182.

Ji, D., Kim, Y., 2004. An entomopathogenic bacterium, *Xenorhabdus nematophila*, inhibits the expression of an antibacterial peptide, cecropin, of the beet armyworm, *Spodoptera exigua*. J. Insect Physiol. 50, 489–496.

Jiang, H., Vilcinskas, A., Kanost, M.R., 2010. Immunity in lepidopteran insects. Adv. Exp. Med. Biol. 708, 181–204.

Jiang, L., Wang, G., Cheng, T., Yang, Q., Jin, S., Lu, G., Wu, F., Xiao, Y., Xu, H., Xia, Q., 2012a. Resistance to *Bombyx mori* nucleopolyhedrovirus via overexpression of an endogenous antiviral gene in transgenic silkworms. Arch. Virol. 157, 1323–1328.

Jiang, S., Mao, L., Liu, X., Yao, Q., Chen, K., 2012b. The natural barrier in insects—peritrophic membrane. J. Biol. 29, 70–73.

Jiang, L., Zhao, P., Cheng, T., Sun, Q., Peng, Z., Dang, Y., Wu, X., Wang, G., Jin, S., Lin, P., Xia, Q., 2013. A transgenic animal with antiviral properties that might inhibit multiple stages of infection. Antiviral Res. 98, 171–173.

Kanost, M.R., Jiang, H., Yu, X.Q., 2004. Innate immune responses of a lepidopteran insect, *Manduca sexta*. Immunol. Rev. 198, 97–105.

Kanost, M.R., Arrese, E.L., Cao, X., Chen, Y.R., Chellapilla, S., Goldsmith, M.R., Grosse-Wilde, E., Heckel, D.G., Herndon, N., Jiang, H., Papanicolaou, A., Qu, J., Soulages, J.L., Vogel, H., Walters, J., Waterhouse, R.M., Ahn, S.J., Almeida, F.C., An, C., Aqrawi, P., Bretschneider, A., Bryant, W.B., Bucks, S., Chao, H., Chevignon, G., Christen, J.M., Clarke, D.F., Dittmer, N.T., Ferguson, L.C., Garavelou, S., Gordon, K.H., Gunaratna, R.T., Han, Y., Hauser, F., He, Y., Heidel-Fischer, H., Hirsh, A., Hu, Y., Jiang, H., Kalra, D., Klinner, C., Konig, C., Kovar, C., Kroll, A.R., Kuwar, S.S., Lee, S.L., Lehman, R., Li, K., Li, Z., Liang, H., Lovelace, S., Lu, Z., Mansfield, J.H., McCulloch, K.J., Mathew, T., Morton, B., Muzny, D.M., Neunemann, D., Ongeri, F., Pauchet, Y., Pu, L.L., Pyrousis, I., Rao, X.J., Redding, A., Roesel, C., Sanchez-Gracia, A., Schaack, S., Shukla, A., Tetreau, G., Wang, Y., Xiong, G.H., Traut, W., Walsh, T.K., Worley, K.C., Wu, D., Wu, W., Wu, Y.Q., Zhang, X., Zou, Z., Zucker, H., Briscoe, A.D., Burmester, T., Clem, R.J., Feyereisen, R., Grimmelikhuijzen, C.J., Hamodrakas, S.J., Hansson, B.S., Huguet, E., Jermiin, L.S., Lan, Q., Lehman, H.K., Lorenzen, M., Merzendorfer, H., Michalopoulos, I., Morton, D.B., Muthukrishnan, S., Oakeshott, J.G., Palmer, W., Park, Y., Passarelli, A.L., Rozas, J., Schwartz, L.M., Smith, W., Southgate, A., Vilcinskas, A., Vogt, R., Wang, P., Werren, J., Yu, X.Q., Zhou, J.J., Brown, S.J., Scherer, S.E., Richards, S., Blissard, G.W., 2016. Multifaceted biological insights from a draft genome sequence of the tobacco hornworm moth, *Manduca sexta*. Insect Biochem. Mol. Biol. 76, 118–147.

Kershaw, M.J., Moorhouse, E.R., Bateman, R., Reynolds, S.E., Charnley, A.K., 1999. The role of destruxins in the pathogenicity of *metarhizium anisopliae* for three species of insect. J. Invertebr. Pathol. 74, 213–223.

Kim, S.H., Lee, W.J., 2014. Role of DUOX in gut inflammation: lessons from *Drosophila* model of gut–microbiota interactions. Front. Cell. Infect. Microbiol. 3, 116.

Kogan, M., 1998. Integrated pest management: historical perspectives and contemporary developments. Annu. Rev. Entomol. 43, 243–270.

Kolliopoulou, A., Van Nieuwerburgh, F., Stravopodis, D.J., Deforce, D., Swevers, L., Smagghe, G., 2015. Transcriptome analysis of *Bombyx mori* larval midgut during persistent and pathogenic cytoplasmic polyhedrosis virus infection. PLoS One 10. e0121447

Kounatidis, I., Ligoxygakis, P., 2012. *Drosophila* as a model system to unravel the layers of innate immunity to infection. Open Biol. 2, 120075.

Kryukov, V.Y., Khodyrev, V.P., Yaroslavtseva, O.N., Kamenova, A.S., Duisembekov, B.A., Glupov, V.V., 2009. Synergistic action of entomopathogenic hyphomycetes and the bacteria *Bacillus thuringiensis* ssp. *morrisoni* in the infection of Colorado potato beetle *Leptinotarsa decemlineata*. Appl. Biochem. Microbiol. 45, 511–516.

Kuraishi, T., Binggeli, O., Opota, O., Buchon, N., Lemaitre, B., 2011. Genetic evidence for a protective role of the peritrophic matrix against intestinal bacterial infection in *Drosophila melanogaster*. Proc. Natl. Acad. Sci. U.S.A. 108, 15966–15971.

Lacey, L.A., Frutos, R., Kaya, H.K., Vail, P., 2001. Insect pathogens as biological control agents: do they have a future? Biol. Control 21, 230–248.

Lacey, L.A., Grzywacz, D., Shapiro-Ilan, D.I., Frutos, R., Brownbridge, M., Goettel, M.S., 2015. Insect pathogens as biological control agents: Back to the future. J. Invertebr. Pathol. 132, 1–41.

Lavine, M.D., Strand, M.R., 2002. Insect hemocytes and their role in immunity. Insect Biochem. Mol. Biol. 32, 1295–1309.
Lee, K.A., Kim, S.H., Kim, E.K., Ha, E.M., You, H., Kim, B., Kim, M.J., Kwon, Y., Ryu, J.H., Lee, W.J., 2013. Bacterial-derived uracil as a modulator of mucosal immunity and gut-microbe homeostasis in *Drosophila*. Cell 153, 797–811.
Lee, K.A., Kim, B., Bhin, J., Kim do, H., You, H., Kim, E.K., Kim, S.H., Ryu, J.H., Hwang, D., Lee, W.J., 2015a. Bacterial uracil modulates *Drosophila* DUOX-dependent gut immunity via Hedgehog-induced signaling endosomes. Cell Host Microbe 17, 191–204.
Lee, K.A., Kim, B., You, H., Lee, W.J., 2015b. Uracil-induced signaling pathways for DUOX-dependent gut immunity. Fly (Austin) 9, 115–120.
Lehane, M.J., 1997. Peritrophic matrix structure and function. Annu. Rev. Entomol. 42, 525–550.
Lehane, M.J., Billingsley, P.F., 1996. Biology of the Insect Midgut. Chapman & Hall, London.
Lemaitre, B., Hoffmann, J., 2007. The host defense of *Drosophila melanogaster*. Annu. Rev. Immunol. 25, 697–743.
Lemaitre, B., Miguel-Aliaga, I., 2013. The digestive tract of *Drosophila melanogaster*. Annu. Rev. Genet. 47, 377–404.
Lemaitre, B., Kromer-Metzger, E., Michaut, L., Nicolas, E., Meister, M., Georgel, P., Reichhart, J.M., Hoffmann, J.A., 1995. A recessive mutation, immune deficiency (*imd*), defines two distinct control pathways in the *Drosophila* host defense. Proc. Natl. Acad. Sci. U.S.A. 92, 9465–9469.
Lemaitre, B., Nicolas, E., Michaut, L., Reichhart, J.-M., Hoffmann, J.A., 1996. The dorsoventral regulatory gene cassette spätzle/Toll/cactus controls the potent antifungal response in *Drosophila* adults. Cell 86, 973–983.
Levy, S.M., Falleiros, A.M., Moscardi, F., Gregorio, E.A., 2011. The role of peritrophic membrane in the resistance of *Anticarsia gemmatalis* larvae (Lepidoptera: Noctuidae) during the infection by its nucleopolyhedrovirus (AgMNPV). Arthropod Struct. Dev. 40, 429–434.
Liu, H., Jiravanichpaisal, P., Cerenius, L., Lee, B.L., Söderhäll, I., Söderhäll, K., 2007. Phenoloxidase is an important component of the defense against *Aeromonas hydrophila* infection in a Crustacean, *Pacifastacus leniusculus*. J. Biol. Chem. 282, 33593–33598.
Liu, F., Ling, E., Wu, S., 2009. Gene expression profiling during early response to injury and microbial challenges in the silkworm, *Bombyx mori*. Arch. Insect Biochem. Physiol. 72, 16–33.
Liu, F., Xu, Q., Zhang, Q., Lu, A., Beerntsen, B., Ling, E., 2013. Hemocytes and hematopoiesis in the silkworm, *Bombyx mori*. Invert. Surviv. J. 10, 102–109.
Lloyd, A., 1996. HIV infection and AIDS. P. N. G. Med. J. 39, 174–180.
Lomer, C.J., Bateman, R.P., Johnson, D.L., Langewald, J., Thomas, M., 2001. Biological control of locusts and grasshoppers. Annu. Rev. Entomol. 46, 667–702.
Lu, A., Zhang, Q., Zhang, J., Yang, B., Wu, K., Xie, W., Luan, Y.X., Ling, E., 2014. Insect prophenoloxidase: the view beyond immunity. Front. Physiol. 5, 252.
Mahadav, A., Gerling, D., Gottlieb, Y., Czosnek, H., Ghanim, M., 2008. Parasitization by the wasp *Eretmocerus mundus* induces transcription of genes related to immune response and symbiotic bacteria proliferation in the whitefly *Bemisia tabaci*. BMC Genomics 9, 342.
Matova, N., Anderson, K.V., 2006. Rel/NF-κB double mutants reveal that cellular immunity is central to *Drosophila* host defense. Proc. Natl. Acad. Sci. U.S.A. 103, 16424–16429.
Mistry, R., Kounatidis, I., Ligoxygakis, P., 2016. Exploring interactions between pathogens and the *Drosophila* gut. Dev. Comp. Immunol. 64, 3–10.

Moreau, S.J., Guillot, S., 2005. Advances and prospects on biosynthesis, structures and functions of venom proteins from parasitic wasps. Insect Biochem. Mol. Biol. 35, 1209–1223.
Morse, R.A., Calderone, N.W., 1999. The value of honey bees as pollinators of U.S. crops in 2000. Am. Bee J. 129, 411–423. 477–486.
Nakamoto, M., Moy, R.H., Xu, J., Bambina, S., Yasunaga, A., Shelly, S.S., Gold, B., Cherry, S., 2012. Virus recognition by Toll-7 activates antiviral autophagy in *Drosophila*. Immunity 36, 658–667.
Nakazawa, H., Tsuneishi, E., Ponnuvel, K.M., Furukawa, S., Asaoka, A., Tanaka, H., Ishibashi, J., Yamakawa, M., 2004. Antiviral activity of a serine protease from the digestive juice of *Bombyx mori* larvae against nucleopolyhedrovirus. Virology 321, 154–162.
Nene, V., Wortman, J.R., Lawson, D., Haas, B., Kodira, C., Tu, Z.J., Loftus, B., Xi, Z., Megy, K., Grabherr, M., Ren, Q., Zdobnov, E.M., Lobo, N.F., Campbell, K.S., Brown, S.E., Bonaldo, M.F., Zhu, J., Sinkins, S.P., Hogenkamp, D.G., Amedeo, P., Arensburger, P., Atkinson, P.W., Bidwell, S., Biedler, J., Birney, E., Bruggner, R.V., Costas, J., Coy, M.R., Crabtree, J., Crawford, M., Debruyn, B., Decaprio, D., Eiglmeier, K., Eisenstadt, E., El-Dorry, H., Gelbart, W.M., Gomes, S.L., Hammond, M., Hannick, L.I., Hogan, J.R., Holmes, M.H., Jaffe, D., Johnston, J.S., Kennedy, R.C., Koo, H., Kravitz, S., Kriventseva, E.V., Kulp, D., Labutti, K., Lee, E., Li, S., Lovin, D.D., Mao, C., Mauceli, E., Menck, C.F., Miller, J.R., Montgomery, P., Mori, A., Nascimento, A.L., Naveira, H.F., Nusbaum, C., O'Leary, S., Orvis, J., Pertea, M., Quesneville, H., Reidenbach, K.R., Rogers, Y.H., Roth, C.W., Schneider, J.R., Schatz, M., Shumway, M., Stanke, M., Stinson, E.O., Tubio, J.M., Vanzee, J.P., Verjovski-Almeida, S., Werner, D., White, O., Wyder, S., Zeng, Q., Zhao, Q., Zhao, Y., Hill, C.A., Raikhel, A.S., Soares, M.B., Knudson, D.L., Lee, N.H., Galagan, J., Salzberg, S.L., Paulsen, I.T., Dimopoulos, G., Collins, F.H., Birren, B., Fraser-Liggett, C.M., Severson, D.W., 2007. Genome sequence of *Aedes aegypti*, a major arbovirus vector. Science 316, 1718–1723.
Neyen, C., Bretscher, A.J., Binggeli, O., Lemaitre, B., 2014. Methods to study *Drosophila* immunity. Methods 68, 116–128.
Nguyen, T.T., Magnoli, I., Cloutier, C., Michaud, D., Muratori, F., Hance, T., 2013. Early presence of an enolase in the oviposition injecta of the aphid parasitoid *Aphidius ervi* analyzed with chitosan beads as artificial hosts. J. Insect Physiol. 59, 11–18.
Notarangelo, L.D., 2010. Primary immunodeficiencies. J. Allergy Clin. Immunol. 125, S182–S194.
Pal, S., St Leger, R.J., Wu, L.P., 2007. Fungal peptide Destruxin A plays a specific role in suppressing the innate immune response in *Drosophila melanogaster*. J. Biol. Chem. 282, 8969–8977.
Palmer, W.J., Jiggins, F.M., 2015. Comparative genomics reveals the origins and diversity of arthropod immune systems. Mol. Biol. Evol. 32, 2111–2129.
Paria, A., Greeshma, S.S., Chaudhari, A., Makesh, M., Purushothaman, C.S., Rajendran, K.V., 2013. Nonspecific effect of double-stranded (ds) RNA on prophenoloxidase (proPO) expression in *Penaeus monodon*. Appl. Biochem. Biotechnol. 169, 281–289.
Park, J., Kim, Y., 2011. Benzylideneacetone suppresses both cellular and humoral immune responses of *Spodoptera exigua* and enhances fungal pathogenicity. J. Asia-Pacif. Entomol. 14, 423–427.
Ponnuvel, K.M., Nakazawa, H., Furukawa, S., Asaoka, A., Ishibashi, J., Tanaka, H., Yamakawa, M., 2003. A lipase isolated from the silkworm *Bombyx mori* shows antiviral activity against nucleopolyhedrovirus. J. Virol. 77, 10725–10729.
Posada-Flórez, F.J., 2008. Production of *Beauveria bassiana* fungal spores on rice to control the coffee berry borer, *Hypothenemus hampei*, in Colombia. J. Insect Sci. 8, 1–13.

Quesada-Moraga, E., Vey, A., 2004. Bassiacridin, a protein toxic for locusts secreted by the entomopathogenic fungus *Beauveria bassiana*. Mycol. Res. 108, 441–452.

Rao, R., Fiandra, L., Giordana, B., de Eguileor, M., Congiu, T., Burlini, N., Arciello, S., Corrado, G., Pennacchio, F., 2004. AcMNPV ChiA protein disrupts the peritrophic membrane and alters midgut physiology of *Bombyx mori* larvae. Insect Biochem. Mol. Biol. 34, 1205–1213.

Richards, E.H., Dani, M.P., 2010. A recombinant immunosuppressive protein from *Pimpla hypochondriaca* (rVPr1) increases the susceptibility of *Lacanobia oleracea* and *Mamestra brassicae* larvae to *Bacillus thuringiensis*. J. Invertebr. Pathol. 104, 51–57.

Richards, S., Gibbs, R.A., Weinstock, G.M., Brown, S.J., Denell, R., Beeman, R.W., Gibbs, R., Beeman, R.W., Brown, S.J., Bucher, G., Friedrich, M., Grimmelikhuijzen, C.J., Klingler, M., Lorenzen, M., Richards, S., Roth, S., Schroder, R., Tautz, D., Zdobnov, E.M., Muzny, D., Gibbs, R.A., Weinstock, G.M., Attaway, T., Bell, S., Buhay, C.J., Chandrabose, M.N., Chavez, D., Clerk-Blankenburg, K.P., Cree, A., Dao, M., Davis, C., Chacko, J., Dinh, H., Dugan-Rocha, S., Fowler, G., Garner, T.T., Garnes, J., Gnirke, A., Hawes, A., Hernandez, J., Hines, S., Holder, M., Hume, J., Jhangiani, S.N., Joshi, V., Khan, Z.M., Jackson, L., Kovar, C., Kowis, A., Lee, S., Lewis, L.R., Margolis, J., Morgan, M., Nazareth, L.V., Nguyen, N., Okwuonu, G., Parker, D., Richards, S., Ruiz, S.J., Santibanez, J., Savard, J., Scherer, S.E., Schneider, B., Sodergren, E., Tautz, D., Vattahil, S., Villasana, D., White, C.S., Wright, R., Park, Y., Beeman, R.W., Lord, J., Oppert, B., Lorenzen, M., Brown, S., Wang, L., Savard, J., Tautz, D., Richards, S., Weinstock, G., Gibbs, R.A., Liu, Y., Worley, K., Weinstock, G., Elsik, C.G., Reese, J.T., Elhaik, E., Landan, G., Graur, D., Arensburger, P., Atkinson, P., Beeman, R.W., Beidler, J., Brown, S.J., Demuth, J.P., Drury, D.W., Du, Y.Z., Fujiwara, H., Lorenzen, M., Maselli, V., Osanai, M., Park, Y., Robertson, H.M., Tu, Z., Wang, J.J., Wang, S., Richards, S., Song, H., Zhang, L., Sodergren, E., Werner, D., Stanke, M., Morgenstern, B., Solovyev, V., Kosarev, P., Brown, G., Chen, H.C., Ermolaeva, O., Hlavina, W., Kapustin, Y., Kiryutin, B., Kitts, P., Maglott, D., Pruitt, K., Sapojnikov, V., Souvorov, A., Mackey, A.J., Waterhouse, R.M., Wyder, S., Zdobnov, E.M., Zdobnov, E.M., Wyder, S., Kriventseva, E.V., Kadowaki, T., Bork, P., Aranda, M., Bao, R., Beermann, A., Berns, N., Bolognesi, R., Bonneton, F., Bopp, D., Brown, S.J., Bucher, G., Butts, T., Chaumot, A., Denell, R.E., Ferrier, D.E., Friedrich, M., Gordon, C.M., Jindra, M., Klingler, M., Lan, Q., Lattorff, H.M., Laudet, V., von Levetsow, C., Liu, Z., Lutz, R., Lynch, J.A., da Fonseca, R.N., Posnien, N., Reuter, R., Roth, S., Savard, J., Schinko, J.B., Schmitt, C., Schoppmeier, M., Schroder, R., Shippy, T.D., Simonnet, F., Marques-Souza, H., Tautz, D., Tomoyasu, Y., Trauner, J., Van der Zee, M., Vervoort, M., Wittkopp, N., Wimmer, E.A., Yang, X., Jones, A.K., Sattelle, D.B., Ebert, P.R., Nelson, D., Scott, J.G., Beeman, R.W., Muthukrishnan, S., Kramer, K.J., Arakane, Y., Beeman, R.W., Zhu, Q., Hogenkamp, D., Dixit, R., Oppert, B., Jiang, H., Zou, Z., Marshall, J., Elpidina, E., Vinokurov, K., Oppert, C., Zou, Z., Evans, J., Lu, Z., Zhao, P., Sumathipala, N., Altincicek, B., Vilcinskas, A., Williams, M., Hultmark, D., Hetru, C., Jiang, H., Grimmelikhuijzen, C.J., Hauser, F., Cazzamali, G., Williamson, M., Park, Y., Li, B., Tanaka, Y., Predel, R., Neupert, S., Schachtner, J., Verleyen, P., Raible, F., Bork, P., Friedrich, M., Walden, K.K., Robertson, H.M., Angeli, S., Foret, S., Bucher, G., Schuetz, S., Maleszka, R., Wimmer, E.A., Beeman, R.W., Lorenzen, M., Tomoyasu, Y., Miller, S.C., Grossmann, D., Bucher, G., 2008. The genome of the model beetle and pest *Tribolium castaneum*. Nature 452, 949–955.

Richards, E.H., Bradish, H., Dani, M.P., Pietravalle, S., Lawson, A., 2011. Recombinant immunosuppressive protein from *Pimpla hypochondrica* venom (rVPr1) increases the susceptibility of *Mamestra brassicae* larvae to the fungal biological control agent, *Beauveria bassiana*. Arch. Insect Biochem. Physiol. 78, 119–131.

Richards, E.H., Dani, M.P., Bradish, H., 2013. Immunosuppressive properties of a protein (rVPr1) from the venom of the endoparasitic wasp, *Pimpla hypochondriaca*: mechanism of action and potential use for improving biological control strategies. J. Insect Physiol. 59, 213–222.

Rosen, F.S., Cooper, M.D., Wedgwood, R.J.P., 1995. The primary immunodeficiencies. N. Engl. J. Med. 333, 431–440.

Salminen, J.-P., Lempa, K., 2002. Effects of hydrolysable tannins on a herbivorous insect: fate of individual tannins in insect digestive tract. Chemoecology 12, 203–211.

Schlenke, T.A., Morales, J., Govind, S., Clark, A.G., 2007. Contrasting infection strategies in generalist and specialist wasp parasitoids of *Drosophila melanogaster*. PLoS Pathog. 3, 1486–1501.

Schnitger, A.K., Kafatos, F.C., Osta, M.A., 2007. The melanization reaction is not required for survival of *Anopheles gambiae* mosquitoes after bacterial infections. J. Biol. Chem. 282, 21884–21888.

Shao, Q., Yang, B., Xu, Q., Li, X., Lu, Z., Wang, C., Huang, Y., Söderhäll, K., Ling, E., 2012. Hindgut innate immunity and regulation of fecal microbiota through melanization in insects. J. Biol. Chem. 287, 14270–14279.

Shapiro-Ilan, D.I., Jackson, M., Reilly, C.C., Hotchkiss, M.W., 2004. Effects of combining an entomopathogenic fungi or bacterium with entomopathogenic nematodes on mortality of *Curculio caryae* (Coleoptera: Curculionidae). Biol. Control 30, 119–126.

Sparks, M.E., Blackburn, M.B., Kuhar, D., Gundersen-Rindal, D.E., 2013. Transcriptome of the *Lymantria dispar* (gypsy moth) larval midgut in response to infection by *Bacillus thuringiensis*. PLoS One 8, e61190.

Stettler, P., Trenczek, T., Wyler, T., Pfister-Wilhelm, R., Lanzrein, B., 1998. Overview of parasitism associated effects on host haemocytes in larval parasitoids and comparison with effects of the egg-larval parasitoid *Chelonus inanitus* on its host *Spodoptera littoralis*. J. Insect Physiol. 44, 817–831.

Strand, M.R., 2008. The insect cellular immune response. Insect Sci. 15, 1–14.

Tanaka, H., Yamakawa, M., 2011. Regulation of the innate immune responses in the silkworm, *Bombyx mori*. Invert. Surviv. J. 8, 59–69.

Tanaka, H., Ishibashi, J., Fujita, K., Nakajima, Y., Sagisaka, A., Tomimoto, K., Suzuki, N., Yoshiyama, M., Kaneko, Y., Iwasaki, T., Sunagawa, T., Yamaji, K., Asaoka, A., Mita, K., Yamakawa, M., 2008. A genome-wide analysis of genes and gene families involved in innate immunity of *Bombyx mori*. Insect Biochem. Mol. Biol. 38, 1087–1110.

Tang, H., 2009. Regulation and function of the melanization reaction in *Drosophila*. Fly (Austin) 3, 105–111.

The Honeybee Genome Sequencing C, 2006. Insights into social insects from the genome of the honeybee *Apis mellifera*. Nature 443, 931–949.

Usha Rani, P., Pratyusha, S., 2014. Role of castor plant phenolics on performance of its two herbivores and their impact on egg parasitoid behaviour. BioControl 59, 513–524.

Van't Hof, A.E., Campagne, P., Rigden, D.J., Yung, C.J., Lingley, J., Quail, M.A., Hall, N., Darby, A.C., Saccheri, I.J., 2016. The industrial melanism mutation in British peppered moths is a transposable element. Nature 534, 102–105.

Vey, A., Matha, V., Dumas, C., 2002. Effects of the peptide mycotoxin destruxin E on insect haemocytes and on dynamics and efficiency of the multicellular immune reaction. J. Invertebr. Pathol. 80, 177–187.

Vilcinskas, A., Matha, V., Götz, P., 1997. Effects of the entomopathogenic fungus *Metarhizium anisopliae* and its secondary metabolites on morphology and cytoskeleton of plasmatocytes isolated from the greater wax moth, *Galleria mellonella*. J. Insect Physiol. 43, 1149–1159.

Wang, S., Jacobs-Lorena, M., 2013. Genetic approaches to interfere with malaria transmission by vector mosquitoes. Trends Biotechnol. 31, 185–193.

Wang, L., Ligoxygakis, P., 2006. Pathogen recognition and signalling in the *Drosophila* innate immune response. Immunobiology 211, 251–261.

Wang, C., St Leger, R.J., 2006. A collagenous protective coat enables *Metarhizium anisopliae* to evade insect immune responses. Proc. Natl. Acad. Sci. U.S.A. 103, 6647–6652.

Wang, B., Kang, Q., Lu, Y., Bai, L., Wang, C., 2012. Unveiling the biosynthetic puzzle of destruxins in *Metarhizium* species. Proc. Natl. Acad. Sci. U.S.A. 109, 1287–1292.

Waterhouse, R.M., Kriventseva, E.V., Meister, S., Xi, Z., Alvarez, K.S., Bartholomay, L.C., Barillas-Mury, C., Bian, G., Blandin, S., Christensen, B.M., Dong, Y., Jiang, H., Kanost, M.R., Koutsos, A.C., Levashina, E.A., Li, J., Ligoxygakis, P., MacCallum, R.M., Mayhew, G.F., Mendes, A., Michel, K., Osta, M.A., Paskewitz, S., Shin, S.W., Vlachou, D., Wang, L., Wei, W., Zheng, L., Zou, Z., Severson, D.W., Raikhel, A.S., Kafatos, F.C., Dimopoulos, G., Zdobnov, E.M., Christophides, G.K., 2007. Evolutionary dynamics of immune-related genes and pathways in disease-vector mosquitoes. Science 316, 1738–1743.

Webb, B.A., Strand, M.R., 2005. The biology and genomics of polydnaviruses. In: Gilbert, L. (Ed.), Comprehensive Molecular Insect Science. Elsevier, Amsterdam, pp. 323–360.

Wertheim, B., Kraaijeveld, A.R., Schuster, E., Blanc, E., Hopkins, M., Pletcher, S.D., Strand, M.R., Partridge, L., Godfray, H.C., 2005. Genome-wide gene expression in response to parasitoid attack in *Drosophila*. Genome Biol. 6, R94.

Wraight, S.P., Ramos, M.E., 2005. Synergistic interaction between *Beauveria bassiana*- and *Bacillus thuringiensis tenebrionis*-based biopesticides applied against field populations of Colorado potato beetle larvae. J. Invertebr. Pathol. 90, 139–150.

Wu, P., Han, S., Chen, T., Qin, G., Li, L., Guo, X., 2013. Involvement of microRNAs in infection of silkworm with *Bombyx mori* cytoplasmic polyhedrosis virus (BmCPV). PLoS One 8, e68209.

Wu, K., Zhang, J., Zhang, Q., Zhu, S., Shao, Q., Clark, K.D., Liu, Y., Ling, E., 2015. Plant phenolics are detoxified by prophenoloxidase in the insect gut. Sci. Rep. 5, 16823.

Wu, K., Yang, B., Huang, W., Dobens, L., Song, H., Ling, E., 2016. Gut immunity in Lepidopteran insects. Dev. Comp. Immunol. 64, 65–74.

Xia, Q., Zhou, Z., Lu, C., Cheng, D., Dai, F., Li, B., Zhao, P., Zha, X., Cheng, T., Chai, C., Pan, G., Xu, J., Liu, C., Lin, Y., Qian, J., Hou, Y., Wu, Z., Li, G., Pan, M., Li, C., Shen, Y., Lan, X., Yuan, L., Li, T., Xu, H., Yang, G., Wan, Y., Zhu, Y., Yu, M., Shen, W., Wu, D., Xiang, Z., Yu, J., Wang, J., Li, R., Shi, J., Li, H., Li, G., Su, J., Wang, X., Li, G., Zhang, Z., Wu, Q., Li, J., Zhang, Q., Wei, N., Xu, J., Sun, H., Dong, L., Liu, D., Zhao, S., Zhao, X., Meng, Q., Lan, F., Huang, X., Li, Y., Fang, L., Li, C., Li, D., Sun, Y., Zhang, Z., Yang, Z., Huang, Y., Xi, Y., Qi, Q., He, D., Huang, H., Zhang, X., Wang, Z., Li, W., Cao, Y., Yu, Y., Yu, H., Li, J., Ye, J., Chen, H., Zhou, Y., Liu, B., Wang, J., Ye, J., Ji, H., Li, S., Ni, P., Zhang, J., Zhang, Y., Zheng, H., Mao, B., Wang, W., Ye, C., Li, S., Wang, J., Wong, G.K., Yang, H., 2004. A draft sequence for the genome of the domesticated silkworm (*Bombyx mori*). Science 306, 1937–1940.

Xie, J., Liu, X., Zhang, Z., Xu, W., Li, Q., 2012. Current status and prospect of researches on mycoinsecticides. J. Jilin Agric. Sci. 37, 49–53.

Yu, R.X., Chen, Y.F., Chen, X.X., Huang, F., Lou, Y.G., Liu, S.S., 2007. Effects of venom/calyx fluid from the endoparasitic wasp *Cotesia plutellae* on the hemocytes of its host *Plutella xylostella* in vitro. J. Insect Physiol. 53, 22–29.

Zambon, R.A., Nandakumar, M., Vakharia, V.N., Wu, L.P., 2005. The Toll pathway is important for an antiviral response in *Drosophila*. Proc. Natl. Acad. Sci. U.S.A. 102, 7257–7262.

Zarubin, T., Han, J., 2005. Activation and signaling of the p38 MAP kinase pathway. Cell Res. 15, 11–18.

Zhang, L., Lu, Z., 2015. Expression, purification and characterization of an atypical 2-Cys peroxiredoxin from the silkworm, *Bombyx mori*. Insect Mol. Biol. 24, 203–212.

Zhang, G., Schmidt, O., Asgari, S., 2006. A calreticulin-like protein from endoparasitoid venom fluid is involved in host hemocyte inactivation. Dev. Comp. Immunol. 30, 756–764.

Zhang, Q., Wang, F., Fang, Q., Ye, G., 2011. *In vitro* cellular responses of *Pieris rapae* (Lepidoptera: Pieridae) hemocytes and the effects of *Pteromalus puparum* venom. Acta Entomol. Sin. 54, 1264–1273.

Zhang, X., He, Y., Cao, X., Gunaratna, R.T., Chen, Y.R., Blissard, G., Kanost, M.R., Jiang, H., 2015. Phylogenetic analysis and expression profiling of the pattern recognition receptors: Insights into molecular recognition of invading pathogens in *Manduca sexta*. Insect Biochem. Mol. Biol. 62, 38–50.

Zhu, Y., Lu, Z., 2013. Identification of differentially expressed genes in the larval midgut of the silkworm (*Bombyx mori*) after bacterial infection. Acta Entomol. Sin. 56, 505–511.

Zou, Z., Evans, J.D., Lu, Z., Zhao, P., Williams, M., Sumathipala, N., Hetru, C., Hultmark, D., Jiang, H., 2007. Comparative genomic analysis of the *Tribolium* immune system. Genome Biol. 8, R177.

Zou, Z., Shin, S.W., Alvarez, K.S., Kokoza, V., Raikhel, A.S., 2010. Distinct melanization pathways in the mosquito *Aedes aegypti*. Immunity 32, 41–53.

CHAPTER EIGHT

Immunology of Insect Vectors: Midgut Interactions of Sandflies and Tsetse with Kinetoplastid Parasites as a Paradigm for Establishing Infection

Megan A. Sloan, Petros Ligoxygakis
Laboratory of Cell Biology, Development and Genetics, University of Oxford, Oxford, United Kingdom

Contents

1. Introduction 232
2. Midgut Establishment 233
3. The Peritrophic Matrix: A Structural Barrier 234
4. Digestive Enzymes 235
5. Tsetse EP Proteins 236
6. Pathogen Recognition 237
7. Antimicrobial Peptides 238
8. Reactive Oxygen Species 239
9. Gut Microbiota and Symbionts 240
   9.1 *Wigglesworthia glossinidia* 241
   9.2 *S. glossinidius* 242
   9.3 *Wolbachia* 242
10. Using the Microbiota as a Transmission Blocking Strategy 243
Acknowledgements 244
References 244

Abstract

For the vast majority of vector borne parasites the ability to overcome the insect midgut defences is central to transmission. However, for many such diseases we know virtually nothing about the molecular mechanisms involved. For vectors such as tsetse flies and sandflies the prospects for rapidly improving our understanding of key interactions occurring in the midgut when challenged by parasites are difficult. This is because the 'tool box' required untangling the interactions is very unlikely to be rapidly developed. For example, there is no realistic prospect of producing transgenic technology for

Advances in Insect Physiology, Volume 52
ISSN 0065-2806
http://dx.doi.org/10.1016/bs.aiip.2017.04.003

tsetse flies because eggs are inaccessible due to intrauterine development of larvae; maintenance of multiple lines of either sand or tsetse flies permitting genetic studies is impossible because of the cost and complexity of culturing colonies; bioinformatics resources are still in their infancy. Nevertheless, through a combination of genomics, transcriptomics, RNAi experiments and work on endosymbionts, researchers in these difficult systems have placed the general framework of parasite establishment in the midgut. Here, we will review the major immune pathways by which tsetse and sandflies respond to kinetoplastid challenge in the midgut and the role of endosymbionts as well as the gut microflora in determining vectorial capacity.

1. INTRODUCTION

Neglected tropical diseases (NTDs) like sleeping sickness, leishmaniasis, hookworm infections, river blindness and elephantiasis are the most common infections of the world's 1.4 billion poorest people and the leading causes of chronic disability and poverty in low- and middle-income countries (Hotez, 2008, 2009). In all, there are 17 NTDs prioritized by the world health organization (WHO), roughly a third of which are transmitted by Diptera or 'true flies' (Table 1). This demonstrates the importance of the Diptera and how understanding their role in these diseases could prove critical to improving global health. Dipterans have a single pair of wings and are holometabolous (undergoing complete metamorphosis). The Diptera is divided into three suborders: Nematocera, Brachycera and Cyclorrhapha. The vectors for many NTDs, including mosquitoes, black flies and sandflies,

Table 1 The WHO's Prioritized Neglected Tropical Diseases With Dipteran Vectors and Their Causative Agents

Disease	Causative Agent(s)	Vector(s)
Lymphatic filariasis	*Wuchereria bancrofti, Brugia malayi, Brugia timori*	*Aedes* spp., *Anopheles* spp., *Culex* spp., *Mansonia* spp.
Onchocerciasis (river blindness)	*Onchocerca volvulus*	*Simulium* spp.
Human African trypanosomiasis (HAT)	*Trypanosoma brucei rhodesiense, Trypanosoma brucei gambiense*	*Glossina* spp.
Leishmaniases	*Leishmania* spp.	*Phleobotomine* sandflies
Dengue and Chikungunya	*DEN-1, DEN-2, DEN-3 and DEN-4,* Chikungunya virus	*Aedes aegypti*

are blood-feeders belonging to the Nematocera, whereas Tsetse (*Glossina* spp.) belong to the Brachycera. This review will focus on two important vectors, sandflies and tsetse, and their interactions with the parasites they transmit: *Leishmania* spp. and *Trypanosoma brucei*. Two sandfly genera are of medical importance in the context of *Leishmania* transmission: *Phlebotomus* and *Lutzomyia*. Sandflies are widely distributed in the earth's subtropical and temperate regions. In contrast, Tsetse flies are found only in sub-Saharan Africa.

2. MIDGUT ESTABLISHMENT

In both insects, the ability of the invader to overcome the insect vector's midgut defences is absolutely central to transmission. This is clearly illustrated as follows. African trypanosomes responsible for sleeping sickness and nagana encounter a severe barrier to their establishment in the midgut of their tsetse fly vectors (reviewed in Lehane et al., 2004; Haines, 2013). Even when tsetse is at its most susceptible, at the first blood meal following emergence from the puparium, the tsetse fly only permits about 50% of *T. brucei* spp. to become established. From the third blood meal onwards (and the fly may take 40–60 blood meals in its life) less than 10% of challenged flies become infected (Lehane et al., 2004). The vast majority of infections fail at the midgut level. Paradoxically therefore—given their importance as vectors—tsetse fly populations are overwhelmingly resistant to trypanosome infection, and the resistance mechanisms are manifested largely in the fly midgut (Lehane, 1997; Lehane and Msangi, 1991). However, our understanding of the molecular events underpinning this midgut-mediated resistance is poor. *Leishmania* parasites seem to have successfully overcome barriers to establishment in their sandfly hosts as they develop in the midgut of challenged laboratory strains of those flies. Again, we understand neither how the parasite establishes itself in the gut nor why the insect tolerates large numbers of parasites (reviewed in Bates, 2008). Nevertheless, in the wild there is only 1% of caught sandflies infected with *Leishmania*. Tsetse flies are obligatory hematophagous insects, whereas sandflies are also plant feeders. These significant lifestyle differences notwithstanding both an indigenous flora in their gut, which may play an important role in the establishment of infection and might explain the largely resistant nature of these insects in the field (see below).

3. THE PERITROPHIC MATRIX: A STRUCTURAL BARRIER

The fully formed peritrophic matrix (PM) is a semipermeable 'mesh' sac comprised of proteins (including peritrophins), chitin and proteoglycans which partitions off gut lumen (and therefore the blood meal) from the epithelium (see Lehane, 1997 as a comprehensive review for PM structure). Keehnen et al. in chapter "Insect antimicrobial defenses: A brief history, recent findings, biases, and a way forward in evolutionary studies" refer to the PM as a barrier in insect immunity; therefore, this review will focus on the role of the PM in NTD vectors. As in mosquitos, the sandfly PM is secreted by the midgut epithelium after being triggered by gut distention upon ingestion of a blood meal. In contrast, the tsetse continuously secretes PM from cells in the proventriculus (Rose et al., 2014; Sádlová and Volf, 2009). Tsetse endosymbionts (see below) have been implicated in PM formation because aposymbiotic flies have no or a severely compromised PM (Weiss et al., 2013).

PM composition is beginning to be elucidated in sandflies and tsetse. Three peritrophins from *Phlebotomus papatasi* predictably contained chitin-binding domains and one, PpPer3, contained a mucin-like domain, suggesting the PM is functionally analogous to mammalian gut's mucus layer (Coutinho-Abreu et al., 2013). The same study also showed PpPer1 is upregulated in the first 24 h of *Leishmania major* infection though it is unclear if the vector mediated this change or perhaps *L. major* modifies vector gene expression in order to protect itself from digestive enzymes and immune effectors diffusing through the PM from the midgut epithelium.

C-type lectins have been identified in the tsetse fly PM leading to speculation that it could have a role in agglutinating invading trypanosomes which once 'trapped' would be expelled upon defecation; however, this is yet to be fully investigated (Weiss et al., 2013). The same study also identified CD38 and a Hemomucin with a strictosidine synthase in the PM. In *Caenorhabditis elegans*, CD38 functions as a scavenger receptor stimulating cytokine production. In other insects strictodine synthase is an important enzyme for synthesis of alkaloids which have been shown to have antimicrobial roles in other species, including plants (de A Gonzaga et al., 2003). Further investigation into their role in trypanomsome infection is needed.

Leishmania escape the PM through a vector-mediated posterior opening (Sádlová and Volf, 2009) and not using a parasite chitinase as previously speculated; as timing of the PM opening in *Phlebotomus duboscqi* is the same in both *L. major* infected and uninfected females. RNAi knockdown of vector chitinase PpChit1 results in a thickened PM and reduced parasite load possibly due to the inability of 'trapped' *L. major* to escape digestive enzymes/immune effector molecules (Coutinho-Abreu et al., 2013). PpChit1 has been highlighted as a potential candidate for a transmission blocking vaccine. A similar story is likely in tsetse as trypanosomes do not appear to express chitinase in the gut (Rose et al., 2014).

4. DIGESTIVE ENZYMES

Once the parasites have escaped the PM lumen, they encounter a new layer of the vector defence—the digestive enzymes which can cause proteolytic damage to the parasites. The PM probably provides the parasites with some protection from these enzymes initially; however, their concentration in the space between the gut epithelium and the PM is much higher.

Depending on the species of fly, the levels of proteases in the midgut reach their peak 18–48 h after ingestion of the blood meal. The most abundant are the trypsin-like enzymes (Telleria et al., 2010), though chymotrypsons, metallocarboxypeptidases, serine proteases and alanyl aminopeptidase have also been reported (Dillon and Lane, 1993; Dillon et al., 2006).

It has long been known that digestive enzymes inhibit parasite establishment in insect vectors. Addition of trypsin inhibitors with the blood meal allows more *Leishmania donovani* to survive during the early stages of infection of *P. papatasi* (Borovsky and Schlein, 1987). Similarly inhibition of *P. duboscqi* lectins lead to enhanced establishment of *L. major* (Volf et al., 2001). Furthermore, RNAi knockdown of LlTryp1, a feeding-induced trypsin, in *Lutzomyia longipalpis* lead to greater survival of *Leishmania mexicana* upon infection (Sant'Anna et al., 2009).

Digestion in the tsetse fly has been studied intensively since the seminal paper of Wigglesworth (1929). There, PM function was discussed to show that it is freely permeable to a whole armoury of host digestive enzymes and to haemoglobin. It was found later that these enzymes included the cysteine-type proteinase Cathepsin B, a zinc-metalloprotease and a zinc-carboxypeptidase (Yan et al., 2002). These enzymes (especially Cathepsin B) have been shown to be activated following a blood meal and hypothesized to be involved in the resistance of the fly to parasite establishment (Yan et al., 2002). Moreover, the *T. brucei* procyclic stage surface is covered in glycoproteins such as EP and GREET. These proteins are recognized and cleaved by tsetse proteases (Liniger et al., 2003).

In this context, therefore, an important question remained unanswered: How does the parasite transverse from the gut lumen to the ecto-peritrophic space avoiding this digestive activity? Using RNA sequencing, Aksoy et al. first determined gene expression in midgut cells at 48 and 72h after feeding mature adults on trypanosome-infected or normal blood. They found that trypanosomes reduced the expression of peritrophins, structural proteins that bind chitin, and digestive enzymes, both associated with PM formation in sandflies (Coutinho-Abreu et al., 2013; Dinglasan et al., 2009). These authors implicate the variant surface glycoprotein (VSG) of the blood stream form of the parasite in humans mediated the disruption of the fly's PM and the downregulation of the digestive enzymes, revealing a dual role for VSG in the life cycle of trypanosomes in the mammalian as well as the insect host (Aksoy et al., 2016).

To assist survival in an environment rich in proteases *Leishmania* parasites have evolved similar strategies. It has been shown that there are changes in digestive enzyme levels in *L. major* and *Leishmania infantum* infections (Dostálová et al., 2011; Jochim et al., 2008; Ramalho-Ortigao et al., 2007; Telleria et al., 2010). *L. major* has serine protease inhibitors in the genome but appears to lack the target enzymes. These inhibitors have been shown to work against insect proteases in vitro (Morrison et al., 2012).

5. TSETSE EP PROTEINS

The tsetse EP proteins have extensive glutamic acid-proline dipeptide repeats for over two-fifths of their sequence and are very similar to the trypanosome EP proteins. Despite this, it is not believed they are the target of molecular mimicry. Their expression is upregulated in both bacterial and trypanosome infections, and they appear to di-/trimerise in aggregates (Haines et al., 2005). EP proteins are expressed throughout the tsetse midgut and have been identified in the PM. The proteins have a lectin domain which may interact with pathogen surface glycans trapping them in the PM (Rose et al., 2014). RNAi knockdown of tsetse EP in the haemocoel increases the trypanosome midgut infection rate, but the mechanism is unknown at present (Haines et al., 2010).

6. PATHOGEN RECOGNITION

There have been several pattern recognition receptor proteins (PRRs) identified in the Diptera, which recognize microbe-associated molecular patterns; however, their interactions with parasites are poorly understood. Comparisons of cDNA libraries produced from infected and uninfected (whole) sandflies identified a number of possible PRRs relevant to *L. infantum/L. mexicana* infection in *Lu. longipalpis* (Dillon et al., 2006). These included lipophosphoglycan LPG-binding galectins, peptidoglycan receptor proteins (PGRP), a Gram-negative-binding protein (GNBP) (Roxström-Lindquist et al., 2004) and a putative scavenger receptor with complement control protein domains/adhesive MAM domain. PGRPs have also been implicated in recognition of trypanosomes in tsetse flies as susceptible tsetse flies had lower levels of PCRP-LB than those of 'self-curing' tsetse flies (Wang et al., 2009). It has also been shown that PGRP-LB has trypanocidal activity (Wang and Aksoy, 2012).

Pattern recognition receptors activate signalling cascades which result in expression of effector molecules such as antimicrobial peptides (AMPs) or reactive oxygen species. In insects there are two major pathways which can be activated: the Toll pathway and the immunodeficiency (IMD) pathway. Results indicate that in trypanosome and *Leishmania* infections the IMD pathway is most relevant. Telleria et al. (2012) showed that Caspar, a negative regulator of the IMD pathway, was downregulated following ingestion of *L. major*, suggesting that the IMD pathway and its effector molecules may negatively effective *Leishmania* development in sandflies (Telleria et al., 2012) (Fig. 1).

Tsetse flies exhibit different immune-related gene expression patterns when challenged with trypanosomes or bacteria. Therefore, the flies may specifically recognize and respond to the invading organism (Beschin et al., 2014). Similar to *Leishmania* infection of sandflies, infection of tsetse with trypanosomes causes higher expression of genes that are transcriptional targets of IMD (Attacin and PGRP-LB) (Weiss et al., 2013) and RNA interference knockdown of IMD genes results increased infection severity (Hu and Aksoy, 2006).

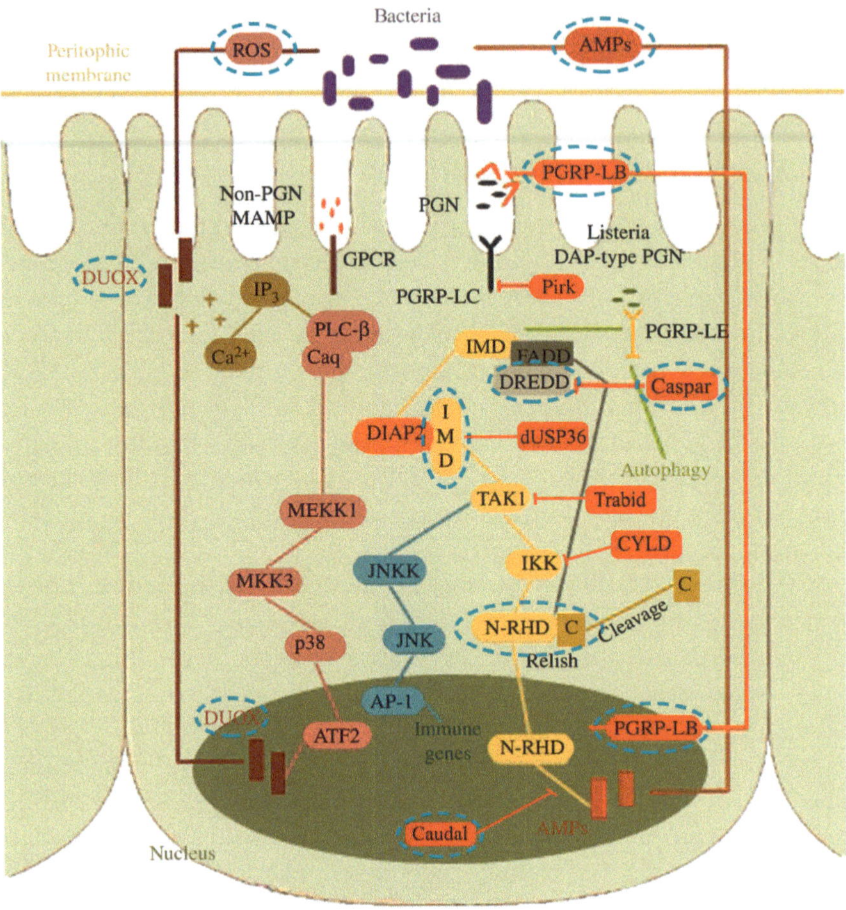

Fig. 1 The IMD pathway in insects. The IMD pathway has been worked out in detail in *Drosophila* and has been connected to both the DUOX and JNK pathways. *Circled* in *red* are the genes known in tsetse and sandflies. For sandflies, see Telleria et al. (2012) and Diaz-Albiter et al. (2012). For tsetse, see Hu and Aksoy (2006), Wang and Aksoy (2012), and Hamidou Soumana et al. (2014). *Modified from Kounatidis I., Ligoxygakis, P., 2012. Drosophila as a model system to unravel the layers of innte immunity to infection. Open. Biol. 2, 120075.*

7. ANTIMICROBIAL PEPTIDES

Diptera have a broad range of, highly conserved, AMPs which can be used to directly attack and eliminate pathogens. Employment of multiple AMPs, often with overlapping functions, can provide a formidable defence. Dipteran AMPs are primarily synthesized in the fat body then secreted into the hemolymph but can also be produced by hemocytes and midgut epithelia. Many are constitutively expressed at a basal level, e.g., Diptericin, but their expression can also be induced by invading parasites. The Dipteran AMPs demonstrated to have antiparasitic activity, cecropins, stomoxyn, defensins, attacin and diptericin, and appear to cause disruption of the plasma membrane and pathogen lysis (Boulanger et al., 2006).

Despite the availability of AMPs in the sandfly midgut *Leishmania* are able to survive. A recent study by Telleria et al. (2013) may provide some explanation. These researchers showed that oral infection of sandflies with *L. mexicana* did not significantly alter defensin expression when compared to uninfected controls nor was defensin expression increased upon on injection of *L. mexicana* into the hemocoel. This suggests *Leishmania* are able to avoid activating the response or directly modulate gene expression to prevent a response occurring (Telleria et al., 2013).

In contrast, infection of *Glossina morsitans* with *T. brucei* was shown to upregulate expression of attacin and cecropin. Subsequent knockdown of the two AMPs resulted in an increase in infection intensity demonstrating the importance of AMPs in management of *T. brucei* infection in tsetse (Hu and Aksoy, 2006). Perhaps the inability to prevent AMP upregulation in the vector is responsible for the high levels of 'self-curing' tsetse flies.

As in mammals, AMPs can also have other roles in the Dipteran immune response. There is evidence for a chemotactic role for defensin to lead hemocytes to the infection site (Bartholomay et al., 2004). Further studies of this alternative role for AMPs in Diptera are necessary and important for understanding vector–parasite interactions.

Such is the importance of AMPs that they are the basis of the paratransgenesis strategy to control *Trypanosoma cruzi* and *Leishmania*. Bacterial flora are isolated from wild vectors, genetically modified to highly express molecules (e.g., AMPs) thus reducing the ability for the parasites to survive in the vector and then reintroduced into wild populations (Hurwitz et al., 2014).

8. REACTIVE OXYGEN SPECIES

Oxygen-derived radical species including the superoxide anion (O2–) and hydroxyl radical (OH) are formed during respiration; as such they're produced at a basal level in the Diptera. Higher levels can be induced by the gut microbiota (see below) and are also induced upon infection with pathogenic bacteria (Moule et al., 2010; Wu et al., 2012). Studies in

Drosophila have revealed the role of the bacterial-induced dual oxidase (DUOX) system, involved in microbial clearance, intestinal epithelial cell renewal, redox-dependent modulation of signalling pathways, cross-linking of biomolecules and discrimination between symbionts and pathogens (reviewed in chapter "Microbiota, Gut Physiology, and Insect Immunity" by Lee et al. of this volume).

Leishmania do not appear to induce reactive oxygen species (ROS) on their own. However, the gut microbiota were able to cause a ROS response in sandflies (Sant'Anna et al., 2014). In addition, ROS kill *Leishmania* in vitro; orally feeding flies hydrogen peroxide (a source of ROS) negatively affected *Leishmania* survival in the insect's midgut. Conversely, silencing the sandfly's detoxifying genes reduced parasite load. However, *Leishmania* infections did not significantly increase hydrogen peroxide levels, suggesting the parasites can avoid causing a ROS response and probably cope with basal ROS levels using antioxidant enzymes, already known to be employed to allow *Leishmania* survival in human macrophages (Sardar et al., 2013). Indeed, peroxidoxin, an antioxidant protein (Levick et al., 1998), is upregulated in *Leishmania* promasitgotes (Diaz-Albiter et al., 2012).

DUOX produces ROS in tsetse flies. ROS are trypanocidal but also act as messenger molecules which trigger Nitric Oxide (NO) production. NO, like the ROS, can cause DNA damage, but it can also trigger AMP production in hemocytes and the fat body (Wu et al., 2012). However, NO inhibitors added to the blood meal compromised trypanosome development in the salivary glands (Macloed et al., 2007). It is possible that at this stage of their lifecycle trypanosomes could use NO to their advantage as was recently shown in malaria, but there is no other experimental evidence for this (Bahia et al., 2013). Unlike *Leishmania*, trypanosome infection does induce increases in hydrogen peroxide levels (Beschin et al., 2014) and supplementation of the blood meal with antioxidants increased the parasite infection rate, demonstrating the importance of ROS in trypanosome infection of tsetse flies (Macloed et al., 2007).

9. GUT MICROBIOTA AND SYMBIONTS

Insects have a less diverse gut microbiome than humans (Geiger et al., 2013). The gut microflora of tsetse has been studied by both culture dependent (Geiger et al., 2009, 2011; Lindh and Lehane, 2011) and culture independent (deep sequencing of 16S rRNA) methods (Aksoy et al., 2014). The identified microbiota are species specific and differ in relation to the location studied. Common gut bacteria found are members of the *Enterobacter*, *Enterococcus* and *Acinetobacter* genera with members of the *Chryseobacterium*, *Sphingobacterium*, *Providencia*, *Lactococcus*, *Pseudomonas* and *Staphylococcus* also being found (Hamidou Soumana et al., 2014). The effects of these gut flora on *T. brucei* infections are not well understood and require further study. However, studies in other insects infer that the Tsetse microflora may have a role in vector competence. From the triatomine bug *Rhodnius prolixus*, *Serratia marcescens* and *Pseudomonas flourescens* strains have both been shown to be able to lyse *T. cruzi* (Azambuja et al., 2004; Mercado and Colón-Whitt, 1982).

In addition to the aforementioned bacteria, laboratory strains of Tsetse have been shown to have three endosymbionts: *Wigglesworthia glossinidia*, *Sodalis glossinidius* and *Wolbachia* spp. The relationship between insects and their endosymbionts is an important one, and more details on concepts and current issues are included in chapter "Insect symbiosis and immunity" by Kim and Lee of this volume. Work done by Weiss et al. (2013) showed that tsetse without symbionts have a compromised PM, which results in earlier interaction between epithelia and parasites during infection. This causes an earlier immune response compared to wild-type infection (Weiss et al., 2013). Of note, that the earlier response did not confer trypanosome resistance. Nevertheless, flies in the wild have been shown to have additional associated microbes (Lindh and Lehane, 2011).

9.1 *Wigglesworthia glossinidia*

W. glossinidia is the Tsetse's dominant and essential endosymbiont and is found intracellularly in the anterior midgut bacteriocytes. It provides the fly with B vitamins (Rio et al., 2016) in exchange for a protected niche (Akman et al., 2002). Provision of these compounds (B1, B6, B9) is important as the fly lacks these vitamins, due to its hematophagous diet. There are also populations in the maternal milk glands which are thought to assist in priming of the immune system and also allow transmission to developing larvae (Weiss et al., 2011). Tsetse lacking *Wigglesworthia* are more susceptible to trypanosome infection than their wild-type counterparts (Pais et al., 2008).

9.2 *S. glossinidius*

S. glossinidius is an evolutionarily more recent acquisition to the Tsetse microbiota and is a nonessential symbiont. It is mostly found in the midgut, although it has been also isolated from a diverse range of tissues (Cheng and Aksoy, 1999). Tsetse with *Sodalis* are three times more likely to be infected with trypanosomes than tsetse without *Sodalis* (Farikou et al., 2010). It has been proposed that in young (teneral) flies *Sodalis* inhibits the fly's anti-trypanosomal lectins and releases *N*-acetyle-D-glucosamine (GlcNAc) through chitinolytic activity, which would assist in the survival of infecting trypanosomes (Rio et al., 2016; Welburn et al., 1993). Both *Wigglesworthia* and *Sodalis* belong to the Enterobacteriaceae family and are vertically transmitted to the intrauterine-developing larvae via milk gland secretions (reviewed in Wang et al., 2013).

9.3 *Wolbachia*

Wolbachia is a nonessential bacterium that infects many invertebrates. The presence of this bacterium is restricted to the reproductive organs of the tsetse fly and is transmitted transovarially (Wang et al., 2013). Of note that in wild populations, *Wolbachia* prevalence is variable in contrast to the high incidence of infection in laboratory flies (Doudoumis et al., 2012). When *Wolbachia*-infected tsetse males mate with uninfected females, a strong incompatibility is induced resulting to no progeny (Alam et al., 2011). The implication here is that female-infected flies will have an advantage ultimately replacing noninfected population. Some *Wolbachia* strains have been shown to reduce lifespan in *Aedes* spp. This is intriguing given that older adults are responsible for more disease transmissions than younger mosquitos; however, further study in this area is needed (Cook et al., 2008).

In contrast to Tsetse, who is strictly haematophagus, sandflies have a more diverse diet, females take regular sugar meals from plants as well as blood and so have a more diverse gut flora. Similarly to tsetse, however, there is growing evidence that the gut microbiome affects the vector competence of sandflies. Groups have demonstrated that *P. papatasi* infected with fungi are significantly more resistant to secondary infection with *L. major* (Adler and Theodor, 1929; Schlein et al., 1985).

In a more recent study, which sequenced the 16S RNA of the inhabitants of *P. papatasi* in 3 geographical locations, 170 bacterial isolates from 40 species were found (Maleki-Ravasan et al., 2015). The only common bacterium in all different geographical location was *Staphylococcus aureus*.

This underlined the potential of sandflies may function as putative vectors of bacterial pathogens in addition to *Leishmania*. Of the most populous bacterial species found belonged to the family of Enterobacteriaceae (a diverse group of Gram-negative bacteria). In addition, another study detected the presence of the Gram-positive bacteria *Bacillus flexus* and *Bacillus pumilus* (Mukhopadhyay et al., 2012). Finally, gradient gel electrophoresis of 16S rRNA gene fragments from adult female *Lutzomyia* guts (*Lu. longipalpis* and *Lu. cruzi)* from three geographically separated regions in Brazil were analyzed by Dillon and coworkers (Santa'Anna et al., 2012). Their analysis showed many plant-associated bacteria of the *Erwinia* and *Ralstonia* spp., in keeping with the ecology of sandflies as plant feeders as well as opportunistic human pathogens of the *Burkhoderia* genus (Santa'Anna et al., 2012).

In addition to bacterial species, Akhoundi et al. (2012) found fungal species belonging to the genera *Aspergillus, Penicillium, Geotrichum, Fusarium, Acremonium and Candida* in the midgut of five species of flies from the *Phlebotomus* genera collected in North-Western Iran (Akhoundi et al., 2012). Interestingly, it seems that *Leishmania* can also provide protection for the sandfly against bacterial infection. Sant'Anna et al. (2014) showed that *L. mexicana* protects *Lu. longipalpis* from *S. marcescens* infection (Sant'Anna et al., 2014). Other than the aforementioned study very little is known about interactions between fungi and *Leishmania* species. Further study into the interactions of Leishmainia species could yet reveal important and interesting information about parasite–vector biology.

10. USING THE MICROBIOTA AS A TRANSMISSION BLOCKING STRATEGY

Studies on the interactions between gut microbiota and vector are only just coming to the forefront of the field and much is still poorly understood. However, there is great potential for transmission blocking strategies in this research which would greatly reduce the burden of insect vector-borne diseases. These include novel vector control methods such as paratransgenesis, where gut microbiota could be genetically modified to produce antiparasite compounds. To this end in vitro culture of *Sodalis*, introduction of an exogenous DNA and repopulation of the insect with the modified bacterium has been achieved (Aksoy et al., 2008; Pontes and Dale, 2011). New environment-friendly tools for disease control could be developed also by harnessing *Wolbachia* symbiosis (for review, see Doudoumis et al., 2013).

ACKNOWLEDGEMENTS

We would like to thank Dr Ilias Kounatidis for drawing Fig. 1. Work in the Ligoxygakis laboratory is supported by the European Research Council (Consolidator Grant no 310912 '*Droso-parasite*') and the BBSRC (Grant no BB/K003569/1) both to P.L. M.S. is supported by a doctoral scholarship from the Wellcome Trust (Infection, Immunity and Translational Medicine Program administered by the Dunn School of Pathology, University of Oxford).

REFERENCES

Adler, S., Theodor, O., 1929. Attempts to transmit Leishmania tropica by bite: the transmission of L. tropica by Phlebotomus sergenti. Ann. Trop. Med. Parasitol. 23 (1), 1–18.

Akhoundi, M., Bakhtiari, R., Guillard, T., Baghaei, A., Tolouei, R., Sereno, D., Toubas, D., Depaquit, J., Abyaneh, M.R., 2012. Diversity of the bacterial and fungal microflora from the midgut and cuticle of phlebotomine sand flies collected in North-Western Iran. PLoS One 7 (11), e50259.

Akman, L., Yamashita, A., Watanabe, H., Oshima, K., Shiba, T., Hattori, M., Aksoy, S., 2002. Genome sequence of the endocellular obligate symbiont of tsetse flies, Wigglesworthia glossinidia. Nat. Genet. 32 (3), 402–407.

Aksoy, S., Weiss, B., Attardo, G., 2008. Paratransgenesis applied for control of tsetse transmitted sleeping sickness. Adv. Exp. Med. Biol. 627, 35–48.

Aksoy, E., Telleria, E.L., Echodu, R., Wu, Y., Okedi, L.M., et al., 2014. Analysis of multiple tsetse fly populations in Uganda reveals limited diversity and species-specific gut microbiota. Appl. Environ. Microbiol. 80, 4301–4312.

Aksoy, E., Vigneron, A., Bing, X., Zhao, X., O'Neill, M., Wu, Y.N., Bangs, J.D., Weiss, B.L., Aksoy, S., 2016. Mammalian African trypanosome VSG coat enhances tsetse's vector competence. Proc. Natl. Acad. Sci. USA 113 (25), 6961–6966. http://dx.doi.org/10.1073/pnas.1600304113.

Alam, U., Medlock, J., Brelsfoard, C., Pais, R., Lohs, C., et al., 2011. Wolbachia symbiont infections induce strong cytoplasmic incompatibility in the tsetse fly *Glossina morsitans*. PLoS Pathog. 7 (12), e1002415.

Azambuja, P., Feder, D., Garcia, E.S., 2004. Isolation of Serratia marcescens in the midgut of Rhodnius prolixus: impact on the establishment of the parasite Trypanosoma cruzi in the vector. Exp. Parasitol. 107 (1–2), 89–96.

Bahia, A.C., Oliveira, J.H.M., Kubota, M.S., Araújo, H.R.C., Lima, J.B.P., Ríos-Velásquez, C.M., Lacerda, M.V.G., Oliveira, P.L., Traub-Csekö, Y.M., Pimenta, P.F.P., 2013. The role of reactive oxygen species in Anopheles aquasalis response to Plasmodium vivax infection. PLoS One 8 (2), e57014.

Bartholomay, L.C., Fuchs, J.F., Cheng, L.L., Beck, E.T., Vizioli, J., Lowenberger, C., Christensen, B.M., 2004. Reassessing the role of defensin in the innate immune response of the mosquito, Aedes aegypti. Insect Mol. Biol. 13 (2), 125–132.

Bates, P.A., 2008. Leishmania sand fly interaction: progress and challenges. Curr. Opin. Microbiol. 11 (4), 340–344. http://dx.doi.org/10.1016/j.mib.2008.06.003.

Beschin, A., Van Den Abbeele, J., De Baetselier, P., Pays, E., 2014. African trypanosome control in the insect vector and mammalian host. Trends Parasitol. 30 (11), 538–547.

Borovsky, D., Schlein, Y., 1987. Trypsin and chymotrypsin-like enzymes of the sandfly Phlebotomus papatasi infected with Leishmania and their possible role in vector competence. Med. Vet. Entomol. 1 (3), 235–242.

Boulanger, N., Bulet, P., Lowenberger, C., 2006. Antimicrobial peptides in the interactions between insects and flagellate parasites. Trends Parasitol. 22 (6), 262–268.

Cheng, Q., Aksoy, S., 1999. Tissue tropism, transmission and expression of foreign genes in vivo in midgut symbionts of tsetse flies. Insect Mol. Biol. 8 (1), 125–132.

Cook, P.E., McMeniman, C.J., O'Neill, S.L., 2008. Modifying insect population age structure to control vector-borne disease. In: Aksoy, S. (Ed.), Transgenesis and the Management of Vector-Borne Disease. Springer New York, New York, NY, pp. 126–140.

Coutinho-Abreu, I.V., Sharma, N.K., Robles-Murguia, M., Ramalho-Ortigao, M., 2013. Characterization of Phlebotomus papatasi peritrophins, and the role of PpPer1 in Leishmania major survival in its natural vector. PLoS Negl. Trop. Dis. 7 (3), e2132.

de A Gonzaga, W., Weber, A.D., Giacomelli, S.R., Dalcol, I.I., Hoelzel, S.C.S., Morel, A.F., 2003. Antibacterial alkaloids from Zanthoxylum rhoifolium. Planta Med. 69 (04), 371–374.

Diaz-Albiter, H., Sant'Anna, M.R.V., Genta, F.A., Dillon, R.J., 2012. Reactive oxygen species-mediated immunity against Leishmania mexicana and Serratia marcescens in the phlebotomine sand fly Lutzomyia longipalpis. J. Biol. Chem. 287 (28), 23995–24003.

Dillon, R.J., Lane, P., 1993. Bloodmeal digestion in the midgut of Phlebotomus papatasi and Phlebotomus langeroni. Med. Vet. Entomol. 7 (3), 225–232.

Dillon, R.J., Ivens, A.C., Churcher, C., Holroyd, N., Quail, M.A., Rogers, M.E., Soares, M.B., Bonaldo, M.F., Casavant, T.L., Lehane, M.J., Bates, P.A., 2006. Analysis of ESTs from Lutzomyia longipalpis sand flies and their contribution toward understanding the insect–parasite relationship. Genomics 88 (6), 831–840.

Dinglasan, R.R., Devenport, M., Florens, L., Johnson, J.R., McHugh, C.A., Donnelly-Doman, M., Carucci, D.J., Yates III, J.R., Jacobs-Lorena, M., 2009. The Anopheles gambiae adult midgut peritrophic matrix proteome. Insect Biochem. Mol. Biol. 39 (2), 125–134. http://dx.doi.org/10.1016/j.ibmb.2008.10.010.

Dostálová, A., Votýpka, J., Favreau, A.J., Barbian, K.D., Volf, P., Valenzuela, J.G., Jochim, R.C., 2011. The midgut transcriptome of Phlebotomus (Larroussius) perniciosus, a vector of Leishmania infantum: comparison of sugar fed and blood fed sand flies. BMC Genomics 12 (1), 223.

Doudoumis, V., Tsiamis, G., Wamwiri, F., Brelsfoard, C., Alam, U., et al., 2012. Detection and characterisation of Wolbachia infections in laboratory and natural populations of different species of tsetse flies (genus Glossina). BMC Microbiol. 12 (Suppl. 1), S3.

Doudoumis, V., Alam, U., Aksoy, E., Abd-Alla, E.M., Tsiamis, G., et al., 2013. Tsetse-Wolbachia symbiosis: comes of age and has great potential for pest and disease control. J. Invertebr. Pathol. 112 (Suppl), S94–103.

Farikou, O., Njiokou, F., Mbida Mbida, J.A., Njitchouang, G.R., Djeunga, H.N., Asonganyi, T., Simarro, P.P., Cuny, G., Geiger, A., 2010. Tripartite interactions between tsetse flies, Sodalis glossinidius and trypanosomes—an epidemiological approach in two historical human African trypanosomiasis foci in Cameroon. Infect. Genet. Evol. 10 (1), 115–121.

Geiger, A., Fardeau, M.-L., Njiokou, F., Ollivier, B., 2013. Glossina spp. gut bacterial flora and their putative role in fly-hosted trypanosome development. Front. Cell. Infect. Microbiol. 3, 34.

Haines, L.R., 2013. Examining the tsetse teneral phenomenon and permissiveness to trypanosome infection. Front. Cell. Infect. Microbiol. 3, 84.

Haines, L.R., Jackson, A.M., Lehane, M.J., Thomas, J.M., Yamaguchi, A.Y., Haddow, J.D., Pearson, T.W., 2005. Increased expression of unusual EP repeat-containing proteins in the midgut of the tsetse fly (Glossina) after bacterial challenge. Insect Biochem. Mol. Biol. 35 (5), 413–423.

Haines, L.R., Lehane, S.M., Pearson, T.W., Lehane, M.J., 2010. Tsetse EP protein protects the fly midgut from trypanosome establishment. PLoS Pathog. 6 (3), e1000793.

Hamidou Soumana, I., Tchicaya, B., Chuchana, P., Geiger, A., 2014. Midgut expression of immune-related genes in Glossina palpalis gambiensis challenged with trypanosoma brucei gambiense. Front. Microbiol. 5, 609.

Hotez, P.J., 2008. Stigma: the stealth weapon of NTD. PLoS Negl. Trop. Dis. 2 (4), e230. http://dx.doi.org/10.1371/journal.pntd.0000230.

Hotez, P.J., 2009. One world health: neglected tropical diseases in a flat world. PLoS Negl. Trop. Dis. 3 (4), e405. http://dx.doi.org/10.1371/journal.pntd.0000405.

Hu, C., Aksoy, S., 2006. Innate immune responses regulate trypanosome parasite infection of the tsetse fly Glossina morsitans morsitans. Mol. Microbiol. 60 (5), 1194–1204.

Hurwitz, I., Forshaw, A., Yacisin, K., Durvasula, R., 2014. Paratransgenic control of Leishmaniasis: new developments. In: Satoskar, A., Durvasula, R. (Eds.), Pathogenesis of Leishmaniasis: New Developments in Research. Springer Science and Business, New York, pp. 25–43.

Jochim, R.C., Teixeira, C.R., Laughinghouse, A., Mu, J., Oliveira, F., Gomes, R.B., Elnaiem, D.-E., Valenzuela, J.G., 2008. The midgut transcriptome of Lutzomyia longipalpis: comparative analysis of cDNA libraries from sugar-fed, blood-fed, post-digested and Leishmania infantum chagasi-infected sand flies. BMC Genomics 9, 15.

Lehane, M.J., 1997. Peritrophic matrix structure and function. Annu. Rev. Entomol. 42, 525–550.

Lehane, M.J., Msangi, A.R., 1991. Lectin and peritrophic membrane development in the gut of *Glossina m. morsitans* and a discussion of their role in protecting the fly against trypanosome infection. Med. Vet. Entomol. 5, 495–501.

Lehane, M.J., Aksoy, S., Levashina, E., 2004. Immune responses and parasite transmission in blood-feeding insects. Trends Parasitol. 20 (9), 433–439.

Levick, M.P., Tetaud, E., Fairlamb, A.H., Blackwell, J.M., 1998. Identification and characterisation of a functional peroxidoxin from Leishmania major1. Mol. Biochem. Parasitol. 96 (1–2), 125–137.

Lindh, J.M., Lehane, M.J., 2011. The tsetse fly Glossina fuscipes fuscipes (Diptera: Glossina) harbours a surprising diversity of bacteria other than symbionts. Antonie Van Leeuwenhoek 99 (3), 711–720.

Liniger, M., Acosta-Serrano, A., Van Den Abbeele, J., Kunz Renggli, C., Brun, R., Englund, P.T., Roditi, I., 2003. Cleavage of trypanosome surface glycoproteins by alkaline trypsin-like enzyme(s) in the midgut of Glossina morsitans. Int. J. Parasitol. 33 (12), 1319–1328.

Macloed, E.T., Maudlin, I., Darby, A.C., Welburn, S.C., 2007. Antioxidants promote establishment of trypanosome infections in tsetse. Parasitology 134 (06), 827–831.

Maleki-Ravasan, N., Oshaghi, M.A., Afshar, D., Arandian, M.H., Hajikhani, S., Akhavan, A.A., Yakhchali, B., Shirazi, M.H., Rassi, Y., Jafari, R., Aminian, K., Fazeli-Varzaneh, R.A., Durvasula, R., 2015. Aerobic bacterial flora of biotic and abiotic compartments of a hyperendemic Zoonotic Cutaneous Leishmaniasis (ZCL) focus. Parasit. Vectors 8 (1), 63.

Mercado, T.I., Colón-Whitt, A., 1982. Lysis of Trypanosoma cruzi by Pseudomonas fluorescens. Antimicrob. Agents Chemother. 22 (6), 1051–1057.

Morrison, L.S., Goundry, A., Faria, M.S., Tetley, L., Eschenlauer, S.C., Westrop, G.D., Dostalova, A., Volf, P., Coombs, G.H., Lima, A.P.C.A., Mottram, J.C., 2012. Ecotin-like serine peptidase inhibitor ISP1 of Leishmania major plays a role in flagellar pocket dynamics and promastigote differentiation. Cell. Microbiol. 14 (8), 1271–1286.

Moule, M.G., Monack, D.M., Schneider, D.S., 2010. Reciprocal analysis of Francisella novicida infections of a Drosophila melanogaster model reveal host-pathogen conflicts mediated by reactive oxygen and imd-regulated innate immune response. PLoS Pathog. 6 (8), e1001065.

Mukhopadhyay, J., Braig, H.R., Rowton, E.D., Ghosh, K., 2012. Naturally occurring culturable aerobic gut flora of adult Phlebotomus papatasi, vector of Leishmania major in the Old World. PLoS One 7 (5), e35748.
Pais, R., Lohs, C., Wu, Y., Wang, J., Aksoy, S., 2008. The obligate mutualist Wigglesworthia glossinidia influences reproduction, digestion, and immunity processes of its host, the tsetse fly. Appl. Environ. Microbiol. 74 (19), 5965–5974.
Pontes, M.H., Dale, C., 2011. Lamda-red mediated genetic modification of the insect endosymbiont *Sodalis glossinidius*. Appl. Environ. Microbiol. 77, 1918–1920.
Ramalho-Ortigao, M., Jochim, R.C., Anderson, J.M., Lawyer, P.G., Pham, V.M., Kamhawi, S., Valenzuela, J.G., 2007. Exploring the midgut transcriptome of Phlebotomus papatasi: comparative analysis of expression profiles of sugarfed, bloodfed and Leishmania major-infected sandflies. BMC Genomics 8, 300. http://dx.doi.org/10.1186/1471-2164-8-300.
Rio, R.V.M., Attardo, G.M., Weiss, B.L., 2016. Grandeur alliances: symbiont metabolic integration and obligate arthropod hematophagy. Trends Parasitol. 32 (9), 739–749.
Rose, C., Belmonte, R., Armstrong, S.D., Molyneux, G., Haines, L.R., Lehane, M.J., Wastling, J., Acosta-Serrano, A., 2014. An investigation into the protein composition of the teneral Glossina morsitans morsitans peritrophic matrix. PLoS Negl. Trop. Dis. 8 (4), e2691.
Roxström-Lindquist, K., Terenius, O., Faye, I., 2004. Parasite-specific immune response in adult Drosophila melanogaster: a genomic study. EMBO Rep. 5 (2), 207–212.
Sádlová, J., Volf, P., 2009. Peritrophic matrix of Phlebotomus duboscqi and its kinetics during Leishmania major development. Cell Tissue Res. 337 (2), 313–325.
Santa'Anna, M.R., Darby, A.C., Brazil, R.P., Montoya-Lemma, J., Dillon, V.M., et al., 2012. Investigation of the bacterial communities associated with females of *Lutzomyia* sand fly species from South America. PLoS One 7, e42531.
Sant'Anna, M.R., Diaz-Albiter, H., Mubaraki, M., Dillon, R.J., Bates, P.A., 2009. Inhibition of trypsin expression in Lutzomyia longipalpis using RNAi enhances the survival of Leishmania. Parasit. Vectors 2 (1), 62.
Sant'Anna, M., Diaz-Albiter, H., Aguiar-Martins, K., Al Salem, W., Cavalcante, R., Dillon, V., Bates, P., Genta, F., Dillon, R., 2014. Colonisation resistance in the sand fly gut: Leishmania protects Lutzomyia longipalpis from bacterial infection. Parasit. Vectors 7 (1), 329.
Sardar, A.H., Kumar, S., Kumar, A., Purkait, B., Das, S., Sen, A., Kumar, M., Sinha, K.K., Singh, D., Equbal, A., Ali, V., Das, P., 2013. Proteome changes associated with Leishmania donovani promastigote adaptation to oxidative and nitrosative stresses. J. Proteomics 81 (0), 185–199.
Schlein, Y., Polacheck, I., Yuval, B., 1985. Mycoses, bacterial infections and antibacterial activity in sandifies (Psychodidae) and their possible role in the transmission of leishmaniasis. Parasitology 90 (01), 57–66.
Telleria, E.L., de Araujo, A.P., Secundino, N.F., d'Avila-Levy, C.M., Traub-Cseko, Y.M., 2010. Trypsin-like serine proteases in Lutzomyia longipalpis—expression, activity and possible modulation by Leishmania infantum chagasi. PLoS One 5 (5), e10697.
Telleria, E.L., Sant'Anna, M.R., Ortigao-Farias, J.R., Pitaluga, A.N., Dillon, V.M., Bates, P.A., Traub-Cseko, Y.M., Dillon, R.J., 2012. Caspar-like gene depletion reduces Leishmania infection in sand fly host Lutzomyia longipalpis. J. Biol. Chem. 287 (16), 12985–12993.
Telleria, E.L., Sant'Anna, M.R., Alkurbi, M.O., Pitaluga, A.N., Dillon, R.J., Traub-Cseko, Y.M., 2013. Bacterial feeding, Leishmania infection and distinct infection routes induce differential defensin expression in Lutzomyia longipalpis. Parasit. Vectors 6, 12.

Volf, P., Svobodová, M., Dvoráková, E., 2001. Bloodmeal digestion and Leishmania major infections in Phlebotomus duboscqi: effect of carbohydrates inhibiting midgut lectin activity. Med. Vet. Entomol. 15 (3), 281–286.

Wang, J., Aksoy, S., 2012. PGRP-LB is a maternally transmitted immune milk protein that influences symbiosis and parasitism in tsetse's offspring. Proc. Natl. Acad. Sci. U.S.A. 109 (26), 10552–10557.

Wang, J., Wu, Y., Yang, G., Aksoy, S., 2009. Interactions between mutualist Wigglesworthia and tsetse peptidoglycan recognition protein (PGRP-LB) influence trypanosome transmission. Proc. Natl. Acad. Sci. U.S.A. 106 (29), 12133–12138.

Wang, J., Weiss, B.L., Aksoy, S., 2013. Tsetse fly microbiota: form and function. Front. Cell. Infect. Microbiol. 3, 69.

Weiss, B.L., Wang, J., Aksoy, S., 2011. Tsetse immune system maturation requires the presence of obligate symbionts in larvae. PLoS Biol. 9 (5), e1000619.

Weiss, B.L., Wang, J., Maltz, M.A., Wu, Y., Aksoy, S., 2013. Trypanosome infection establishment in the tsetse fly gut is influenced by microbiome-regulated host immune barriers. PLoS Pathog. 9 (4), e1003318.

Welburn, S.C., Arnold, K., Maudlin, I., Gooday, G.W., 1993. Rickettsia-like organisms and chitinase production in relation to transmission of trypanosomes by tsetse flies. Parasitology 107 (2), 141–145.

Wigglesworth, V.B., 1929. Digestion in the tsetse-fly: a study of structure and function. Parasitology 21, 288–321.

Wu, S.-C., Liao, C.-W., Pan, R.-L., Juang, J.-L., 2012. Infection-induced intestinal oxidative stress triggers organ-to-organ immunological communication in Drosophila. Cell Host Microbe 11 (4), 410–417.

Yan, J., Cheng, Q., Narashimhan, S., Li, C.-B., Aksoy, S., 2002. Cloning and functional expression of a fat body-specific chitinase cDNA from the tsetse fly, *Glossina morsitans morsitans*. Insect Biochem. Mol. Biol. 32 (9), 979–989.

FSC
www.fsc.org
MIX
Papier | Fördert gute Waldnutzung
FSC® C083411

Kolibri 360 GmbH
Ferdinand-Jühlke-Straße 7
99095 Erfurt, Deutschland
produktsicherheit@kolibri360.de